Introduction to *n*MOS and CMOS VLSI Systems Design

AMAR MUKHERJEE

Department of Computer Science
University of Central Florida

Prentice-Hall, Englewood Cliffs, N.J. 07632

Library of Congress Cataloging-in-Publication Data

MUKHERJEE, AMAR.
    Introduction to nMOS and CMOS VLSI systems
design.

    Bibliography: p.
    Includes index.
    1. Integrated circuits—Very large scale
integration—Design and construction.   2. Metal
oxide semiconductors.   3. Metal oxide semiconductors,
Complementary.   I. Title.
TK7874.M86 1985   621.395   85-19161
ISBN   0-13-490947-X

Editorial/production supervision and
   interior design: Diana Drew
Cover design: Whitman Studio, Inc.
Manufacturing buyer: Gordon Osbourne

Printed in the United States of America

10   9   8   7   6

ISBN 0-13-490947-X   01

Prentice-Hall International, Inc., *London*
Prentice-Hall of Australia Pty. Limited, *Sydney*
Editora Prentice-Hall do Brasil, Ltda., *Rio de Janeiro*
Prentice-Hall Canada Inc., *Toronto*
Prentice-Hall Hispanoamericana, S.A., *Mexico*
Prentice-Hall of India Private Limited, *New Delhi*
Prentice-Hall of Japan, Inc., *Tokyo*
Prentice-Hall of Southeast Asia Pte. Ltd., *Singapore*
Whitehall Books Limited, *Wellington, New Zealand*

To the Memory
of My Parents

Contents

4 The Technology of Semiconductors 118

5 Design—Fabrication Interface 141

6   Delay and Power   182

7   Systems Design   211

Appendix A: The Physics of Semiconductor Devices   336

Appendix B: MOSIS Scalable and Generic CMOS Design Rules   349

Index   365

Preface

Until a decade ago, chip design expertise was the sole province of a handful of designers who had the prerequisite knowledge that spanned the diverse disciplines of physics, electrical engineering, systems and computer science. In recent years, a systematic design methodology has evolved which allows the entire design process to be broken down into a number of design layers with abstractions that hide the details of the lower layers. This has led to the development of powerful automated design tools without which the chip revolution would not have taken place. This technology is now available in the classroom because of the efforts of a large number of researchers and teachers in the U.S. and abroad.

The present text adopts a hierarchical design methodology for VLSI and presents both nMOS and CMOS design in an organized fashion. The emphasis is on fundamental design concepts. The text has evolved over the last four years from lecture notes by the author for a VLSI design course in the Computer Science Department at the University of Central Florida. The organization of the book roughly follows the organization of this course. Chapter 1 gives an overview of the VLSI design process. The following two chapters deal with basic logical elements and logic design methods in nMOS and CMOS. The physics of MOS devices is included in Appendix A. Chapter 4 gives a brief introduction to fabrication processes so the reader gains a basic understanding of semiconductor technology. Chapter 5 is concerned with design rules and computation of circuit parameters from layout. Chapter 6 treats delay and power computation with a simplified model. Chapter 7 presents systems-level design and includes a number of student design projects. These projects have been created using the Berkeley VLSI design software and fabricated by MOSIS (MOS Implementation System) facilities located at the

USC Information Sciences Institute in California. Chapter 8 focuses on the design principles of memory circuits. Chapter 9 contains descriptions of several algorithms used in the implementation of computer-aided design tools for VLSI. Appendix B contains the most recent MOSIS scalable CMOS design rules.

The text is suitable for a senior-level undergraduate or a first-year graduate course in Computer Science/Computer Engineering or Electrical Engineering departments.

I wish to acknowledge a large number of students and colleagues who have contributed to this book. Alan Eustace has participated in many technical discussions and has made several contributions to the book. At the early stages of the development of the VLSI laboratories at the University of Central Florida, both Alan Eustace and Timothy Curry contributed a great deal. The design projects of Kenneth Donovan, Phil Gatt, and Sam Richie have been included in the text, and I gratefully acknowledge their contributions. Other noteworthy contributions were made by N. Ranganathan, Eduardo Diaz, Mike Eisler, Larry Lawrence, Don Harper, and all the students who took the VLSI design course at the University of Central Florida.

I would like to especially thank Professor Randy H. Katz of the University of California at Berkeley, Professor Christopher J. Terman of the Massachusetts Institute of Technology and James R. Sutton of Harris Semiconductors in Melbourne, Florida, for their critical comments on an earlier version of the manuscript. I would also like to thank Professors Terman, Ron Rivest of the Massachusetts Institute of Technology, and John Ousterhout and Clark Thompson of the University of California at Berkeley for granting permission to include some material based on their work in this text. I have derived materials from several publications and doctoral dissertations; these are properly cited in the text and are gratefully acknowledged. Special thanks also goes to George Lewicki and Kathleen Fry of MOSIS for their support and their permission to include the CMOS design rules in Appendix B.

I would like to acknowledge the Computer Science Department of the University of Central Florida for the resources made available to me for developing a set of lecture notes which formed the basis for the manuscript. I would also like to thank the faculty of the Computer Science Department at the University of Central Florida for their moral support.

Karl Karlstrom and Diana Drew of Prentice-Hall deserve special thanks for keeping an impossible production schedule.

My wife, Pampa, and our daughters, Mita and Paula, deserve very special thanks for their continued support, understanding, and encouragement throughout this project.

Finally, a note regarding my name. Many of my earlier publications cited in the book have my name as Amar Mukhopadhyay. "Mukhopadhyay" is simply a long Sanskrit form of "Mukherjee."

—Amar Mukherjee

1

Overview of the Design Methodology

1.1 Introduction

The term *very large scale integration* (VLSI) reflects the capabilities of the semiconductor industry to fabricate a complex electronic circuit consisting of thousands of components on a single silicon substrate. The growth of semiconductor technology in recent years has been described by "Moore's law," enunciated in the late 1960s, which projected quadrupling of component density in a chip every three to four years. Several factors contributed to this tremendous growth: reduction of line width of the basic device and interconnection wires due to the development of high-resolution lithographic techniques and improved processing capabilities; increase in the size of the silicon wafer due to improved reliability of processing, growth of the accumulated circuit, and layout design experience; better understanding of system-level design issues, leading to improved architectures that can exploit the technology; and the availability of better automated design tools for circuit layout, simulation, verification, and testing.

Until a decade ago, chip design expertise was the sole province of a handful of designers who had the prerequisite knowledge that spanned over system, logic, circuit, layout, and processing. In many instances, the system-level issues got the least attention because of the lack of adequate abstraction of design levels. The book by Mead and Conway (1980) showed how to deal with design concepts and system issues at levels divorced from device and processing details. The authors developed a set of simplified design rules to abstract the *n*MOS processing details, a set of ratio rules to produce conservative circuits, a two-phase clocking discipline to simplify the logic design and timing problems, and a hierarchic design methodology that allowed the system designer to easily translate the digital system into an *n*MOS chip. Having

1

an operational multiproject chip assembly system and silicon foundries directly linked with the textual material gave students exciting classroom "hands-on" experience with custom VLSI (Conway et al., 1980). The mystique about chip design was removed and the floodgates were opened for a VLSI design course at the college level. New enthusiasm and awareness of VLSI were generated among researchers in the university, industry, and government. The rapid explosion of publications, conferences, and research funding in the VLSI area can be taken as indicators of this movement.

1.2 Layers of Abstraction

The VLSI design process spans a diversified spectrum of disciplines in physics, chemical engineering, electrical engineering, and computer science (Clark, 1980). Because of the diversity of tasks and design issues, a systematic approach to breaking the process into a number of design layers and subtasks is essential.

1.2.1 The Physical Layer

At the lowest level lies the physical world of semiconductor conduction. A complex fabrication technology has evolved to control and exploit the physical processes for the purpose of building useful devices. As these devices are put together to form interesting circuits, the flow of currents and level of voltages become the carrier of information in the circuit. Electrical engineers thus use an abstraction of the physical world represented in the form of transistors, resistors, capacitors, and connectors. The design issues that are of concern at this stage are the speed of the circuit as determined by the circuit parameters; signal degradation in connecting wires; the total power consumption of the circuit, which determines heat buildup; and a precise sequencing of the voltage and current waveforms that control the flow of information within the circuit. A conservative design methodology such as the one adopted in this book takes care of these issues by sticking to a set of electrical rules and timing conventions that guarantee correct operation of the circuit. In a typical industrial environment, these issues get the most attention from electrical engineers who are attempting to design optimum custom circuits.

1.2.2 The Layout Layer

The interface between the physical world and the electrical world is the layout layer, which serves as the link between the circuit and the fabrication process that builds the circuit. An abstract description of the layout model that hides the underlying physical processes is as follows. A circuit is built on the surface of a silicon substrate by interconnecting pieces of material in three primary layers: a conducting layer in *metal* used for pure electrical connection; and two layers of semiconductors (*poly-silicon* and *diffusion*) used for building such devices as switches, inverters, and gates as well as for electrical connection. These layers are normally insulated from each other by insulating material. Electrical connection can be established between the layers by special structures called *contact cuts* (to connect the metal layer with the polysilicon or diffusion layer) and *buried* or *butting contact* (to connect the polysilicon with the diffusion layer). The semiconductor material can be of *n-type*, meaning that *electrons* carrying a negative charge are available to support electrical conduction if a voltage gradient can be maintained, or of *p-type*, if holes carrying positive charge to support the electrical conduction are more frequent. To enhance the availability of the charge carriers, the semiconductor layers can be *implanted* with additional charge-carrying ions by specialized processes. Whenever a piece of polysilicon material overlays two opposite sides with two pieces of diffusion material, an *enhancement-mode* transistor is formed. If the silicon surface underneath the polysilicon material is implanted, a *depletion-mode* transistor is formed. The top view of the layout of a typical simple circuit is shown in Fig. 1.1. The design issue that is of primary concern at the layout level is to optimize the area of the layout subject to a set of *design rules* that permit overlapping and extension of the various structures such that the circuit implemented by the fabrication process will work under conservative assumptions of process variations and the resolution of a typical lithographic step (explained in Chapter 5). Issues that are pertinent at the layout level come from a consideration of the design at higher levels, such as the logic and systems level, and the interaction of these levels at the electrical level. The layout of a circuit must take into account the context surrounding the layout: whether the circuit has to be used in the regular structure of an iterative network; whether certain lines in the circuit are global, carrying power, ground, data, or control; and so on. The effect on the electrical design comes from the accumulated contribution of resistance and capacitance of the global wires. Since the wires that carry the signals in a chip could be metal, polysilicon, or diffusion, one must be aware of the effect of layout on the electrical properties of the network. Metal has low resistance and low capacitance. Next to metal is polysilicon, which has low to moderate capacitance but rather high resistance. Diffusion has the highest capacitance and moderate resistance. Finally, there are concerns with respect to the minimum width of metal wires that carry current to the various devices, due to a phenomenon called *metal migration*.

1.2.3 The Circuit and the Logic Layers

The logic level of the design process uses an abstraction of the underlying electrical circuit in which the currents and voltages are limited to discrete levels. In digital circuits, the two levels of signals that are permissible are those that represent logic 0 and logic 1, and the circuit is represented by diagrams such as those shown in Fig. 1.2 and Fig. 1.3, corresponding to the layout of Fig. 1.1. Symbols are used here to

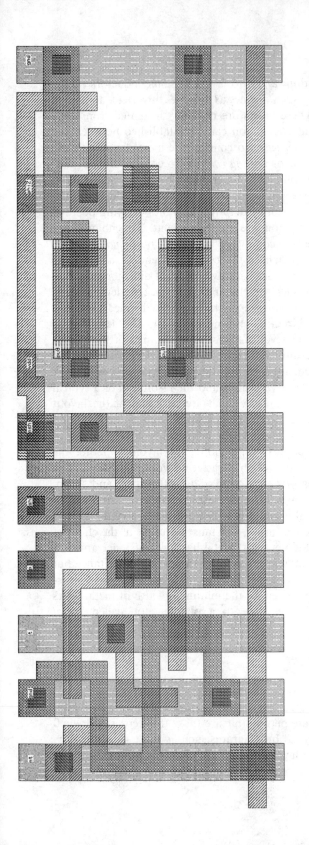

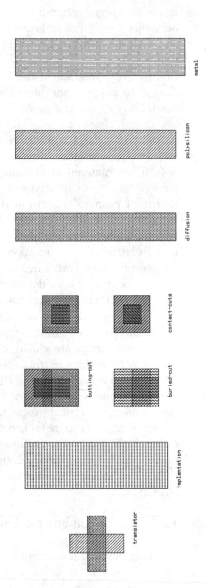

Figure 1.1 Layout of a logic circuit.

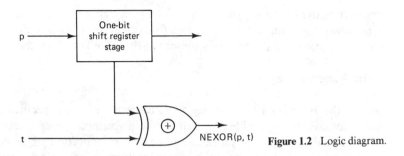

Figure 1.2 Logic diagram.

represent switches, inverters, and gates which manipulate logic signals. Quite often, refinements of a logic diagram of a network such as the one shown in Fig. 1.3 spelled out in terms of the elementary transistors and augmented by associated electrical parameters (ratios representing the resistances and capacitance of connecting wires) are good enough representations to capture the circuit design phase of the design. The logic-level representation hides the details of delay and timing considerations by making the circuits work synchronously. In synchronous networks the information flow between different parts of the network is controlled so as to happen at discrete time instants determined by the occurrence of "clock pulses" which serve as the time marker inside the circuit. The major design issues at the logic-level stage are the correctness of the overall logical operations of the circuit in terms of the elementary operations performed by the logical elements in the specified timing sequence. The

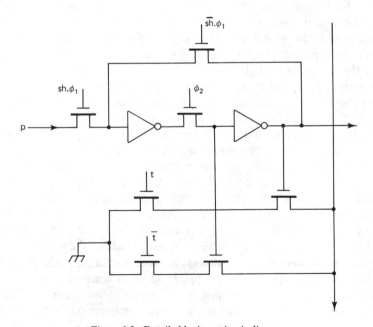

Figure 1.3 Detailed logic or circuit diagram.

other concern is the simplicity and elegance of the design, so that it blends well with the higher-level algorithmic process at the functional level, on the one hand, and with the flow of data and the control structure at the layout level, on the other.

1.2.4 The Function Layer

The task at the functional level of design is to come up with specifications for a conceptual solution to the problem in terms of an algorithmic process, and specifications for the major functional blocks and their interconnection to implement the heart of the algorithm. Different alternative algorithms must be considered at this point to evaluate the regularity of the structure, data flow organization, cell types, geometric placements, communication needs between the blocks, and so on. If the function is too complex, it should be decomposed into subsystems and the process iterated hierarchically. Since the functional decomposition induces a geometrical decomposition of the layout and the functional hierarchy maps into a geometrical embedding process with well-defined boundaries, the process of design is separated into smaller subtasks that can be executed independently. In practice, however, complete independence of subtasks is not achievable because of design factors that are common to all subtasks, which force the total design to go through several iterations.

1.2.5 The Systems Layer

At the highest level of design, called the systems level, one is concerned with connecting the major subsystems, together with communication interfaces with the external world; global wiring strategies; selecting layers for carrying global control, data, and power; placement of major subsystems; and routing strategies. This level of the design is strongly linked to the process of *floor plan designing*, which is an attempt to combine systems design with placement and routing of major subsystems on the basis of initial estimates of module size and the locations of input/output points. The final floor plan is the floor plan of the implemented chip. We discuss the design of complete chips in Chapter 7, but will take this opportunity to make some observations regarding certain important features of system-level design. The heart of the design is usually located in the middle of the chip, which consists of a few basic cell types. This means that the chips have a high *regularization factor* (Lattin et al., 1981), which is roughly the ratio of the number of transistors in the chip to the number of transistors laid out by the designer. A high regularization factor also implies that the global signals can possibly be distributed to different parts of the chip using regular wiring strategies. A major bulk of the wiring consists of horizontal metal wires that carry the power, ground, and data lines. Since metal has a low resistance and capacitance, it is preferred for all long- to medium-distance communication within the chip. All power and ground lines are invariably carried in metal because of low heat dissipation and low voltage drops along the lines due to low resistance. Metal also provides good speed because of its low capacitance. The only disadvantage is that the metal wires take up more space than do other materials. Next to metal is polysilicon, which has low to moderate capacitance but rather high resistance. Polysilicon lines are generally used to run control lines in a direction perpendicular to the direction of metal wires. The power and ground lines are never run in polysilicon since it has a large resistance which causes considerable voltage drops. Diffusion wires are used mostly for "local computation." They are never used to carry signals over a long distance because of their relatively large capacitance. The power and ground routing also has to meet the special requirement that most active components of the chip must be fed by them, both of them must be in metal, and the width of the lines must be such that it does not carry excessive currents, causing metal migration, which will thin out portions of the wires, and ultimately create an open circuit. A common strategy is to use *interdigitation*. Note also that the width of the wiring for ground and power increases as it goes out toward the periphery of the chip, to carry a larger amount of current. The interdigitated pattern is also a feasible scheme for distributing control signals.

1.3 Technology Trends and Design Styles

The technology of semiconductor devices is advancing at a rapid rate. The history of its growth has been documented in great detail in several texts and articles. (See the References at the end of the chapter.) In this section we give a very brief and broad overview of the trends in MOS technology, which will be our primary concern in this text. For most high-speed and low-noise application, such as the large mainframe processor, *bipolar* technology is used. The heat dissipation in bipolar circuits is rather large and requires an elaborate cooling arrangement for its operation. The *gallium arsenide* (GaAs) technology shows great promise in space applications since the devices perform remarkably well in environments with radiation and varying temperature. The MOS technology is growing in leaps and bounds because of high-volume commercial applications. The minimum feature size of MOS technology is decreasing rapidly, resulting in high-density complex custom chips with proven reliability of the two MOS technologies (nMOS and CMOS processes are described in detail in Chapter 4). CMOS has the leading edge and is projected by industry analysts as being the dominant technology for the next decade. The greatest advantage of CMOS over nMOS is its static power consumption, which is an order of magnitude smaller than nMOS static power requirements. The nMOS technology has had some advantages in terms of size or area of silicon needed to produce equivalent functionality, but with decreasing feature size for CMOS, that advantage is rapidly evaporating. In 1985, the minimum feature size for CMOS stands at 1.5 microns (μm); the projected dimensions in the 1990s is 0.5 μm. The maximum number of gates per chip in 1985 is about 65,000 (approximately four transistors per gate); it will be 200,000 in 1988 and will reach more than 800,000 in the 1990s. The gate delay is about 2.5 nanoseconds (ns) in 1985; it will be 1.5 ns in 1988 and less than 1 ns in the 1990s. The assembly of the chip is also improving drastically;

80 pins per package is now commonplace, 120-to-250-pin packages will be available soon. Current projections are for 500 pins in 1988 and over 1000 pins in the 1990s. The masking technology (see Chapter 4 for more details) is currently using electron-beam and x-rays. In the 1990s, photolithography will use x-rays exclusively.

The design styles and design tools are also undergoing rapid evolution. The *gate array* method provides a standard cell library with predefined placement on the chip. The cell library contains a set of basic functions such as flip-flops and NOR or NAND gates. The interconnection of these cells is customized for specific applications. The *standard cell* approach provides only the cell library; the customer retains the flexibility of both placement and interconnect. The *standard custom* approach is an extension of the standard cell approach, with the additional feature that customers can add their own local cell libraries. Finally, the *custom* design style incorporates maximum flexibility; the designers have at their disposal the entire array of design tools (logic capture, layout, design verification, simulation, timing verification, placement, and routing; see Chapter 9 for a detailed discussion of these topics), which they use to design custom chips. The design tools sometimes referred to as computer-aided design (CAD) tools, are also undergoing rapid evolution. Good logic and timing simulators were not available until about 1982. It is now projected that custom workstations incorporating all the latest design tools will be available by about 1988. A restricted class of "silicon compilers" has been developed and is in the research and development stages. Similarly, ideas for "wafer-level-integration soft-configurable silicon systems" may be implemented in the next decade. In summary, we are in the midst of a technological revolution in microelectronic technology. There seems to be no limit to what can be done or what will be done.

1.4 Plan for the Remainder of the Book

In Chapter 2 the two basic logical elements of an MOS digital network—the switch and the inverter—are discussed. The operation of the MOSFET (metal-oxide-semiconductor field-effect transistor) or MOS transistor is presented and a simplified abstract model of the device as a switching element is developed.

We then discuss the nMOS and the CMOS inverter circuits. A digital network can be built by interconnecting inverters and other primitive elements by following a set of rules of interconnection. We then present rules that guarantee correct operation of the networks. The physical process of electron–hole conduction in semiconductors is presented in detail in the Appendix. In Chapter 2 we refer to the current–voltage characteristics of these devices qualitatively, leaving the quantitative treatment for the Appendix.

In Chapter 3 we discuss how an arbitrary digital network can be built using inverters and other elements, such as NOR and NAND gates. We present different classes of switch networks, NOR–NAND networks, and programmable logic arrays

(PLAs). We then discuss clocked nMOS and CMOS networks, one- and two-phase clocking disciplines, precharge logic, and the "domino logic" for CMOS. We then present a few elementary sequential networks and storage devices which form the building blocks for subsystem design.

Chapter 4 describes basic concepts and processes used in the production of semiconductor devices. This chapter provides the reader with only a glimpse of semiconductor production, leaving the engineering details to books specifically written to describe these processes. The following are described: metal-gate nMOS and CMOS processes, silicon-gate nMOS and bulk CMOS processes, oxide-isolation CMOS, and silicon-on-sapphire CMOS.

In Chapter 5 we discuss the design–fabrication interface that hides the lower-level physical design details from the system designer. This interface takes the form of a set of design rules and a set of procedures to compute the electrical parameters of a layout which can be used to validate the correctness of the design and to provide a basis for estimating the performance of the system by simulation tools. The design rules for nMOS and bulk CMOS are presented. The Magic design rules incorporated in the most recent Berkeley VLSI design tools are also discussed; these rules correspond to either n-well or p-well processes, with two levels of metal interconnect. We then discuss procedures to compute electrical parameters—capacitance and resistance—of a layout. The chapter ends with a discussion of sample layouts.

Chapter 6 is concerned with speed and power estimates of logic gates and circuits. The proposed estimates are based on simplified models and are intended to provide the designer with an understanding of the fundamental parameters that affect the performance of the circuit. The analysis will provide the designer with rough figures. The utility of such an analysis lies in providing insight into the physical parameters affecting performance, which can then be incorporated into models of simulation programs written to handle large and complex circuits.

Chapter 7 is concerned with functional subsystem and system design. The approach is to present several interesting circuits and systems as test cases to illustrate the fundamental design issues. We present several design examples of interconnection networks, nonnumeric processors, and arithmetic processors.

Chapter 8 is concerned with memory systems. We discuss typical nMOS and CMOS memory circuits and general principles to organize these circuits to build systems. We discuss trade-offs with respect to power, speed, and silicon areas. The aim of this chapter is to provide a broad overview of the principles and technology of memory systems.

The implementation of an integrated-circuit chip involves three major steps: design, mask making, and fabrication. For each step, a whole world of technology and tools has evolved. The architecture of the chip originates at the "design house." Here, a system-level specification undergoes several levels of translations via the function, logic, and gate/switch levels to produce a layout description of the chip. The tools applicable to carrying on this process are discussed in Chapter 9. A large number and variety of design tools are presently being used and developed for integrated-circuit chips. In this chapter we discuss some of the basic tools and algorithms used for the development of an integrated design automation system. These include circuit layout tools, layout verification tools such as design rule checking and circuit extraction, simulation tools for functional and timing verification, and routing tools for interconnection. Since the design tools are constantly evolving, the life cycle of a specific tool may not be more than three or four years. Therefore, the descriptions of some of the intersecting algorithms have been emphasized in the chapter.

REFERENCES

Clark, W. A., "From Electron Mobility to Logical Structure," *Comput. Surv.*, Vol. 12, No. 3, 1980.

Conway, L., A. Bell, and M. E. Newell, "MPC 79: A Large Scale Demonstration of a New Way to Create Systems in Silicon," *Lambda*, 2nd quarter 1980.

Lattin, W. W., J. A. Bayliss, D. L. Budde, J. R. Rattner, and W. S. Richardson, "A Methodology for VLSI Chip Design," *Lambda*, 2nd quarter 1981, pp. 34–44.

Mead, C., and L. Conway (Eds.), *Introduction to VLSI Systems*. Reading, Mass.: Addison-Wesley, 1980.

2

The Switch and the Inverter

2.1 Introduction

In this chapter we discuss the two basic logical elements of the MOS digital network: the switch and the inverter. We discuss briefly the operation of the MOSFET (metal-oxide-semiconductor field-effect transistor) or MOS transistor and present a simplified abstract model of the device as a switching element. We then discuss the nMOS and the CMOS inverter circuits. A digital network can be built by interconnecting these primitive elements by following a set of rules of interconnection. We present three rules that guarantee logically correct operation of the networks. These rules are the switch interconnection rule, the ratio rule for the nMOS inverter with restored input and degraded or unrestored input, and a rule which ensures that every node in the circuit is driven. These rules have been derived to hide from the logical designer the lower-level details regarding the underlying physical processes of conduction of electrons and holes in the semiconductors, which are presented in detail in the Appendix. We will also refer to the current–voltage characteristics of these devices qualitatively, leaving a quantitative treatment for the Appendix.

2.2 MOS Transistor

There are two basic types of MOS transistors: the n-channel and the p-channel. The structure of an *n-channel* MOS transistor is shown in Fig. 2.1. It consists of two islands of n-type diffusions embedded in a p-type substrate, which are connected via metal or polysilicon to external conductors called *source* and *drain*. On the surface,

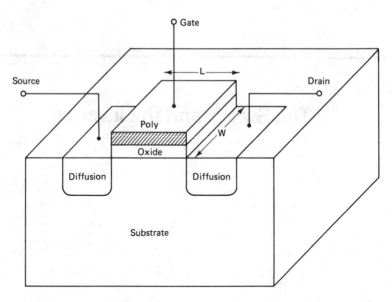

Figure 2.1 Structure of a MOS transistor.

a thin layer of silicon dioxide (SiO_2) is formed and on top of this a conducting material made of polysilicon called a *gate* is deposited. If the substrate material is *n*-type and the diffused islands are *p*-type, a similar structure will represent a *p-channel* MOS transistor. The region between the two diffused islands under the oxide layer is called the *channel* region. A layout view from the top of the transistor is shown in Fig. 2.2(a), where L and W denote the length and width of the channel, respectively. The minimum size of the MOS transistor is extremely process dependent and is subject to rapid changes with the progress of technology. Current technology is in the range of $W = 3$ μm and $L = 1.5$ μm effective. [One micron (or micrometer, μm) is one-thousandth of a millimeter.] Note there is a difference between the *drawn* dimension and the *effective* dimension; these could differ considerably depending on the process. Schematic circuit symbols are shown for *n*- and *p*-channel transistors in Fig. 2.2(b) and (c), respectively. In the most common mode of operation of the transistor, the source and substrate are grounded and the drain is connected to a supply voltage V_{dd} through a load resistor or a switching device, which is positive for an *n*-channel transistor and negative for a *p*-channel transistor.

The terminal characteristics of the device are given by a plot of drain-to-source current I_{ds} against drain-to-source voltage V_{ds} for different values of gate-to-source voltage V_{gs}. All voltages are referenced with respect to the source voltage, which is assumed to be at ground potential. The source and substrate are assumed to be connected together. The characteristic curves are shown in Fig. 2.3 for both *n*- and *p*-channel transistors. The curves are very approximate and depict typical trends under current technology, which is undergoing rapid changes. The curves reveal the following properties: The drain-to-source current flows only when the magnitude of

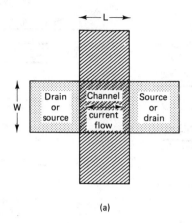

(a)

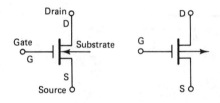

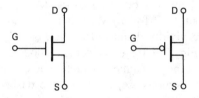

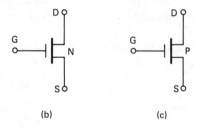

(b)   (c)

Figure 2.2   (a) Layout of a MOS transistor; (b) circuit symbols for an *n*-channel transistor; (c) circuit symbols of a *p*-channel transistor.

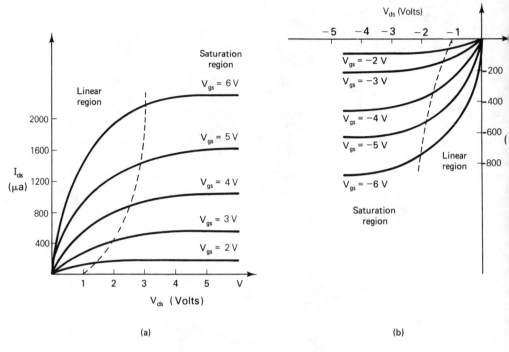

Figure 2.3 $I_{ds}-V_{ds}$ characteristics for (a) n-channel transistor and (b) p-channel transistor.

the gate-to-source voltage exceeds a minimum value called the *threshold voltage*, V_{th}; that is, $|V_{gs}| > |V_{th}|$ for enhancement mode transistors. (A typical value for V_{th} for an nMOS 5-V process is about 1 V; for a pMOS it is also about -1 V, and it is denoted as V_{thp}.) This is more vividly displayed in the $I_{ds}-V_{gs}$ characteristic curves for a given V_{ds} as shown in Fig. 2.4 for enhancement mode transistors. Note that both I_{ds} and V_{ds} are negative for the p-channel characteristic curve. The characteristic curve can be divided into two regions, separated by the dashed *saturation curve*, which corresponds to the value of I_{ds} for $|V_{ds}| = |V_{gs}| - |V_{th}|$. The region to the left (right for p-channel) of this curve is called the *linear region*, where the device behaves like a voltage-controlled resistor. The region to the right (left for p-channel) of the curve is called the *saturation region*, where I_{ds} remains practically constant with increasing magnitude of V_{ds}.

The equations giving the drain-to-source current I_{ds} for a nMOS transistor are derived in the Appendix and are given below.

$$I_{ds} = 0 \qquad\qquad\qquad\qquad V_{gs} - V_{th} < 0 \quad \text{(off)}$$

$$I_{ds} = \frac{\beta}{2}\left(V_{gs} - V_{th}\right)^2 \qquad\qquad 0 \le V_{gs} - V_{th} \le V_{ds} \quad \text{(saturated)}$$

$$I_{ds} = \beta\left(V_{gs} - V_{th} - \frac{V_{ds}}{2}\right)V_{ds} \qquad V_{gs} - V_{th} > V_{ds} \quad \text{(linear)}$$

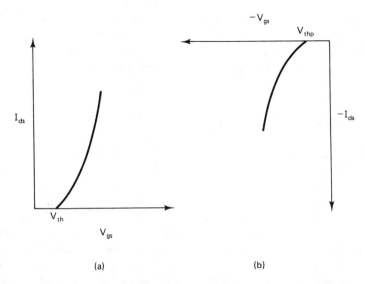

Figure 2.4 $I_{ds}-V_{gs}$ characteristics for (a) n-channel and (b) p-channel transistor.

where $\beta = (\mu\varepsilon/T_{ox})(W/L)$; μ is the average mobility of the charge carrier (electron for n-channel, hole for p-channel), ε is the permittivity of the oxide material, and T_{ox} is the thickness of the oxide. The mobility μ of the n-channel transistor is about twice that of the p-channel transistor and the threshold voltage V_{th} for the n-channel transistor is about half that of the p-channel transistor. This accounts for the fact that the amount of current flowing through the transistor is much greater for the n-channel transistor than for the p-channel transistor for given $|V_{ds}|$ and $|V_{gs}|$. Also note that the polarities of the currents and voltages of a n-channel transistor are just the opposite of those for a p-channel transistor.

The three terminals of the device are held at voltage levels such that the gate-induced charge is the dominant phenomenon for the device. This type of MOS device is called an *enhancement-mode* device since the formation of the channel has been enhanced by the presence of the gate voltage. Another type of n-channel MOS device, called a *depletion-mode* device, has a thin continuous n-type channel built under the gate by ion implementation. (A p-channel depletion-mode transistor, although conceptually possible, has never been used in practical circuits.) This is accomplished early in the fabrication process by using an ion acceleration gun to shoot phosphorus ions into the channel areas of the substrate, which lodge themselves near the surface up to some accurately controlled depth. The net effect of this is to change the threshold voltage V_{th} to a negative value. The characteristic curves are similar to that of the enhancement-type transistor except for the V_{th} value. A typical value for V_{th} under current technology is -3 to -4 V for nMOS. A source-to-drain current will flow when $V_{gs} = 0$. A negative gate voltage will deplete the electrons away from the surface and will stop the current. The magnitude of the negative voltage must be greater than a minimum voltage in order that the depletion phenomenon can occur. This negative voltage is called the *depletion threshold voltage* V_{dep}. Typically, $V_{dep} = -0.8V_{dd}$ for an *n*MOS. Thus with $V_{dd} = 5$ V, $V_{dep} = -4$ V. These values are again extremely process dependent and are subject to change.

2.3 MOS Transistor as a Switch

When the transistor is operating in the linear region, the device acts as a linear resistance under gate voltage control. In this mode the transistor can be used as an on–off switch, as symbolized in Fig. 2.5. The transistor must also operate in the cutoff region to act as a switch, but when the transistor is on, it operates in the linear region. The switch is turned off by setting $V_{gs} = 0$, the channel disappears, and only a small amount of leakage current flows at the drain end. The switch is turned on by setting $V_{gs} = V_{dd}$ and the current path provides a resistance R_{ch}. For a *p*-channel transistor the switch is turned on if $V_{gs} = -V_{dd}$. If the transistor is connected in series with a high-impedance circuit the total current flowing through the transistor will be very small and the voltage drop $V_{ds} = I_{ds}R_{ch}$ across the channel will also be very small. In the context of its use as a logical element, an input voltage source signal, V_{in}, which is either zero or V_{dd} representing logical 0 or 1, respectively, is applied as shown in Fig. 2.6. The output node V_{out} is either held at its previous logic value if the switch is open or its capacitance C is charged up to V_{in} if the gate voltage allows the transistor path to be closed, forcing a current for a short period of time. The transistor used in this way is called a *pass transistor* or a *transmission gate* and the process of transferring the charge from the input node to the output under the control of gate voltage is called *charge steering*.

An *n*MOS enhancement-mode transistor passes a low 0 signal well and passes a high 1 signal a little degraded, as shown in Fig. 2.7. Although in both of these situations, the transistor is operating in linear region, when passing a 1 signal, the maximum value of the voltage at the output is $V_{dd} - V_{th}$, since beyond this value the

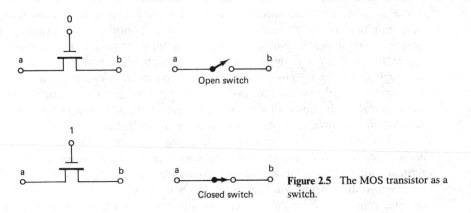

Open switch

Closed switch

Figure 2.5 The MOS transistor as a switch.

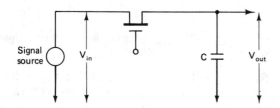

Figure 2.6 Pass transistor.

gate-to-source voltage will drop below threshold and the transistor will be turned off. Note that the roles of the source and drain are interchangeable and the device acts truly as a bilateral switch. The signal at the output of the pass transistor is thus a "weak" 1 signal, but if the input voltage is 0 or any value less than or equal to $V_{dd} - V_{th}$, it will be passed almost unchanged except for a small drop due to the resistance of the channel. A complementary situation exists for the p-channel pass transistor, which passes a signal toward the positive side as shown in Fig. 2.7(b). When passing a 0 signal the minimum value of the output voltage is $|V_{thp}|$, since below this value the gate-to-source voltage will be higher than $-V_{thp}$ and the transistor will be turned off. Thus a p-channel transistor passes a weak 0 signal but a strong 1 signal. These facts lead to one basic rule of interconnection of MOS switches: *The output of a* p- *or* n-*channel switch should not be connected to control the input of another switch of the same type.* This is because the output of a switch could produce a weak signal which might be inadequate to turn on the transistor whose gate is connected to it. This rule is not applicable to transmission gate structures for CMOS, which produce an undegraded signal, as we will see later in Section 2.7.

Another special kind of switch is the "always closed" depletion-mode nMOS transistor, shown in Fig. 2.8. This has sometimes been called a "yellow transistor" since yellow color is used to denote the implantation region. The general use of this device is to provide a conducting path for any signal whenever the gate is 0 or positive. It passes a low signal very well and a high signal slightly weakened since there is a voltage drop across the transistor. Some examples of such circuits are

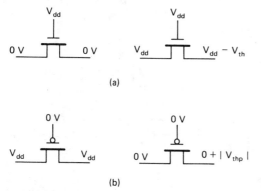

Figure 2.7 Signal degradation in a pass transistor: (a) *n*-channel; (b) *p*-channel.

Figure 2.8 A "yellow" transistor.

presented in later chapters (see Section 3.3). However, use of this type of transistor in circuit design has very limited potential for future technology with two levels of interconnect. Even for nMOS technology, use of this kind of switch does not result in significant area savings.

2.4 Basic *n*MOS Inverter

The simplest form of an nMOS logic circuit is an inverter, as shown in Fig. 2.9. It consists of a "load" resistance R called the *pull-up* resistor, and a *pull-down* transistor T connected in series between supply voltage V_{dd} and ground Gnd. The resistor is sized to limit the pull-up current to some fraction of maximum pull-down current provided by the transistor. A typical value for R in a 5-V process might be 40 kilohms (kΩ), which limits the pull-up current to about 125 μA. The input voltage V_{in} is applied to the gate of T, which provides a very high input impedance via the gate capacitance C_g. In a typical configuration, the output will drive a similar load capacitance C.

If V_{in} is less than the threshold voltage V_{th} of T, it will be turned off and the load capacitance C will be charged to V_{dd}, which equals V_{out}. V_{out} actually never equals V_{dd}, due to a small leakage current flowing through T. If V_{in} is raised beyond V_{th}, T will initially conduct in the saturation region; the capacitance C will start discharging, pulling down V_{out} to a lower voltage; and the current I_{ds} will increase due to the larger value of V_{gs}, which equals V_{in}, until the transistor moves into the linear region of the $I_{ds} - V_{ds}$ curve, as shown in Fig. 2.10 via the points a through e in the dashed curve. When $V_{in} = V_{dd}$, a large current will flow in T and the output voltage will be given by V_{lo}:

$$V_{lo} = V_{dd} \frac{R_{ch}}{R_{ch} + R} \tag{2.1}$$

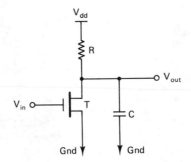

Figure 2.9 The nMOS inverter.

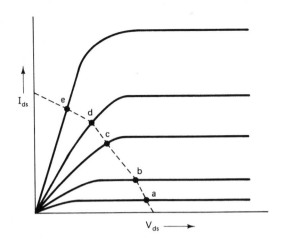

Figure 2.10 Current–voltage characteristics of the inverter.

where R_{ch} is the channel resistance with $V_{gs} = V_{dd}$. Since R_{ch} is very small compared to R, V_{out} will be very near ground potential. In a typical application, the output of the inverter is used to drive another inverter and V_{lo} must be less than V_{th} to make sure that the pull-down transistor of the driven inverter is turned off. Since $V_{th} = 0.2V_{dd}$, it is necessary that

$$R \geq 4R_{ch} \tag{2.2}$$

For satisfactory operation, the low level should be significantly less than V_{th} to provide adequate noise margin. However, we will assume the value of R given by Eqn. (2.2) as an approximation. This implies that the output capacitance C discharges through R_{ch} at least four times faster during the pull-down phase than during its pull-up phase through R. This basic asymmetry of switching times is a fundamental limitation of "*ratioed*" logic which this inverter represents. To have a higher speed of operation, R must be made small. This not only requires a smaller value for R_{ch} (requiring a larger silicon area for the pull-down transistor) to satisfy the ratio criteria, but also entails more power dissipation in the load resistor [$(V_{dd} - V_{lo})^2/R$ watts] due to increased current. An effect equivalent to lowering R can be achieved by lowering the value of V_{th}, which results in a larger source-to-drain current. This means higher speed, but to produce a lower value of the output voltage V_{lo}, R_{ch} should be correspondingly lower. For example, if $V_{th} = 0.1V_{dd}$, the ratio equation becomes $R \geq 9R_{ch}$. Lower channel resistance implies a large silicon area for the pull-down.

Building a large resistive load on the silicon surface takes up quite a bit of area. A better solution is to use a depletion-mode transistor as a pull-up load, as shown in Fig. 2.11, with source and gate connected together so that $V_{gs} = 0$. To determine the transfer characteristics—V_{out} as a function of V_{in}—we do the following analysis.

Since the pull-down transistor is an enhancement-mode n-channel transistor, the source-to-drain current I_{pd} is the same as I_{ds} of a MOS transistor. Also, since the input voltage V_{in} is the same as V_{gs} and the output voltage V_{out} of the inverter is

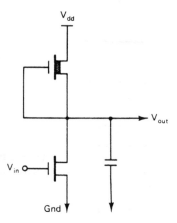

Figure 2.11 The nMOS inverter with a pull-up load.

the same as V_{ds}, we can write the pull-down current equations as

$$I_{pd} = 0 \qquad\qquad\qquad\qquad V_{in} - V_{th} < 0 \quad \text{(off)}$$

$$I_{pd} = \frac{\beta_d}{2}(V_{in} - V_{th})^2 \qquad\qquad 0 \le V_{in} - V_{th} \le V_{out} \quad \text{(saturated)}$$

$$I_{pd} = \beta_d\left(V_{in} - V_{th} - \frac{V_{out}}{2}\right)V_{out} \qquad V_{in} - V_{th} > V_{out} \quad \text{(linear)}$$

where $\beta_d = (\mu\varepsilon/T_{ox})(W/L)$ for the pull-down transistor.

The pull-up depletion-mode transistor has a negative threshold voltage V_{dep} about $-0.6V_{dd}$ to $-0.8V_{dd}$ and its V_{gs} is always 0, so that $V_{gs} - V_{dep} > 0$, and the pull-up always conducts. Its source-to-drain current is given by

$$I_{pu} = \frac{\beta_u}{2}|V_{dep}|^2 \qquad\qquad\qquad |V_{dep}| \le V_{dd} - V_{out} \quad \text{(saturated)}$$

$$I_{pu} = \beta_u\left(|V_{dep}| - \frac{V_{dd} - V_{out}}{2}\right)(V_{dd} - V_{out}) \qquad |V_{dep}| > V_{dd} - V_{out} \quad \text{(linear)}$$

where $\beta_u = (\mu\varepsilon/T_{ox})(W/L)$ for the pull-up transistor.

The plots of I_{pu} and I_{pd} as a function of the output voltage V_{out} are shown in Fig. 2.12(a) and (b), respectively. Note that I_{pu} is independent of V_{in}, whereas I_{pd} increases with increasing V_{in} as depicted by a family of curves. For a particular input voltage V_{in}, the output voltage V_{out} is obtained by intersecting the I_{pd} and I_{pu} plots, since the same current must flow in both the pull-up and pull-down transistors. This

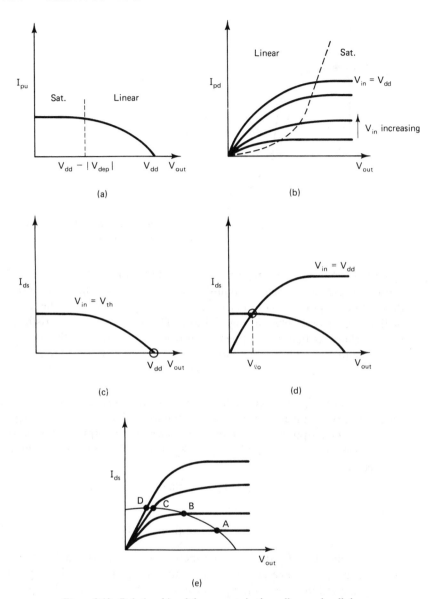

Figure 2.12 Relationship of the currents in the pull-up and pull-down.

is shown in Fig. 2.12(c) and (d) for the two important cases: first, $V_{in} < V_{th}$ and $V_{out} = V_{dd}$, and second, $V_{in} = V_{dd}$ and V_{out} corresponding to the lowest output voltage V_{lo}. To obtain the voltage characteristics the I_{pu} curve is superimposed on the family of I_{pd} curves [Fig. 2.12(e)] and one notes that the operating point moves through points A, B, C, D as V_{in} is increased from 0 to V_{dd}, which yields the voltage transfer curve for an inverter, shown in Fig. 2.13. The transfer curve can be divided

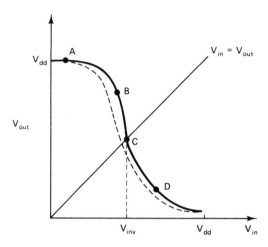

Figure 2.13 Input–output characteristics of the inverter.

approximately into four regions: A (pull-down off, pull-up linear), B (pull-down saturated, pull-up linear), C (pull-down saturated, pull-up saturated), and D (pull-down linear, pull-up saturated), corresponding roughly to the operating points A, B, C, D, respectively, in Fig. 2.12(e). The transfer characteristic has the same general shape as that with a linear load resistor R, but since the depletion transistor makes more current available during the transition from low to high, and particularly when the pull-down is turned off, the circuit provides faster charge and discharge times for the load capacitance.

In many digital circuits, the output of an inverter must be connected to the input of another inverter. Let us define the *logic threshold or switching point* voltage V_{inv} to be the output voltage of an inverter when the output and input of the inverter are connected together; that is, $V_{out} = V_{in} = V_{inv}$. From the static characteristics, we see that if

$$V_{in} \leq V_{inv} \qquad \text{then} \qquad V_{out} \geq V_{inv}$$

and if

$$V_{in} > V_{inv} \qquad \text{then} \qquad V_{out} < V_{inv}$$

which means that the device will function as an inverter as long as a 0 signal is recognized as a voltage less than V_{inv} and a 1 signal as a voltage greater than V_{inv}. Ideally, a good inverter should produce 0 V and V_{dd} as 0 and 1 signals, respectively; that is, $V_{inv} = 0$. When $V_{in} = 0$, the 1 signal is properly generated since $V_{out} = V_{dd}$. When $V_{in} = V_{dd}$, the output V_{out} does not go all the way to 0 V, but it must be sufficiently low, lower than V_{th}, in order to turn the next stage inverter solidly off. The pull-down is at linear region and the pull-up is in saturation at this low output voltage. Equating the pull-up and pull-down currents, we have

$$\beta_d \left(V_{in} - V_{th} - \frac{V_{lo}}{2} \right) V_{lo} = \frac{\beta_u}{2} |V_{dep}|^2$$

Assuming that $V_{in} = V_{dd}$ and ignoring $V_{lo}/2$ in the presence of $V_{in} - V_{th}$, we have

$$V_{lo} = \left(\frac{1\beta_u}{2\beta_d}\right) \frac{(V_{dep})^2}{V_{dd} - V_{th}}$$

$$\approx \left(\frac{1}{2k}\right) \frac{(V_{dep})^2}{V_{dd} - V_{th}}$$

where $k = \beta_d/\beta_u \approx (L/W)_{pu}/(L/W)_{pd}$, assuming that μ, ϵ, and T_{ox} are the same for the pull-up and the pull-down. The quantity k is called the *ratio of the inverter*. The implication of this equation is that the dimensions of the transistors can be so chosen as to ensure that V_{lo} is less than V_{th} so that any transistor connected to this output is turned off. This is equivalent to saying that a logic 0 is actually produced at the output when the input is logic 1. If the value of k is increased, it will have several effects. First, it will lower V_{lo}, making the gap between V_{lo} and V_{th} wider. A smaller fraction of V_{dd} is made available as V_{out} since R_{pd} is relatively smaller than R_{pu}. This implies a steeper voltage transfer curve in Fig. 2.13 (dotted curve). This also implies a smaller value for the logic threshold voltage V_{inv}. This can be seen more analytically by setting $V_{in} = V_{inv}$ and equating the saturation currents of the pull-up and pull-down transistors (assuming that the operating region is close to C in the voltage transfer curve), which yields

$$\left(\frac{W}{L}\right)_{pd} (V_{inv} - V_{th})^2 = \left(\frac{W}{L}\right)_{pu} (-V_{dep})^2$$

or

$$V_{inv} = V_{th} - \frac{V_{dep}}{\sqrt{k}}$$

We therefore have three different parameters, V_{th}, V_{dep}, and k, to control the value of V_{inv}. We could reduce V_{th} to make V_{inv} smaller, but this requires a larger k in order to bring V_{out} smaller than V_{th}. This increases the asymmetry of the inverter with respect to charge and discharge times, decreases the total current, and necessitates more area for the transistors. Making V_{dep} more negative will increase current driving but will necessitate increasing k and hence require more area.

The resistance of a rectangular semiconductor region of length L and width W is given by

$$\text{resistance} = \frac{L}{W} \times \text{sheet resistivity}$$

where sheet resistivity is defined to be the resistance of a unit square of the material. When V_{in} is low, the pull-down is off and the equivalent electrical circuit is a single resistance R_{pu}, the resistance of the pull-up transistor connected to V_{dd}, as shown in Fig. 2.14(a). When V_{in} is high, the inverter can be represented as a voltage divider, shown in Fig. 2.14(b), with R_{pu} connected in series with R_{pd}, the pull-down

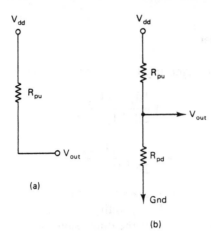

(a)

(b)

Figure 2.14 Equivalent electrical circuits for the inverter.

resistance between V_{dd} and the ground, and V_{out} being read as the output across R_{pd}, given by

$$V_{out} = V_{dd}\frac{R_{pd}}{R_{pd} + R_{pu}}$$

But $V_{out} \leq V_{th}$. Taking $V_{th} = 0.2V_{dd}$, we have

$$R_{pu} \geq 4R_{pd}$$

Since R_{pu} is directly proportional to $(L/W)_{pu}$ and R_{pd} to $(L/W)_{pd}$, we have

$$\left(\frac{L}{W}\right)_{pu} \geq 4\left(\frac{L}{W}\right)_{pd}$$

which means that the ratio $k \geq 4$. The value of R_{pu} in Fig. 2.14(a) and (b) actually depends on V_{ds} because of the non-linear values of the current voltage characteristics. The requirement $k \geq 4$ is still correct because of two compensating errors. One is treating the resistance values as linear and the second is assuming that a low level of 0.2 V_{dd} would be acceptable.

Another idea for a load is to use a saturated enhancement transistor by connecting the gate to V_{dd} so that $V_{gs} = V_{ds}$, as shown in Fig. 2.15. The transistor offers a high resistance by being in the saturated mode. When the pull-down is turned off, V_{out} rises and C gets charged by I_{ds}. When V_{out} reaches $V_{dd} - V_{th}$, the channel disappears since there will be no channel support voltage left. Thus the output never reaches V_{dd}, which is a limitation on this kind of load. To maintain the same logic level, we need higher V_{dd}, resulting in more power dissipation. To get around this difficulty, the load transistor can be separately biased to the linear region by a bias voltage V_{gg} and making sure that $V_{gs} - V_{th} > V_{dd}$, which will set the high to V_{dd}. The low value of the output is set by the resistance ratio.

The actual characteristics of an inverter depend on a large number of processing, geometric, and circuit parameters. For typical nMOS operations with $V_{th} = 0.2V_{dd}$, the ratio $k = 4$ yields satisfactory results. Some typical characteristics of an inverter obtained by SPICE simulation are presented in Chapter 6.

24

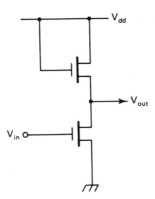

Figure 2.15 Inverter with an
enhancement transistor as the load.

2.5 Steered Input to an nMOS Inverter

Consider the circuit of Fig. 2.16(a). The input signal V_s is steered through a pass transistor to produce an output V'_{in} which is applied to the gate of the inverter. The signal V_s could be derived from an arbitrary gate or switch network. We know that V'_{in} is a degraded signal whose maximum value is $V_{dd} - V_{th}$. Worse even, the source and substrate of the pass transistor steering V_s are not connected together to ground potential. This typically increases the value of V_{th} from $0.2V_{dd}$ to $0.3V_{dd}$, due to a phenomenon called *body effect*. Thus $V_{in} = V_{dd} - V_{thb}$, where V_{thb} is the threshold voltage of the pass transistor, including body effect.

   In order that the output V'_{out} be the same as the output V_{out} of an inverter with ratio k, as shown in Fig. 2.16(b), the circuit of Fig. 2.16(a) must have a different ratio, k'. This can be derived as follows. When the input signal is high, in both configurations the pull-downs operate in the linear region and the pull-ups operate in the saturated region of their characteristic curves. The output voltage equals the

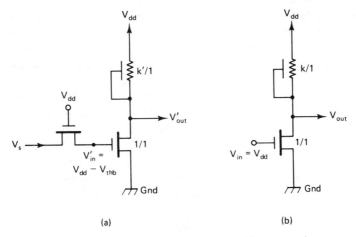

(a)   (b)

Figure 2.16 (a) Steered input to an inverter; (b) restored input to an inverter.

saturated pull-up currents times the linear pull-down channel resistance [Eq. (A.9) in the Appendix]. Equating these two voltages and after simplification, one obtains

$$k' = k \frac{V_{dd} - V_{th}}{V_{dd} - V_{th} - V_{thb}} \tag{2.3}$$

where $k' = (L/W)'_{pu}/(L/W)'_{pd}$ and $k = (L/W)_{pu}/(L/W)_{pd}$. Taking typical values of $V_{th} = 0.2V_{dd}$ and $V_{thb} = 0.3V_{dd}$, we have $k' = 8/5k$. More conservatively,

$$k' = 2k \tag{2.4}$$

This result can be interpreted as follows. The effect of the degraded signal at the input to a pull-down is to double the pull-down resistance, which necessitates doubling the pull-up resistance in order to bring the output of the inverter to the same level as that of a circuit with ratio k. We know that k is 4 with fully restored signal. Therefore, k must be at least 8 with steered input signal. A side effect of this is to reduce the total current in the inverter, which will increase the charging and discharging time of the load capacitance, thereby decreasing the speed and power dissipation of the circuit.

2.6 CMOS Inverter

In CMOS, both p- and n-channel transistors are used. The silicon substrate is an n-type, in which a p-type "well" or "tub" is created by diffusion. The n-channel transistors are created in the p-well region. The p-channel transistors are made in the n-substrate (see Chapters 4 and 5 later). The basic structure of the p-well CMOS inverter is shown in Fig. 2.17. Note the special "p-plugs" used to connect the p-well substrate and the source of the n-channel transistor to ground. Similarly, an "n-plug" connects the n-substrate and the source of the p-type transistor to V_{dd}, a positive supply voltage. Other structures, such as guard rings to prevent formation of parasitic transistors and contact cuts, are not described in this preliminary sketch.

Two schematic circuit representations of the inverter are shown in Fig. 2.18. We will use the circuit symbol shown in Fig. 2.18(a), but the one shown in Fig. 2.18(b) is also useful since it explicitly shows the substrate connections. All voltages are referenced with respect to V_{ss}, the ground potential.

The operation of the circuit as an inverter can be described as follows. When the input voltage $V_{in} = 0$, the gate of the p-channel transistor is at V_{dd} below the source potential, that is, $V_{gs} = -V_{dd}$, which will turn on this transistor, offering a low-resistance path to load capacitance C, which will be charged up to V_{dd}. No current flows through the n-channel transistor, which is turned off since $V_{gs} = 0$ for this transistor. If the input voltage is now increased to its threshold voltage and then

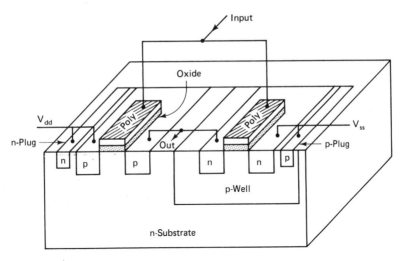

Figure 2.17 Structure of a CMOS inverter.

to V_{dd}, the n-channel transistor will conduct while the p-channel transistor is turned off, discharging the load capacitance C to ground potential. Note that the current flows until the output node reaches V_{dd} (when charging) or ground (when discharging), the transistors to provide either the charge or the discharge current.

A detailed explanation of the operation of the inverter will now be given. Since both the transistors are enhancement-mode transistors, we will use the symbols V_{gs}, V_{th}, and V_{ds} as before with a letter "p" or "n" attached to the end of the subscript to indicate whether the quantity refers to the p-channel or n-channel device. For example, V_{thn} and V_{thp} denote the threshold voltage of the n-channel device and p-channel device, respectively. Referring all the voltages to $V_{ss} = 0$ (the substrate

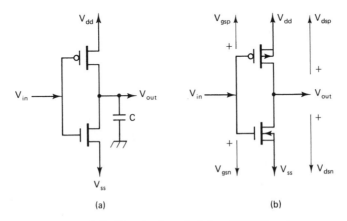

Figure 2.18 Two circuit symbols for a CMOS inverter.

voltage of the *n*-device), we can write

$$V_{gsn} = V_{in}$$

$$V_{dsn} = V_{out}$$

$$V_{gsp} = V_{in} - V_{dd}$$

$$V_{dsp} = V_{out} - V_{dd}$$

The drain-to-source currents for the *n*-channel and *p*-channel transistors are plotted in Fig. 2.19 on a common *x*-axis with the sign of I_{dsp} reversed (dashed curves). The voltage transfer characteristic of the inverter is depicted in Fig. 2.20, which can be divided into five operating regions.

   Region I: $0 \leq V_{in} \leq V_{thn}$.   The *n*-transistor is off and the *p*-transistor is set to operate in the linear region, but there is no actual current flow until V_{in} crosses the threshold V_{thn}. V_{gsp} varies in the range $-V_{dd}$ to a voltage a little over $-V_{dd} + V_{thn}$ and the operating point of the *p*-transistor moves from higher to lower values of currents in the linear region. A typical operating point is A shown in Fig. 2.19, marked as region I in Fig. 2.20. The output voltage V_{out} remains close to V_{dd} as V_{in} increases from 0 to V_{thn}.

   Region II: $V_{thn} \leq V_{in} < V_{inv}$.   The upper limit of V_{in} is given by the logic threshold voltage V_{inv} of the inverter for this region. Recall that V_{inv} is the output voltage at which $V_{in} = V_{out}$. A typical operating point B is shown in Fig. 2.19. The

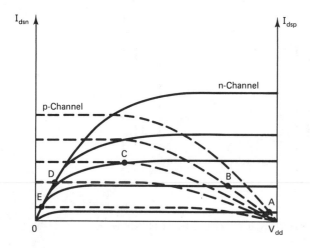

Figure 2.19 Drain-to-source currents with the *n*- and *p*-channel transistors superimposed.

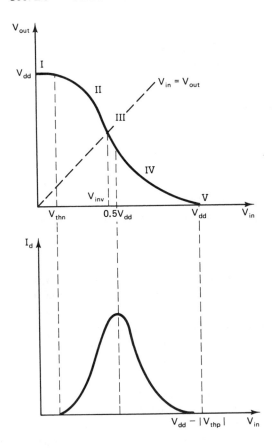

Figure 2.20 Input–output voltage characteristics and the transient current characteristics of a CMOS inverter.

n-transistor moves into the saturation region, whereas the p-transistor remains in the linear region of their I_{ds} curves. The total current flowing through the transistors increases and the output voltage tends to drop fast.

Region III: $V_{in} \approx V_{inv}$.   Both the transistors are in saturation regions of their characteristic curves, the drain current attains a maximum value, and the output voltage falls rapidly. A typical operating point is C shown in Fig. 2.19. By substituting $V_{gsn} = V_{in} = V_{inv}$, and $V_{gsp} = V_{in} - V_{dd} = V_{inv} - V_{dd}$ and equating the saturation currents for both the devices, we get

$$\tfrac{1}{2}\beta_n(V_{inv} - V_{thn})^2 = \tfrac{1}{2}\beta_p(V_{inv} - V_{dd} - V_{thp})^2$$

where β_n and β_p denote the quantity β defined earlier for the n- and p-transistors, respectively. Solving for the logic threshold voltage V_{inv}, we get

$$V_{inv} = \frac{\sqrt{\beta_p}\left(V_{dd} + V_{thp}\right) + \sqrt{\beta_n}\,V_{thn}}{\sqrt{\beta_p} + \sqrt{\beta_n}}$$

$$= \frac{\sqrt{k}\left(V_{dd} - |V_{thp}|\right) + V_{thn}}{1 + \sqrt{k'}}$$

where $k' = \beta_p/\beta_n$.

Region IV: $V_{inv} < V_{in} \leq V_{dd} - |V_{thp}|$.   As the input voltage is increased beyond V_{inv}, the n-transistor leaves the saturation region to enter the linear region, whereas the p-transistor continues to stay in the saturation state. The magnitudes of both the drain current and the output voltage continue to drop. The operating point is now at D on the characteristic curves.

Region V: $V_{dd} - |V_{thp}| \leq V_{in} \leq V_{dd}$.   At this point, the p-channel transistor is turned off and the n-channel transistor is turned hard on in the linear region, drawing a small current (point E in Fig. 2.19), which reduces to zero as V_{in} increases beyond $V_{dd} - |V_{thp}|$ since the p-transistor turns off the current path. The load capacitance C is now completely discharged and the output is connected to the ground potential.

The transient current characteristic is shown together with the voltage transfer curve in Fig. 2.20. The transient current characteristic contrasts with that of an nMOS inverter, which draws a current if the input is held at logic 1, whereas the CMOS inverter draws no current at any stable state of its output (except for a small leakage current). This means that the CMOS circuits have an advantage over the nMOS circuits in terms of power consumption. Since the current flows only during transition, the CMOS circuits have dynamic power consumption, which depends on the frequency of input changes.

Another advantage of the CMOS inverter is that it is *ratioless*, which means that the output does not depend on the relative dimensions of the p- and n-channel transistors, which could be of minimum size. The input to the inverter acts like a control signal steering logic 1 through the p-channel and logic 0 through the n-channel transistors, both signals being transmitted undegraded. This implies that the circuit produces a larger output voltage swing, resulting in higher noise immunity than in nMOS circuits. To make the rise and fall transition time symmetrical, many commercial CMOS circuits have provided a fixed ratio of 2:3 between the width of the p- and n-channel transistors to partially account for the disparity of the electron and hole mobilities which differ by a factor of about 2.5. This ratio is, however, extremely process sensitive. It depends on doping of p versus n, effective channel length differences, effective threshold voltage differences, and so on. In most industrial designs, the channel length is usually kept pretty much fixed and the channel width is varied.

2.7 Steering Input to a CMOS Inverter: The Transmission Gate

We have noted in Section 2.5 that the signal steered to the input of an nMOS inverter is a degraded signal. This was due to the fact that the n-channel switch produces a weak 1 signal, but passes a 0 signal undegraded. We also noted that the complementary situation holds true for a p-channel transistor. These principles can be combined to form a "perfect" steering element from considerations of transmitting high and low values for CMOS circuits, called a *transmission gate*, shown in Fig. 2.21(a). It consists of a parallel connection of a p-channel and an n-channel transistor with their source terminals connected to the input, their drain terminals connected to the output, and their gates connected to a pair of complementary control signals C and $\overline{C}$. When $C = 1$, both transistors are conducting. If the input signal $x = 0$, it will pass undegraded via the n-channel path; if $x = 1$, it will pass undegraded via the p-channel path. When $C = 0$, both transistors will be turned off. It thus behaves like a normally open perfect switch which is closed when the control signal $C = 1$. A symbol for this circuit is shown in Fig. 2.21(b). If the connections to

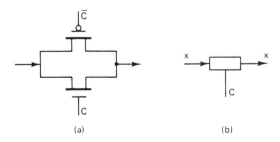

(a)   (b)

Figure 2.21 Transmission gate (normally open).

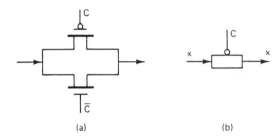

(a)   (b)

Figure 2.22 Transmission gate (normally closed).

the gate terminals are reversed, we will get a normally closed switch as shown in Fig. 2.22(a) and symbolized in Fig. 2.22(b).

2.8 Summary

In this chapter we have discussed the two basic logical elements of nMOS and CMOS digital circuits and explained the rules of interconnecting them to build arbitrary circuits. A detailed discussion of the physical processes, current–voltage equations, and electrical properties is presented in the Appendix. We have discussed some of the advantages of CMOS circuits over nMOS circuits: lower power dissipation; higher noise immunity; ratioless logic, resulting in higher speed; and the ability to construct a transmission gate—a nearly perfect switch. In the following chapter we discuss the logic design methods. The technology that fabricates millions of such circuits in one single chip is discussed in Chapter 4.

REFERENCES

Most of the references cited at the end of the Appendix are also relevant for this chapter. The first three are general references and the last one, by Terman, presents an excellent treatment of inverter operation in the context of developing a simulation model.

Mead, C., and L. Conway (Eds.), *Introduction to VLSI Systems*. Reading, Mass.: Addison-Wesley, 1980.

Meindl, J. D., "Microelectronic Circuit Elements," *Sci. Am.*, Vol. 237, No. 3, 1977, p. 70.

Glasser, A. B., and G. E. Subak-Sharpe, *Integrated Circuit Engineering*. Reading, Mass.: Addison-Wesley, 1979.

Terman, C. J., "Simulation Tools for Digital LSI Design." Ph.D. dissertation, Massachusetts Institute of Technology, Cambridge, Mass. (MIT/LCS/TR-304), Sept. 1983.

EXERCISES

2.1. The inverter circuit is the basic logic circuit symbolizing the structure of a generalized logic gate which consists of a pull-up network and a pull-down network. A NOR and a NAND gate are shown in Fig. E2.1. Note that the pull-up transistor has a different symbol, signifying load resistance. Assume that β_1 and β_2 denote the quantity β for the two transistors in the pull-down network. Write equations describing the linear and saturated currents flowing in the pull-down circuit when both the transistors are connected to the same input voltage V_{in}. From these expressions determine the equivalent β, β_{nor}, and β_{nand} for the NOR and the NAND gates, respectively, for the single pull-down transistor which can replace the pull-down circuit.

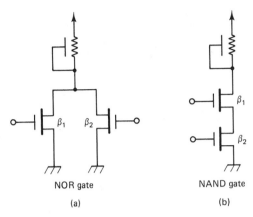

NOR gate   NAND gate

(a)   (b)

2.2. The NOR and NAND gates for CMOS are shown in Figs. 3.38 and 3.39, respectively, in Chapter 3. Derive from the basic principles the voltage transfer characteristics and the current–voltage characteristics assuming that an input voltage V_{in} is applied to all the gate inputs at the same time.

2.3. Derive a curve showing the total current flowing into an nMOS inverter as the input voltage V_{in} is varied from 0 V to V_{dd}. Will the CMOS inverter circuit work if V_{dd} is less than the sum of V_{thn} and $|V_{thp}|$? Explain.

3

Logic Design

3.1 Introduction

In Chapter 2 we discussed several basic elements of logic design: the switch, the transmission gate, and the inverter. In this chapter we discuss how an arbitrary digital network can be built using these and other elements, such as NOR and NAND gates. We present different classes of switch networks reminiscent of relay-contact networks (Caldwell, 1958), NOR–NAND networks, and programmable logic arrays (PLAs). We then discuss clocked nMOS and CMOS networks, one- and two-phase clocking disciplines, precharge logic, and the "domino logic" for CMOS. We then present a few elementary sequential networks and storage devices that form the basic building blocks for subsystem design discussed later in the book.

As we have explained in Chapter 1, the logic design level is an abstraction of the underlying world. The interconnection rules and the clocking discipline hide from the logical designer the details of the electrical world: the circuit parameters, the timing information, and the current–voltage waveforms. To understand this, we first need to describe a notational scheme appropriate for logic-level abstractions.

3.2 Sticks and Mixed Notation

To describe logic networks, it is convenient to use a "stick" notation, which will help us visualize the function as well as the topology of the network. This is particularly true for nMOS circuits. A stick representation hides lower-level circuit details and electrical parameters such as current, speed, and noise, but comes one step closer to actual layout. It views the *n*MOS network as an interconnection of wires in three layers denoted by the colors red, green, and blue, representing the layers polysilicon, diffusion, and metal, respectively, as shown in Fig. 3.1(a). In black-and-white drawings, these are represented by dashed lines, solid lines, and double-stranded lines, respectively. The functions determined by the topology in *n*MOS circuits are as follows:

1. When two wires in the same color intersect or touch, they are electrically connected [Fig. 3.1(b)].
2. Contact between polysilicon and metal or diffusion and metal is represented by crosses called contact cuts. Contact between polysilicon and diffusion is made by "butting" contact or "buried" contact, but we will indicate it by a cross, as shown in Fig. 3.1(c).
3. Whenever a polysilicon wire (red) crosses a diffusion wire (green), a switch is formed in the form of an *n*MOS transistor, as shown in Fig. 3.1(d). The polysilicon wire forms the gate of the transistor and the two sides of the diffusion wire form the source and the drain of the transistor. If the area intersecting the polysilicon and diffusion is surrounded by a yellow region or a dashed square, the transistor is recognized as a depletion-mode transistor or a yellow transistor, as shown in Fig. 3.1(e).

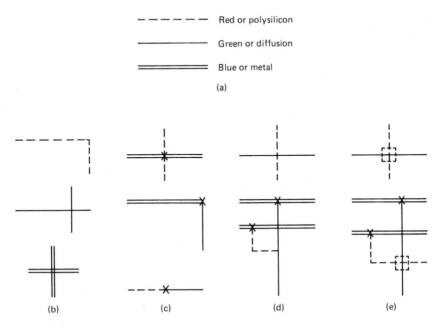

(a)

(b)   (c)   (d)   (e)

Figure 3.1 Stick symbols.

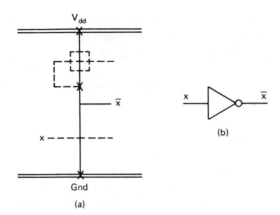

(a)

(b)

Figure 3.2 The nMOS inverter notation.

4. The stick notation for the nMOS inverter is shown in Fig. 3.2(a). Note that the pull-up notation seems a bit peculiar compared with the layout shown in Fig. 5.14. The symbolic notation shown in Fig. 3.2(b) will be used more frequently. The mixed notation illustrated in Fig. 3.3 is also in common use.

Note that the stick notation uses only vertical and horizontal wires to emulate what is known as Manhattan geometry for layout. A curved polysilicon wire intersecting a curved diffusion wire will also form a perfectly acceptable transistor in what is called the Boston geometry for layout.

Developing a stick notation for the CMOS network is complicated by the fact that the relationship between the topology and the layout is far from direct: for example, the fact that the n-channel transistors must be built in a p-well and the p-channel transistors under the n region can be captured by using two different colors (green for n-diffusion, brown for p-diffusion); the contact cut has several variations; the n-diffusion cannot cross the p-well to connect to electrically equivalent nodes and needs contacts and metal connection between two types of diffusions. Furthermore, there are several variations of the CMOS process and the sticks will have completely different interpretations at the layout level, depending on the particular process. We will occasionally use a simplified stick notation for CMOS only as a tool to sketch quickly the functional relations in the circuit. The notations 1, 2, and 3 explained above will be adhered to except that all contacts will be represented

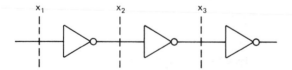

Figure 3.3 Mixed notation for a cascade of three inverters whose inputs (in diffusion) are controlled by signals x_1, x_2, and x_3 (in polysilicon).

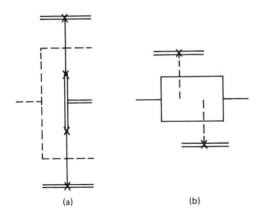

Figure 3.4 (a) Stick notation for a CMOS inverter; (b) stick notation for a transmission gate.

(a)   (b)

by a cross and there will be no depletion-mode transistor. The stick notation to represent the CMOS inverter and the transmission gates is shown in Fig. 3.4. A symbol for the CMOS inverter will be the same as that shown in Fig. 3.2(b) for the purpose of mixed notation.

3.3 nMOS Combinational Networks with Pass Transistors and Inverters

In this section we present several classes of nMOS combinational networks composed of inverters and pass transistors. The structure of these networks is very similar to that of the relay-contact networks presented by Caldwell in his classic text (Caldwell, 1958). The similarities of these networks follow directly from the functional equivalence of an nMOS switch with a bilateral contact, as shown in Fig. 3.5. There are, however, some important differences, as described later. We will discuss several classes of these networks with simple examples.

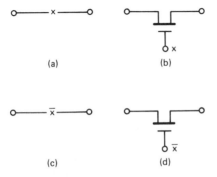

(a)   (b)

(c)   (d)

Figure 3.5 (a) Normally open contact which closes when the control variable x has value 1; (b) pass transistor equivalent to the contact in part (a); (c) normally closed contact which opens when $x = 1$; (d) pass transistor with gate controlled by $\bar{x}$, equivalent to the contact in part (c).

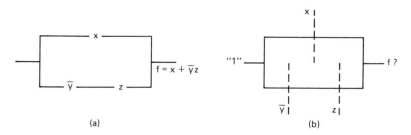

(a)   (b)

Figure 3.6 (a) Relay-contact network for $f = x + \bar{y}z$; (b) wrong nMOS switch network for f.

3.3.1 Networks Derived from Canonic Forms

It is well known that a switching function can be expressed in one of two canonic forms: the sum-of-products or the product-of-sums forms. Consider the function f expressed as a sum-of-products form:

$$f = x + \bar{y}z \tag{3.1}$$

It would seem that the nMOS network shown in Fig. 3.6(b) realizes f since it is "equivalent" to the relay-contact network of Fig. 3.6(a). But there is an important difference: When all the paths are open in a relay-contact network, the "transmission" function f is assumed to have logic value 0. This may not be true for an nMOS network; it may retain the logic value 1 stored at the output capacitance for some previous input, or the output might go to an undefined or intermediate value between 1 and 0 while the capacitance is being discharged. We will say that the output *floats*. This situation will arise in Fig. 3.6(b) for $x = z = 0$, $y = 1$, or for any combination of inputs for which the expression f in Eq. (3.1) becomes 0. A correct nMOS switch network for f is shown in Fig. 3.7 in which the network for the complementary function $\bar{f} = \bar{x}(y + \bar{z})$ is attached to the output node, which drives the output to 0 value whenever $f = 0$.

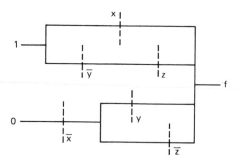

Figure 3.7 Correct nMOS switch network for $f = x + \bar{y}z$.

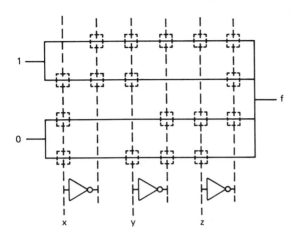

Figure 3.8 Realization of f with regular organization of data and control signals.

The network of Fig. 3.8 also realizes f but it uses a regular structure for organizing the flow of data (constants 1 and 0) and the control information (i.e., x, y, z). Although the circuit uses more switches than are necessary, the area of the layout may be approximately the same as that derived from Fig. 3.7. This does not, of course, imply that the logic minimization has no value in nMOS. As a case in point, the same function f, if realized from the canonic sum-of-products form $f = \bar{x}\bar{y}z + x\bar{y}\bar{z} + xy\bar{z} + \bar{x}y\bar{z} + xyz$, will need at least four times the area needed by the network of Fig. 3.8. An alternative realization of the same function is shown in Fig. 3.9, which uses metal to carry the control signals. The layout based on this network needs more area compared to the area needed in Fig. 3.8 due to design

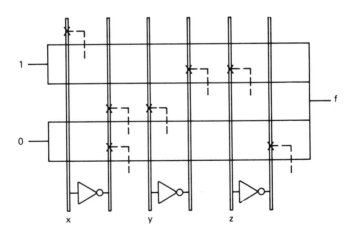

Figure 3.9 Realization of f with metal lines for control signals and no yellow transistors.

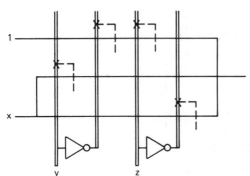

Figure 3.10 Efficient realization based on the Shannon expansion theorem.

rule constraints on metal wires and contact cuts, but will have some advantages in terms of speed, as we will see in a later chapter. This statement is applicable to all circuits that use switch logic.

3.3.2 Networks Derived from the Shannon Expansion Theorem

The fact that the network shown in Fig. 3.10 also realizes $f = x + \bar{y}z$ might raise some eyebrows about the method described in the preceding section. The efficiency of this network in terms of layout area is due to the fact that the role of the input variable has been changed from control to data, at the same time making sure that the output node never floats. Such network structures can be derived using the well-known Shannon expansion theorem (Shannon, 1938; 1949):

$$f(x_1, x_2, \ldots, x_n) = x_1 f(1, x_2, \ldots, x_n) + \bar{x}_1 f(0, x_2, \ldots, x_n) \cdots \qquad (3.2)$$

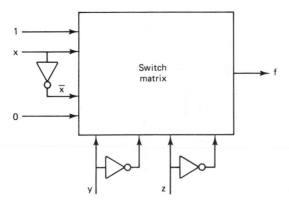

Figure 3.11 General structure for realization of an arbitrary three-variable function.

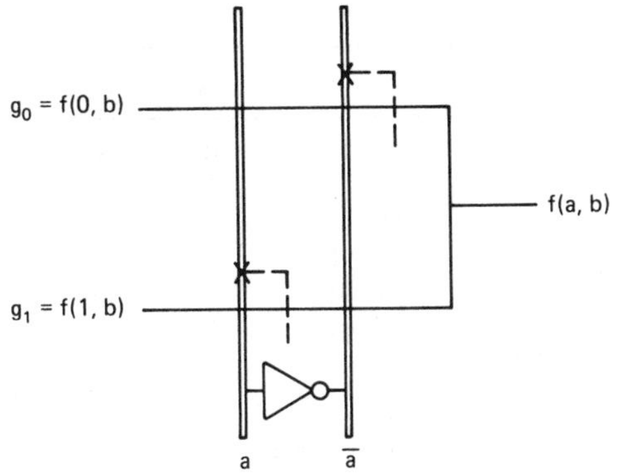

Figure 3.12   Universal logic module for two-variable functions.

Thus for $f = x + \bar{y}z$, we can write

$$f = \bar{y}\bar{z}(x) + \bar{y}z(1) + y\bar{z}(x) + yz(x)$$
$$= \bar{y}z(1) + (y + \bar{z})(x) \tag{3.3}$$

which leads directly to the network shown in Fig. 3.10. An obvious generalization of this network for realizing an arbitrary three-variable function based on the foregoing principle is shown in Fig. 3.11. It can be shown that the maximum size of the matrix for realizing an arbitrary three-variable function is 4×4.

For two-variable functions, a 2×2 switch matrix with a *fixed* structure as shown in Fig. 3.12 is all that is necessary. A two-variable switching function $f_i(a, b)$ can be described by the truth table shown in Table 3.1, where $G_j(0 \leq j \leq 3)$ is 0 or 1 and i denotes the decimal number corresponding to the binary number (G_3, G_2, G_1, G_0). Table 3.2 gives the values of $g_0 = f_i(0, b)$ and $g_1 = f_i(1, b)$ for some of the important two-variable functions. All other two-variable functions can be realized similarly (see the Exercises). Thus the circuit of Fig. 3.12 forms a *universal logic module* (ULM) or a *function block* for all two-variable functions. A somewhat less area-efficient two-variable function block can be derived by using the complete Shannon expansion:

$$f(a, b) = \bar{a}\bar{b}G_0 + \bar{a}bG_1 + a\bar{b}G_2 + abG_3 \tag{3.4}$$

TABLE 3.1

a	b	f_i
0	0	G_0
0	1	G_1
1	0	G_2
1	1	G_3

TABLE 3.2

i	$f_i(a, b)$	Name	g_0	g_1
8	ab	AND	0	b
14	$a + b$	OR	b	1
1	$\bar{a}\bar{b}$	NOR	$\bar{b}$	0
7	$\bar{a} + \bar{b}$	NAND	1	$\bar{b}$
6	$a\bar{b} + \bar{a}b$	EXOR	b	$\bar{b}$
9	$ab + \bar{a}\bar{b}$	NEXOR	$\bar{b}$	b

The circuit is shown in Fig. 3.13. The circuit can also be used as a 4-to-1 *multiplexer*, where G_0, G_1, G_2, and G_3 are considered as separate inputs and the two-bit code (a, b) selects one and only one of the inputs to be connected to the output. By the same reason the circuit in Fig. 3.12 can be used as a 2×1 multiplexer, selecting one of the inputs from (g_0, g_1) by a single-bit code (a).

In principle it is possible to extend the networks described above to synthesize combinational functions of an arbitrarily large number of variables and a multiplexer of an arbitrarily large number of inputs. In practice, the size is limited by the signal delay through the chain of pass transistors in a given path. The maximum number of pass transistors in a cascade to steer a signal should not exceed four for minimum delay (see Section 6.5). Also, remember that the output of a switch network is a degraded signal and must be properly restored before it is used anywhere else.

One final remark: The multiplexer circuit works because the inputs are connected to the output via mutually disjoint paths. This is not necessarily true for an arbitrary switch network. For example, the circuit of Fig. 3.7 provides two transmission paths for 1 for input $(x = 1, y = 0, z = 1)$. But consider the circuit of Fig. 3.14(a). What is f? It is ambiguous; it could be either ab or $\bar{a}$. Since for input

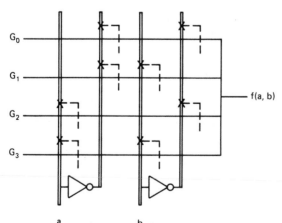

Figure 3.13 Two-variable function block which can also be used as a 4-to-1 multiplexer.

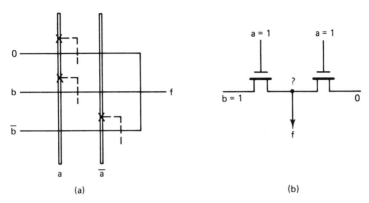

Figure 3.14 (a) Risky circuit; (b) charge sharing.

$(a = 1, b = 1)$, the output is driven by both 0 and 1 via two paths. This results in a *fighting* situation in which the dominant signal, that is, the signal carried by the higher capacitive line or the line with the stronger current source (i.e., stronger driving device) will establish itself at the output or the output might go to an undefined intermediate state, as shown in Fig. 3.14(b). This is risky and should be avoided. This leads to another rule for networks, the consistency rule, which states that a node should not be driven simultaneously by opposite signal values.

3.3.3 Networks Derived from Iterative Structures

An *iterative network* consists of a one-dimensional array of *basic cells*. The concept of a basic cell is important in VLSI design since it exploits regularity of structure to yield dense layout together with other advantages, such as expandability with the size of the input problem, and testability. The theory of design of iterative networks is well established in switching theory literature (Caldwell, 1958; Hennie, 1961; McCluskey, 1965; Mukhopadhyay and Stone, 1971). Iterative networks could be sequential or combinational. In this section we present a few interesting *n*MOS networks which will highlight the general design procedure.

The parity checker.   The parity checker circuit produces a 1 output if there is an odd number of 1's in the input. If the input occurs in a time sequence, a synchronous finite-state machine (FSM) will produce a parity output sequence as illustrated in Fig. 3.15(a). On the other hand, if the input is a *spatial* or simultaneous input string of finite length, an iterative combinational network as shown in Fig. 3.15(b) will produce the same output simultaneously. Hennie (1961) has shown that the output produced by any finite-state machine can also be produced spatially by an iterative network.

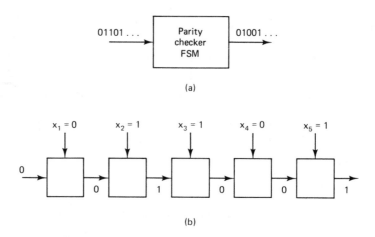

(a)

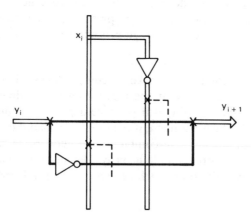

(b)

Figure 3.15 (a) Parity checker FSM; (b) iterative network for parity checking.

The description of a *typical cell* in an iterative network is similar to the description of a state table for the FSM. The intercell signal carries information for two possible states: $y_i = 1$ if there are an *odd* number of 1's in $(x_1, x_2, \ldots, x_{i-1})$ and $y_i = 0$ if there are an *even* number of 1's in $(x_1, x_2, \ldots, x_{i-1})$. The "next" state is given in Table 3.3, which yields $y_{i+1} = y_i \oplus x_i$, or $y_{i+1} = $ EXOR (y_i, x_i). An *n*MOS circuit realizing y_{i+1} is shown in Fig. 3.16. Note that the network is simply a cascade of EXOR cells. Also, the y_i signal has to be restored after every four cells since four consecutive x_i's could possibly have value 0, thus steering y_i through four pass transistors.

A sequence detector.   The problem is to design a basic cell of an iterative network that will detect a single consecutive group of three 1's in an input binary string. A typical cell for a relay-contact network is shown in Fig. 3.17(a). The symbol

Figure 3.16 Typical cell of the parity checker.

TABLE 3.3 CELL TABLE FOR THE PARITY CHECKER

Input State: y_i	x_i	Output State: y_{i+1}
0	0	0
0	1	1
1	0	1
1	1	0

$\bar{0}$ means a single consecutive group of 0's, the symbol $\bar{0}\bar{1}$ means a single consecutive group of 0's followed by a 1, and so on. The solution in Fig. 3.17(a) is rather unusual since it does not use the formalism of developing a cell table, state assignment, next state variable, and so on. The boundary input 1 is applied to terminal 1 of the leftmost cell and 0 to terminals 2, 3, and 4 of the leftmost cell; the output is taken from terminal 4 of the rightmost cell. The four terminals 1, 2, 3, 4 represent the basic four states; the fifth state, corresponding to an input sequence not recognized by

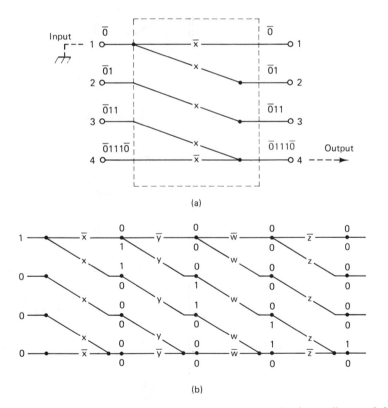

(a)

(b)

Figure 3.17 (a) Typical cell of the sequence detector; (b) three cells cascaded together. (After Caldwell, 1958.)

the circuit, is represented by all four "transmission" functions at the output becoming zero simultaneously. The following encoding (values of terminals 1, 2, 3 and 4) has been used to represent the states:

$\bar{0}$: 1000 (zero or more 0's)
$\bar{0}1$: 0100 (the first 1 of the group)
$\bar{0}11$: 0010 (two consecutive 1's in the group)
$\bar{0}11\bar{1}0$: 0001 (three consecutive 1's in the group)

Any other combination of the input has an encoding of 0000. To handle a variable input sequence of length n, a cascade of n such cells has to be used and one variable per cell is applied from the top. The initial state of the network at the leftmost input side of the cell is set to be 1000. A correct sequence is detected if terminal 4 receives a 1 signal and terminals 1, 2, and 3 receive a 0 signal at the output of the rightmost

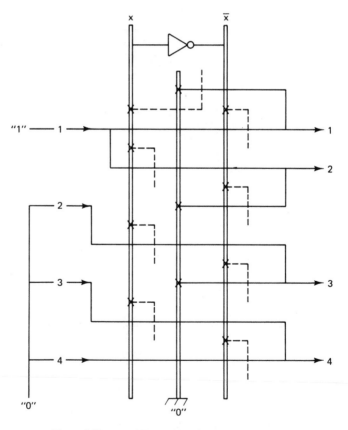

Figure 3.18   An nMOS cell for the sequence detector.

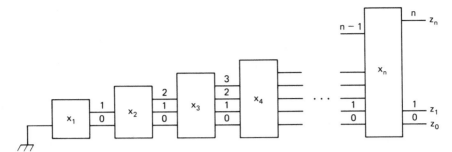

Figure 3.19 Symmetric circuit.

cell. The state transitions induced by the variable x are also depicted in the contact network. Thus, if $x = 1$, the state $\bar{0}1$ gets transformed to the state $\bar{0}11$, and so on. If $x = 0$, states $\bar{0}$ and $\bar{0}111\bar{0}$ stay the same; for any other state, the circuit enters into the state 0000, denoting an illegal input.

A cascade of four basic cells is shown in Fig. 3.17(b). Two sets of outputs are shown for each cell: one for inputs $x = y = w = 1$ and $z = 0$ (shown on top of the output lines), and the other for inputs $x = z = 0$ and $y = w = 1$ (shown below the output line).

In adopting this cell to an nMOS circuit, one has to be cautioned about the possibility of some of the output terminals being floated or not driven. The circuit is shown in Fig. 3.18. The boundary input is applied as a 1 to terminal 1 and as a 0 to terminals 2, 3, and 4 at the leftmost cell. Note that the line connected to 0 in the

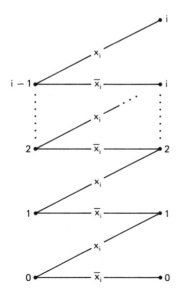

Figure 3.20 The ith cell of the symmetric circuit.

middle and the associated pass transistors are required to ensure that none of the output terminals float.

Symmetric function network.   The symmetric function network illustrates an interesting basic cell definition. A symmetric function circuit has n inputs $(x_1, x_2, \ldots, x_n)$, and has $n + 1$ outputs $(z_0, z_1, \ldots, z_n)$. The output z_a is 1 if and only if exactly a of the inputs assume value 1. The number a is called Shannon's a-number (Shannon, 1938). The nMOS network will again be an interesting adaptation of a relay-contact symmetric circuit as proposed by Caldwell (1958); it is shown in Fig. 3.19 with the basic cell shown in Fig. 3.20. The ith cell has switches controlled by the variable x_i only; it has i inputs and $i + 1$ outputs. The jth output

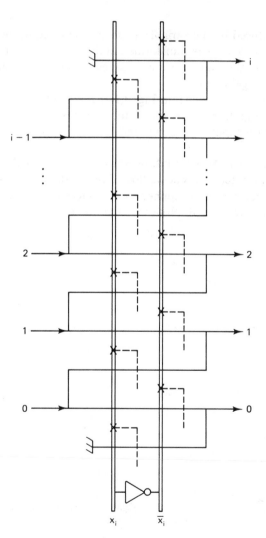

Figure 3.21 Stick diagram of the basic cell.

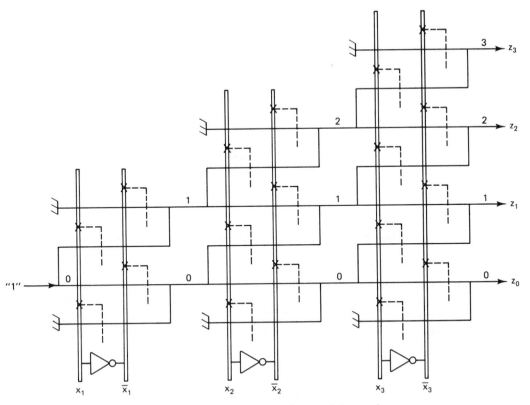

Figure 3.22 Symmetric network for $n = 3$.

$(0 \leqslant j \leqslant i)$ has a transmission function 1 if exactly j of the variables from $(x_1, x_2, \ldots, x_i)$ have value 1. In adopting this cell for an nMOS realization, one has to prevent the situation when the output terminal 0 floats if $x_i = 1$ or the (i th) terminal floats when $x_i = 0$. The solution consists of simply forcing a 0 signal at these nodes with two additional switches. This leads immediately to the basic nMOS cell shown in Fig. 3.21. A symmetric function network for $n = 3$ is shown in Fig. 3.22. Note that the zeroth terminal of the leftmost cell receives a constant 1 input. The symmetric function network was called a *tally* circuit in earlier texts (Mead and Conway, 1980).

3.4 nMOS Combinational Logic Circuits

3.4.1 Logic Gates

The basic NOR and NAND logic functions can be implemented by simply replacing the pull-down transistor of the inverter by parallel and series connection of transistors as shown in Figs. 3.23 and 3.24, respectively. The truth table, the logic circuit

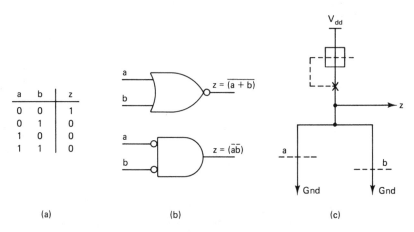

a	b	z
0	0	1
0	1	0
1	0	0
1	1	0

$z = \overline{(a + b)}$

$z = \overline{(ab)}$

(a)   (b)   (c)

Figure 3.23 NOR gate: (a) truth table; (b) logic symbol; (c) stick diagram.

symbols, and stick diagrams are also shown. The inputs a and b are applied via polysilicon lines; the output can be taken in diffusion from the source side of the pull-up or via polysilicon from the gate side. The functions of the circuit are easily explained. For example, for the NAND gate if inputs a and b are both high, the pull-down path will be closed and the output will be pulled low. Note that the gate resistances for the pull-down path are in series and thus the pull-down path resistance is doubled. To maintain the same voltage level as that of a standard $4:1$ inverter, either the pull-up path resistance must be doubled or in the pull-down path the channel resistance must be half the resistance of a standard pull-down. This can be done by doubling the width of the pull-down transistors or by increasing R_{pu} by increasing the length of the pull-up transistor or a combination thereof. The latter option means longer times for charging the output load capacitance and hence slower speed. For a NOR gate, the ratio of each of the pull-down transistors could be equal to that of a standard pull-down transistor in the inverter since V_{out} will be

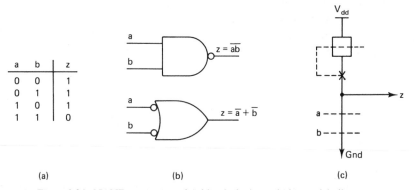

a	b	z
0	0	1
0	1	1
1	0	1
1	1	0

$z = \overline{ab}$

$z = \overline{a} + \overline{b}$

(a)   (b)   (c)

Figure 3.24 NAND gate: (a) truth table; (b) logic symbol; (c) stick diagram.

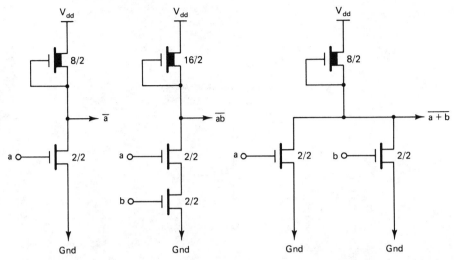

Figure 3.25 Commonly used notations for the ratio L/W of an inverter, a NAND gate, and a NOR gate.

lower than V_{th} when only one of the transistors is on. The output voltage will go down further when both the transistors are on, but this does not affect the logical NOR function of the circuit. The current in the circuit is limited by the saturation current in the pull-down. For this reason, the charge/discharge time of a two-input NOR gate is comparable to that of an inverter if we ignore the slight increase in capacitance due to the increase in wiring area. For this reason, most nMOS logic networks prefer a NOR primitive to a NAND primitive.

A commonly used notation for denoting ratios is illustrated in Fig. 3.25. The numbers denoting the length-to-width ratio L/W are attached beside each transistor. The three circuits shown in Fig. 3.25 have the same $k = 4$. In most commercial CMOS and nMOS schematics, the numbers are always expressed as W/L. The reason for this is that it is very common to specify a nonstandard width (e.g., to increase the drive) while using the standard length. In this case only the W number is given; the length is implied.

The idea of series–parallel combination of pull-down transistors can be used to synthesize any arbitrary Boolean function with an appropriate "pull-down network." For example, the logic circuit for $f = \overline{(ab + c)}$ is shown in Fig. 3.26. Note that the pull-down resistance R_{pd} is the minimum of the resistances of the two transistors in series (ratio 2:4) in parallel with another transistor (ratio 2:2), yielding $R_{pd} = 1$. The ratio k is again 4. Note that we have assumed all the inputs to be restored. If the inputs are steered, the effective resistances of the transistors whose gates receive the inputs have to be doubled for the computation of ratio k.

For arbitrarily large combinational functions with many inputs, the method of using a pull-down network is not very practical since the circuit may need a very

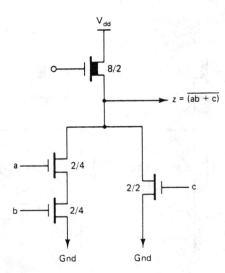

V_{dd}

8/2

$z = \overline{(ab + c)}$

a

2/4

2/2 c

b

2/4

Gnd

Gnd

Figure 3.26 Complex nMOS gate.

large pull-up resistance R_{pu} to produce an effective ratio of $k = 4$. This will reduce the drive current so much that the charging transient will make the circuit excessively slow.

Instead of using a pull-down network, another approach to synthesizing arbitrary functions is to use an interconnection of NOR–NAND networks. This does not always lead to an efficient network unless we take special care to reduce intergate communication and to structure the network in a regular fashion. This is done for programmable logic arrays (PLAs), described in the next section. Also, nMOS circuits can sometimes exploit special transistor configurations to synthesize functions that would normally be rather inefficient using a NOR–NAND network or PLA. A case in point is the exclusive-OR gate (or the equivalence function), shown in Fig. 3.27. The circuit in Fig. 3.27(a) produces the exclusive-OR function, denoted $XOR(a, b)$, and the one shown in Fig. 3.27(b) also produces $XOR(a, b)$ with the indicated ratios, but this circuit needs only one of the variables (and its complement) coming from one side of the circuit. Notice that the same circuit can produce the complementary function if we negate one of the inputs, such as a. These two circuits need both the inputs and their complements and have been used in the design of comparison circuits. The circuit in Fig. 3.27(c) does not need complemented inputs and produces $NEXOR(a, b)$. This can be seen as follows. When $a = b = 1$, both the transistors are on and the output is connected to 1. When $a = b = 0$, the output is connected to V_{dd}; both of the transistors are off. When $a \neq b$, the "on" transistor connects the output to the zero voltage, which provides the pull-down path. The pull-up will have a dimension determined by the resistance of this pull-down path. This path may have a complex structure with other pull-ups connected to it, but for the sake of simplicity the effective resistance of the path to ground is taken as the resistance of the pull-down network. The design imposes a dependence between the design of the gate and the circuit that drives it, a poor policy in general since changes in one may require changes in the other which may get overlooked.

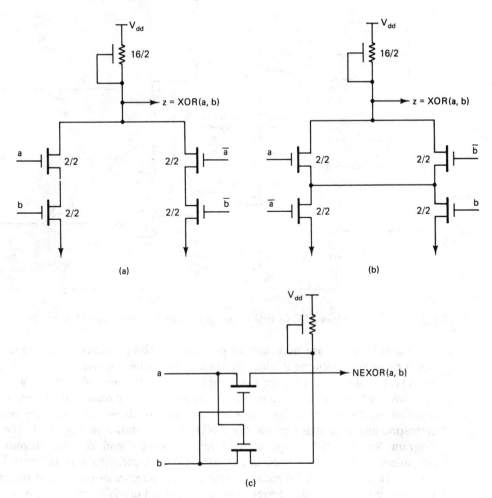

Figure 3.27 (a) Exclusive-OR gate; (b) an alternative equivalence gate; (c) equivalence gate.

3.4.2 nMOS Programmable Logic Array or PLA

Programmable logic arrays (PLAs) provide a flexible and efficient way of synthesizing arbitrary combinational functions in a regular structure. The circuit is based on a representation of Boolean functions as a set of sum-of-products terms. A set of such functions can be mapped into a two-stage NOR network as shown in Fig. 3.28, for the functions

$$z_1 = ab + \bar{a}bc$$
$$z_2 = ab$$
$$z_3 = a + \bar{b}c$$

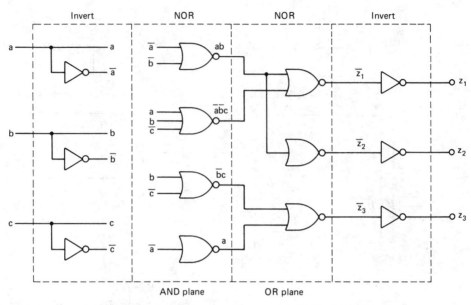

Figure 3.28   General functional structure of a NOR–NOR PLA.

The input inverters are necessary to produce the complements of the inputs. The AND plane produces the distinct product terms in the expressions; the OR plane collects these terms to produce the complements of the desired outputs, which are then inverted to produce the true outputs. An nMOS circuit with a very regular interconnection structure for the same network is shown in Fig. 3.29 and the corresponding stick diagram for the AND plane is shown in Fig. 3.30. The stick diagram for the PLA maps directly into a regular and compact layout with polysilicon and diffusion lines interspersed in the same direction and metal lines running in an orthogonal direction. Note that the number of pull-ups at the end of the rows in the AND plane correspond to the distinct product terms (i.e., a, ab, $\bar{a}\bar{b}c$, and $\bar{b}c$) in the expressions; the pull-down network is a NOR gate consisting of the transistors in parallel formed at the selected intersections of the polysilicon lines carrying the inputs (and their complements) and the lines carrying the ground potential. Similarly, columns in the OR plane compute the complements of the sum of product terms corresponding to distinct output functions (z_1, z_2, and z_3). It is therefore clear that by simply putting appropriate pull-down transistors at different intersections in Fig. 3.29, it is possible to "program" the network to synthesize any arbitrary three-input three-output functions having a maximum of four distinct product terms. In fact, some PLA generator software first generates the network "template" based on the number of inputs, the number of outputs, and the number of terms and then "programs" the structure to customize for a particular set of functions. The input signals are sometimes driven by superbuffers (discussed in Chapter 6), which produce restored signals to the inputs of the pull-down transistors. The pull-ups, therefore, need only have the 8:2 ratio used for a standard inverter.

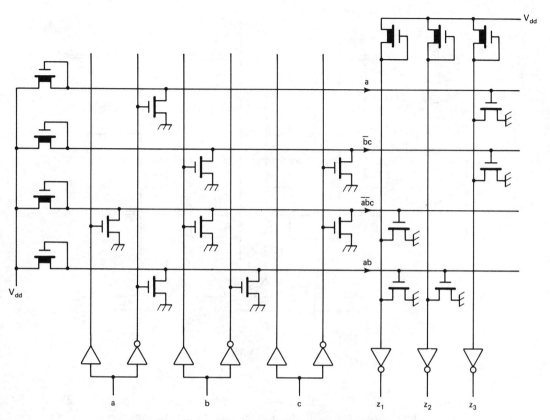

Figure 3.29 Circuit schematic for an *n*MOS PLA.

Since the PLA is one of the most frequently used logic circuits, currently available PLA generators have incorporated several enhancement features and some others may be incorporated for future PLAs. We will discuss a few of them here. First, the size of the AND plane depends on the number of distinct product terms that can be optimized by using logic minimization techniques such as the Quine–McCluskey (McCluskey, 1965) algorithm and its generalization to multioutput combinational functions. Second, being general purpose and programmable, the PLA is not always a dense layout for specific functions. Consider the circuit shown in Fig. 3.31. This is a five-input, six-output, eight-term PLA using an equivalent area for a five-input, four-term, four-output PLA plus six additional pull-ups. The AND plane is situated in the middle and the OR plane is *folded* on both sides of the AND plane. Notice that the horizontal lines and some vertical lines have been *split* to allow for more terms and output functions, thereby yielding a denser PLA. It is conceivable to use a number of such AND and OR planes to generalize the circuit structure. For such a circuit structure to be realizable, the output functions must satisfy certain decomposability properties. For example, z_2 independent of inputs a, b; z_5 is independent of c, d, e; and so on. In order to share the same

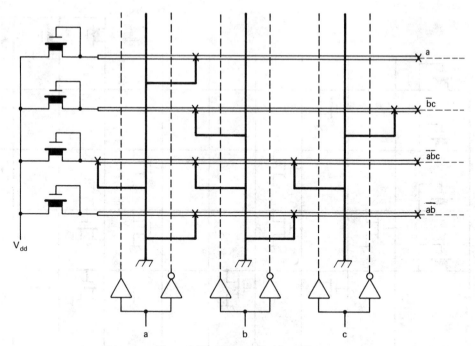

Figure 3.30 Stick notation for a PLA.

column in the output OR plane, the two functions must also have mutually disjoint terms. Furthermore, it should be possible to order the input variables such that splitting of the rows is feasible. Another area-shaving trick for PLAs is to use predecoded inputs. For example, the four input lines a, $\bar{a}$, b, $\bar{b}$ could be replaced by the predecoded terms ab, $\bar{a}b$, $\bar{a}b$, ab. Other sets of predecoded terms could now be combined in the array to produce a complex set of terms with a smaller number of rows. Algorithms to determine decomposability and functional separability (Ashenhurst, 1959) of a set of Boolean functions are extremely complex, and therefore practical PLA optimization tools are not common. Several papers on PLA folding have appeared in the literature.

The area, speed, and power become critical parameters as the size of the PLA increases. The gate capacitances of the inputs carried by long polysilicon lines become the key factor. In moderate to large PLAs, the polysilicon resistance becomes as important a factor as the capacitance. If the lines were purely capacitive, the drivers could simply be increased; but with the large R added to the line, the resultant signal will be seriously degraded no matter how large the drivers are. This has been one of the prime factors in the push to develop double-metal and refractory-polysilicon processes. The *refractory-polysilicon processes* place a thin layer of metal directly on top of the polysilicon, reducing the resistance from about 20 Ω/square to about 1 Ω/square. In the *double-metal process* the polysilicon gate

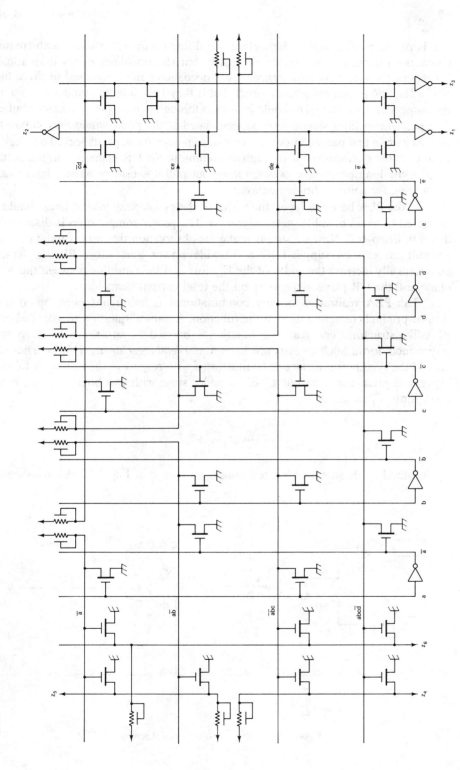

Figure 3.31 Illustration of PLA folding.

line is run in parallel and underneath a metal line on the extra layer, with frequent contacts—the capacitance increases slightly but the resistance drops dramatically. To reduce the effect of capacitance, low impedance buffers are used to drive these lines. The OR planes are usually small, but if they have a large number of columns, the outputs of the AND plane could be driven through similar low impedance buffers. The polysilicon lines should also be terminated at the point where they connect to the last gate in that particular column or row to reduce the capacitances. To safeguard from possible degradation of the signals coming to the OR plane due to the voltage drop in the long polysilicon wires, the output pull-ups usually have a larger ratio, accounting for some additional delay.

If the PLA becomes large, the width of the power and ground lines should be increased to avoid possible metal migration. The power computation is discussed in detail in Chapter 6. Now we simply make the observation that a maximum of about 100 pull-ups can be supplied by a 3λ-wide power line. Most PLA generators automatically increase the width of the V_{dd} line and the Gnd line between the AND plane and the OR plane, depending on the total current demand.

The PLA realization of any combinational function has been based on a sum-of-products representation of the function. A natural question to ask is whether PLA-like structures can realize a function expressed as an arbitrarily deep sum-of-products form. Such circuits are known as *Weinberger arrays* (1967). The basic idea of the Weinberger array can be illustrated with respect to the following Boolean expression given the carry-out C_o of an adder stage with two binary bits a, b and carry input C_i as

$$C_o = ab + C_i(a\bar{b} + \bar{a}b)$$

A multilevel NOR gate network realizing C_o is shown in Fig. 3.32. A stick diagram

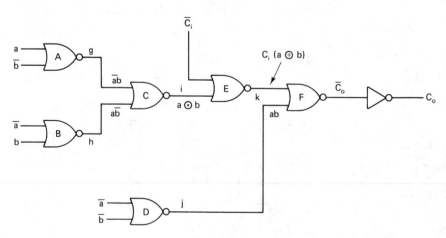

Figure 3.32 Multistate combinational network.

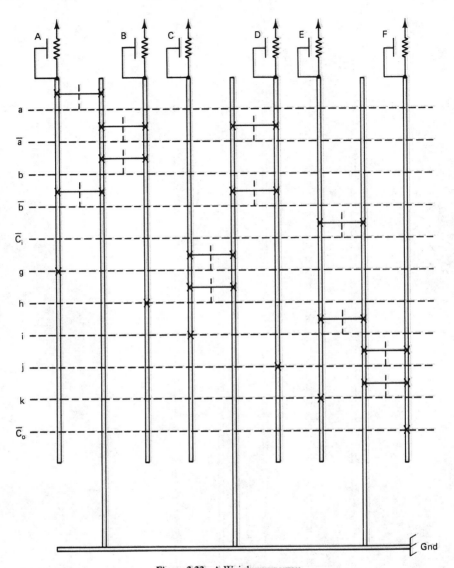

Figure 3.33 A Weinberger array.

of a possible Weinberger array is shown in Fig. 3.33. The circuit consists of a one-dimensional array of "elongated" NOR gates. The inputs to the NOR gates are derived from signal lines running horizontally in polysilicon which are connected to external inputs (a, b, C_i) and internal inputs such as $g = \text{NOR}(a, \bar{b})$, $h = \text{NOR}(\bar{a}, b)$, $i = \text{NOR}(g, h)$, $j = \text{NOR}(\bar{a}, \bar{b})$, and $k = \text{NOR}(i, \bar{C_i})$. The output $\bar{C_o}$ is made available on a horizontal line. Note that a pair of pull-ups is sharing a common ground line. Other variations of the layout are possible, which the reader may wish to explore.

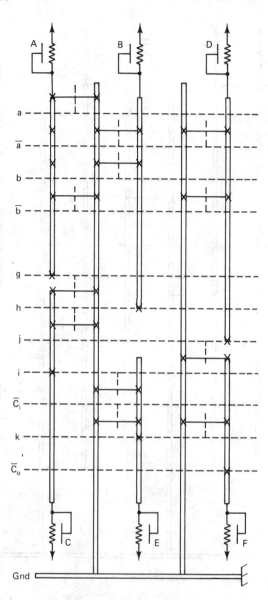

Figure 3.34 Optimized Weinberger array.

The size of the array is determined by the number of NOR gates and the number of internal and external nodes in the circuit. There is room for optimization. Note that the ordering of the rows in the array is arbitrary, which means that a kind of *folding* as illustrated in Fig. 3.34 is possible to increase the transistor density and reduce the area and delay in the signal lines. Other optimizations, such as trimming unused wire length and eliminating pull-ups by directly connecting the output of a

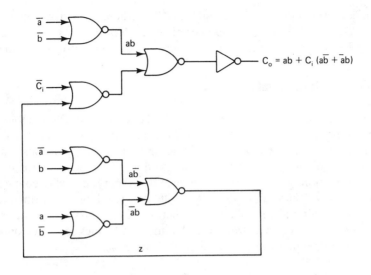

$$C_o = ab + C_i \, (a\bar{b} + \bar{a}b)$$

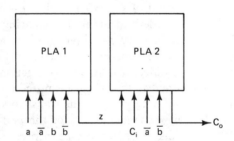

Figure 3.35   Feedback-reduced PLA.

NOR gate to the input of a physically adjacent gate, are possible. It will be very desirable to develop software for a Weinberger generator similar to those available for PLA generators. For further discussion of the problem, see Ullman (1983).

Another approach to designing PLA using multilevel expressions is illustrated in Fig. 3.35, again with respect to C_o. This is a way to use the equation that can be factored into more than two logic levels. The NOR–NOR logic of the PLA is instituted starting with the deepest nested level of the expression. The outputs of each level of nesting are fed back to the next higher level of logic, and so on. The resulting PLA is called a *feedback reduced PLA* (FRPLA). Note that the *feedback* paths do not really form a loop in the circuit, but rather, form a "coil." Depending on the particular function, this could lead to considerable area shaving. For example, an ordinary PLA can use area $O(n \times 2^n)$, whereas FRPLAs have an area complexity of $O(n^2)$, where n is the number of inputs (Glaser, 1982). But one disadvantage of an FRPLA is that the circuit is much slower because of the multiple loops.

3.5 CMOS Combinational Networks

In this section, we present several classes of CMOS combinational networks, composed of transmission gates, inverters, and pass transistors. We will present networks derived from canonic forms and Shannon expansions similar to the nMOS circuits discussed in the previous sections. The equivalence of a switch and a transmission gate allows us to use the theory of relay-contact network rather effectively (Caldwell, 1958). We will also see how the principle of duality and DeMorgan's theorem allow construction of arbitrary combinational functions. We discuss several optimization techniques and rules of interconnection. We then discuss domino logic and precharged CMOS logic.

3.5.1 Networks Derived from Canonic Forms

Consider the nMOS circuit of Fig. 3.7 realizing $f = x + \bar{y}z$. A CMOS network for the same function can be derived simply by replacing the n-channel transistors that transmit 1 to the output f by p-channel transistors and by complementing the inputs to the gates of these transistors as shown in Fig. 3.36. In so doing, we have actually achieved three different results by one action. We ensured that the output never "floats," we guaranteed that 1 is transmitted undegraded to the output by the p-channel paths, and we have grouped the transistors in a p-channel group for the pull-up path and in an n-channel group for the pull-down path, so that a CMOS process (such as the bulk p-well process) can produce these groups in n-substrate and p-well regions, respectively. The ground connection to the pull-down path is also denoted as V_{ss}.

The structure of the network of Fig. 3.36 generalizes to arbitrary combinational networks as shown in Fig. 3.37. The top p-transistor network F provides transmission paths for all input combinations $(x_1 \cdots x_n)$ for which the desired function $f(x_1 \cdots x_n) = 1$; the bottom n-transistor network $\bar{F}$ provides transmission paths for all input combinations for which $f(x_1 \cdots x_n) = 0$. Furthermore, the network structures for F and $\bar{F}$ bear a dual relationship by DeMorgan's theorem, stated in its most general form as

$$f(x_1, x_2, \ldots, x_n, +, .) = \bar{f}(\bar{x}_1, \bar{x}_2, \ldots, \bar{x}_n, ., +)$$

which says that the complement of any function can be obtained by replacing each variable by its complement and by interchanging the OR operation with the AND operation at each level of the expression for f. Thus, if f is specified as a sum-of-products form (i.e., $f = x + \bar{y}z$), we can derive the $\bar{F}$ network from f

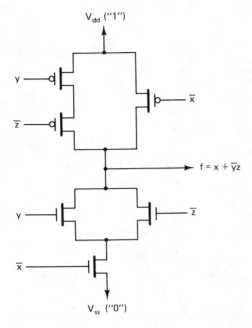

Figure 3.36 CMOS network for $f = x + \bar{y}z$.

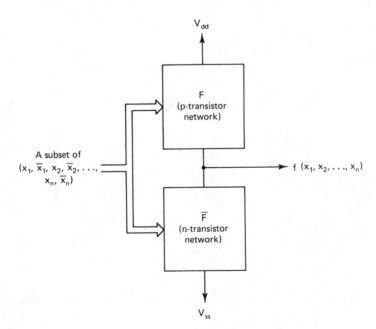

Figure 3.37 CMOS network for arbitrary combinational function.

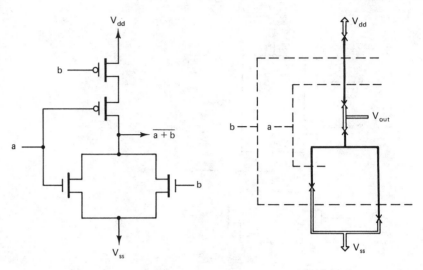

Figure 3.38 CMOS two-input NOR gate and its stick representation.

expressed as a product-of-sums [i.e., $\bar{f} = \bar{x}(y + \bar{z})$] in the form of a series–parallel network of n-channel switches. The dual of this network is used to realize F, except that we do not have to use the complements of the variable since the complementary p-channel switches are used for the F network, as illustrated in Fig. 3.36.

Using the foregoing principle, a CMOS network for arbitrary combinational functions can be derived. The commonly used two-input NOR and NAND gates together with their stick representations are shown in Figs. 3.38 and 3.39,

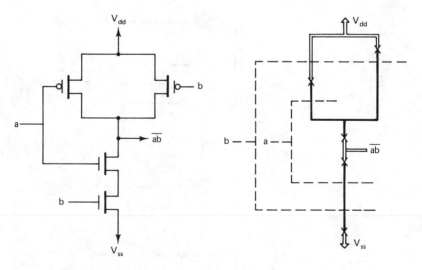

Figure 3.39 CMOS two-input NAND gate and its stick representation.

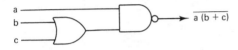

$$\overline{a\,(b + c)}$$

Figure 3.40 Symbol for a complex CMOS gate.

respectively. The network in Fig. 3.36 actually represents a "complex" CMOS gate and is sometimes symbolized as shown in Fig. 3.40. The AND–OR–INVERT gate, expressed as $f = \overline{(ab + cd)}$, is shown in Fig. 3.41, consisting of a parallel connection of series n-transistors for $\overline{F}$ and a series connection of parallel p-transistors for F. Another interesting complex gate is the "majority function,"

$$M(a, b, c) = ab + bc + ca$$

$$= ab + c(a + b)$$

which also represents the "carry function" from an adder stage (M represents carry output bit if c is carry in, and a and b represent the input bits to the stage). Two CMOS networks for $\overline{M}$ based on the expressions above are shown in Fig. 3.42(a) and (b). The function $M(a, b, c)$ also has the interesting property that it is *self-dual* (i.e., complementing the input yields the complementary function). This means that

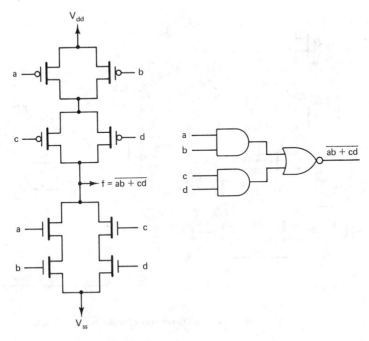

Figure 3.41 CMOS AND–OR–INVERT gate.

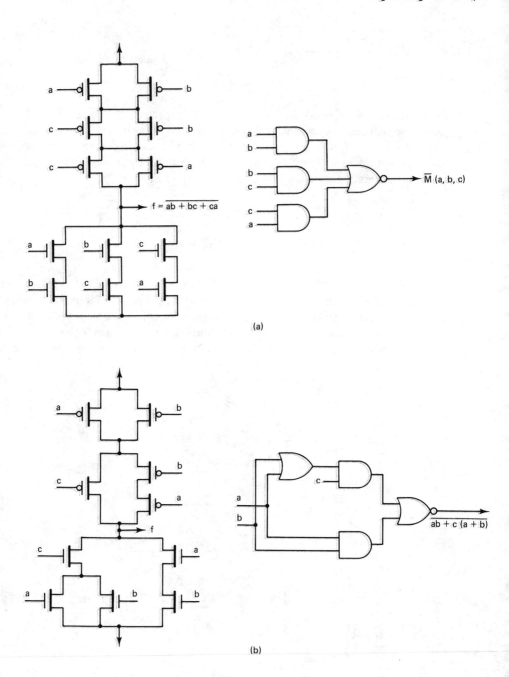

Figure 3.42 CMOS networks for the majority function.

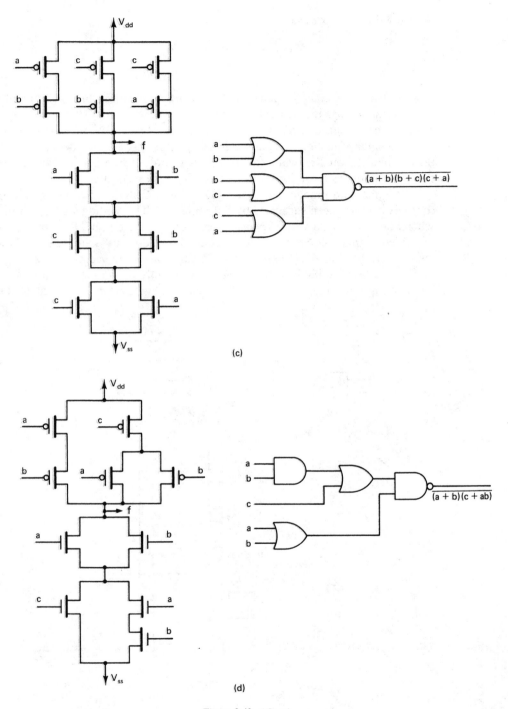

(c)

(d)

Figure 3.42 (*Cont.*)

we can write $M(a, b, c)$ as

$$M(a, b, c) = (a + b)(b + c)(c + a)$$
$$= (a + b)(c + ab)$$

yielding two more complex gates for $\overline{M}(a, b, c)$, as shown in Fig. 3.42(c) and (d).

Summarizing, the general procedure to synthesize a CMOS network for an arbitrary combinational function f is as follows. Start with an expression for the complementary function $\bar{f}$, obtain a series–parallel n-transistor network for $\overline{F}$, and then obtain the dual network structure F with only p-transistors whose gates have the same inputs as those for the corresponding dual elements.

A quick and practical method of deriving a CMOS network for an arbitrary complex gate has been suggested by Sutton (1985). The method starts with the logic schematic of the gate with a *bubble* (negation) at the output. Recognize that any "bubbled-output" gate can be drawn as a "bubbled-input" gate using DeMorgan's theorems (replacing AND gates with OR gates, and vice versa). Draw the gate in both ways. Use the bubbled output form to directly derive the n-channels and the bubbled-input form for the p-channels. Each AND gate symbol represents "stacked"

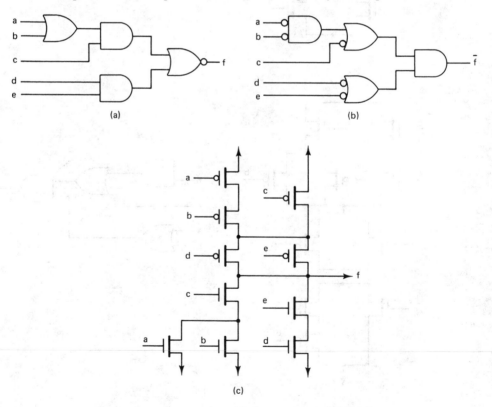

Figure 3.43   Sutton's method of synthesizing complex gates.

transistors (in series) and the OR symbol represents parallel transistors. The method is illustrated in Fig. 3.43. In part (a), the bubbled output gate is shown; part (b) shows the bubbled input gate. Note the duality relations between pairs of gates. The circuit is shown in Fig. 3.43(c). For an n-channel pull-down network, a and b are connected in parallel stacked with c, the whole structure being in parallel with a d-e stack. For a p-channel, a and b are stacked in parallel with c, the whole structure being stacked with d-e in parallel. Because this method is visually/graphically oriented, it is usually faster and less prone to errors than working directly from the equations. Also, this method is exceptionally useful when checking out a layout, when the equations may not be handy.

The network structures discussed in this section use n-channel and p-channel forms in the $\bar{F}$ and F networks, respectively (Krambeck et al., 1982). This means that the circuit needs considerably more area compared to an equivalent nMOS circuit. Also, since each output has to drive both a p-channel and an n-channel transistor, the output capacitive load is considerably higher (at least a factor of 2 higher) than the loads on nMOS circuits. One advantage of the complementary CMOS circuit is that it consumes no static power because, as in a CMOS inverter, the current flows only when the input values change.

3.5.2 Networks Derived from the Shannon Expansion Theorem

In this section we show how CMOS networks similar to nMOS networks derived in Section 3.3.2 can be obtained using Shannon's expansion theorem. These networks might have a degraded signal problem, which can be taken care of by using transmission gates. We then show how a universal logic module can be built using transmission gates and Shannon's expansion principle. We will then discuss some optimization techniques applicable to this class of networks. For the sake of comparison with nMOS, we begin with the function given by Eq. (3.3), reproduced here as

$$f = x + \bar{y}z = \bar{y}z(1) + (y + \bar{z})x$$

which yields the network shown in Fig. 3.44(a). The network consists of a p-channel part to transmit a 1 to the output and an n-channel part to transmit x to the output. Compare this structure with a very similar structure of the circuit in Fig. 3.10. The network is logically correct, but for input combinations $y = x = 1$ or $\bar{z} = x = 1$, the output will be a degraded 1 signal. This can be remedied by replacing each of the n-channel pass transistors in the pull-down path by a suitable transmission gate, as shown in Fig. 3.44(b), requiring both the true and complemented forms of inputs y and z to drive the gates. Although wasteful, the p-channel transistors in the pull-up path can also be replaced by transmission gates. In fact, the CMOS inverter circuit reproduced in Fig. 3.45(a) can be looked upon as a circuit resulting from removing the n-transistor from the pull-up path and the p-transistor from the pull-down path. We could not have done that anyway since the gates of these

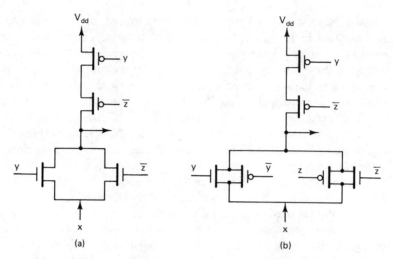

Figure 3.44 (a) Network based on the Shannon expansion theorem; (b) the network in part (a) corrected to remedy degraded signals using transmission gates.

transistors would have to be controlled by $\bar{x}$—the inversion function to be realized by the circuit! On the other hand, if we had $\bar{x}$ available, we could have used the inverter to synthesize a 2×1 multiplexer circuit, as shown in Fig. 3.45(b), where the inputs V_{dd} and V_{ss} are replaced by the signal inputs g_0 and g_1 that we wish to multiplex.

The principle of using transmission gates only in both pull-up and pull-down paths leads to a general technique of synthesis of CMOS networks. Let us first consider a two-variable function $f(a, b)$. Using Shannon's expansion theorem, we can write

$$f(a, b) = ag_1 + \bar{a}g_0$$

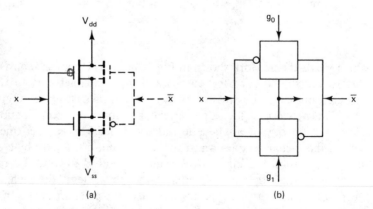

Figure 3.45 A 2×1 multiplexer realized out of an inverter circuit.

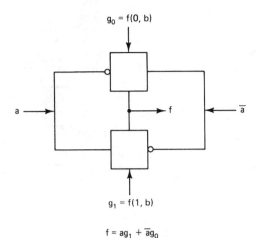

$g_0 = f(0, b)$

a →

f

← $\bar{a}$

$g_1 = f(1, b)$

$f = ag_1 + \bar{a}g_0$

Figure 3.46 CMOS two-input universal logic module.

where $g_1 = f(1, b)$ and $g_0 = f(0, b)$. The network in Fig. 3.46 forms a universal logic module based on the equation above. Specific values of g_0 and g_1 as given in Table 3.2 will yield specific two-variable functions. For example, Fig. 3.47(a) and (b) show EXOR and NOR networks, respectively. Since the pull-down path needs to transmit only a 0, an optimized NOR circuit can be derived by eliminating the p-transistor in the pull-down path, as shown in Fig. 3.47(c). Similar optimization can be done for most two-variable functions.

A somewhat less area-efficient two-variable universal logic module or "function block" can be derived using the complete Shannon expansion of Eq. (3.4). The network is shown in Fig. 3.48(a), where (G_3, G_2, G_1, G_0) are binary constants for a given function, and the network is the CMOS equivalent of the nMOS circuit of Fig. 3.13. Again, optimization is possible depending on the values of the constants to yield a specific network. If we make each of G_3, G_2, G_1, and G_0 arbitrary functions of one variable $(0, 1, c, \bar{c})$, the same network will act as a three-variable function block similar to the network of Fig. 3.11. Alternatively, a larger array based on complete Shannon expansion for three or more variables can be built in a similar fashion.

The circuit of Fig. 3.48(a) also acts as a 4-to-1 multiplexer when G_0, G_1, G_2, G_3 are considered as separate inputs and the 2-bit code (a, b) selects one and only one of the inputs to be connected to the output. The 2×1 multiplexer selects one from (g_0, g_1) depending on the single-bit code (x), as described in Fig. 3.45. A 4-to-1 multiplexer that uses a much smaller number of transmission gates is shown in Fig. 3.48(b). For each input, however, a pair of complementary control signals are needed. Only one pair of the control signals can be asserted at any time.

The consistency rule that was discussed for the nMOS circuit is also applicable to CMOS circuits. For example, the circuit shown in Fig. 3.49 is the CMOS analog

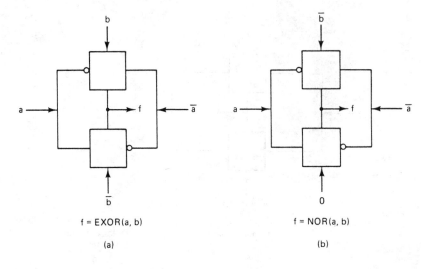

f = EXOR(a, b)   f = NOR(a, b)

(a)   (b)

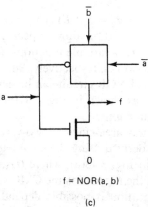

f = NOR(a, b)

(c)

Figure 3.47 (a) Exclusive-OR gate realized from the universal logic module; (b) NOR gate realized from the universal logic module; (c) NOR gate with one transistor in the pull-down part.

of the *n*MOS circuit of Fig. 3.14 in that it will result in an undefined or floating output for $a = b = 1$ due to a fighting situation. We must obey the rule that a node should not be driven simultaneously by opposite signal values.

Summarizing, the method of synthesis of an arbitrary combinational function discussed in this section is general, but may need both true and complemented input. Each of the networks of Section 3.5.1 can be considered to be a special case of a network derived from Shannon's expansion theorem optimized to eliminate the *n*-transistors in the pull-up network and the *p*-transistors in the pull-down network, resulting in subsequent elimination of the need to use the variable in both true and complemented forms. Thus the network of Fig. 3.38 can be thought of as an optimized

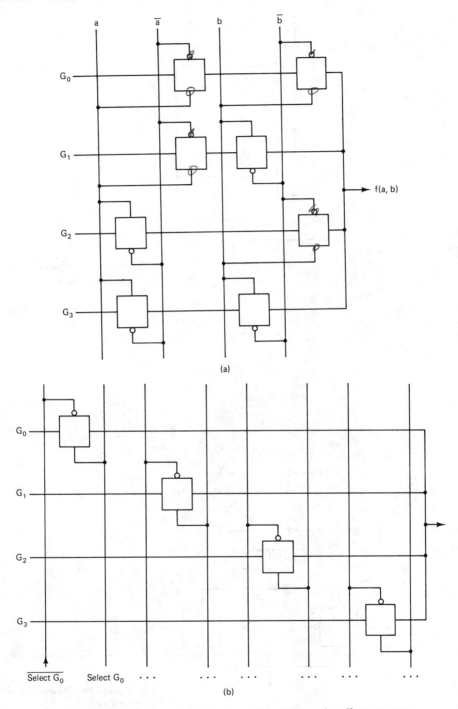

Figure 3.48 Two-variable universal logic module using complete Shannon expansion; (b) a 4 × 1 multiplexer which uses a smaller number of transmission gates.

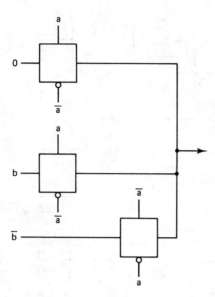

Figure 3.49 CMOS circuit with an undefined output.

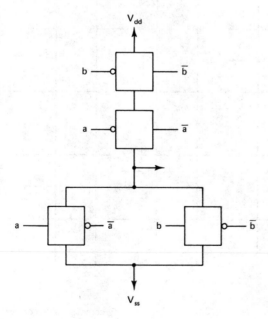

Figure 3.50 Redundant CMOS NOR network.

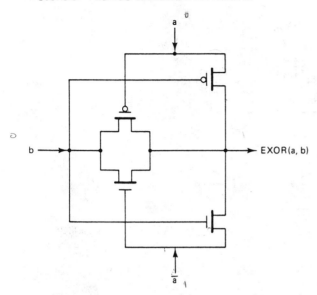

Figure 3.51 Exclusive-OR circuit.

version of the redundant NOR network shown in Fig. 3.50, which is based on the expansion

$$NOR(a, b) = \bar{a}\bar{b}(1) + (a + b)(0)$$

The circuits presented in this section share the disadvantages of complementary CMOS circuits in terms of area and output capacitive loads. The situation is a little worse since both p- and n-channel transistors appear in the pull-down and pull-up networks, and the output and its complement have to be generated to drive secondary stages. The "tricky" circuits such as the one illustrated above may cause unanticipated problems during the design; viz., the circuit may not simulate or the delay may be too long if these circuits are cascaded, or after circuits are fabricated.

Finally, it must be noted that there are circuits that are simply "tricky" and do not follow from any systematic method. We will see several such circuits later in this book, but the one shown in Fig. 3.51 realizing the EXOR function illustrates the point. The circuit consists of an inverter and a transmission gate connected to perform the following functions. When $a = 1$, the circuit behaves like an inverter on input b, cutting off the transmission gate. When $a = 0$, the inverter circuit is cut off and the transmission gate transmits b to the output. This is equivalent to an exclusive-OR function. If inputs b and $\bar{b}$ are interchanged, the same circuit will yield the NEXOR or the equivalence function.

3.5.3 Pseudo-nMOS Networks

If the p-channel F network in a complementary CMOS circuit is replaced by a single p-channel load transistor with its gate connected to V_{ss}, a pseudo-nMOS

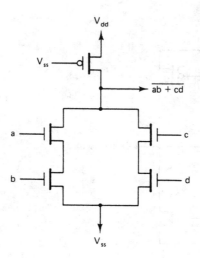

Figure 3.52 Pseudo-nMOS
AND–OR–INVERT gate.

circuit results. For example, a pseudo-nMOS AND–OR–INVERT gate is shown in Fig. 3.52 (which is functionally equivalent to the circuit of Fig. 3.41). The operation of the circuit is as follows. If a path is not closed in the pull-down network, the output will be connected to V_{dd} via the conducting p-channel transistor. If the input combination is such that the pull-down network path is closed, the output becomes V_{ss}, a logic 0.

In comparison to complementary CMOS networks, pseudo-nMOS or nMOS networks yield a considerable savings in area and both of these classes of circuits will be faster than complementary CMOS circuits. In practice, however, this is not the case since the pull-up current always flows and the pull-down circuit will have to sink it, resulting in slower pull-down speed. To offset this, the pull-up current can be made smaller, but this will raise pull-up time.

3.6 Clocked Logic

Capacitance plays an important role in all digital circuits. In our discussion so far we have used logic values 1 and 0 to describe the functions of the circuit, but in practice, these values represent an abstraction of the voltages in the circuit, the transitions between 1 and 0 represent the physical process of charging and discharging of the capacitance of the terminals and of interconnecting wires called the *parasitic capacitance*. The voltages representing the logic values take finite response time to attain their final values contributing to the *delay* in the circuit. This is because a current flows into or out of a capacitance C as long as the voltage V across it is changing as given by the equation

$$I = C\frac{dv}{dt}$$

Thus, to produce a voltage change dv across C, a time dt is needed, given by

$$dt = \frac{C}{I}dv$$

In *combinational* networks of the type discussed so far, capacitance is regarded as a factor contributing to the poor speed of the circuit. But capacitance has two important virtues. First, capacitance means electrical inertia; that is, every change needs finite response time. Without this, circuits will become extremely sensitive to spurious pulses or "glitches" or "hazards." There are two kinds of hazards. If the signal changes its value for a brief time, it is called a *static* hazard, as shown in Fig. 3.53(a). A *dynamic* hazard occurs if the signal during its transition to its value bounces back and forth at least once, as shown in Fig. 3.53(b). Note that the logic representations of the hazards are really abstractions of actual noisy signals, as shown in the top part of Fig. 3.53. Ironically, hazards are caused by stray delays at the input lines to the logic circuits. Second, capacitance holds a charge and a charge can represent information. A digital circuit manipulates the stored charges to realize a logic function. If the charging and discharging processes occur at discrete moments in synchrony with a *clock* pulse, the circuits are said to be *clocked* or *synchronous*. *Asynchronous* circuits are free-running circuits without clock pulse in which the timing relations between signals are determined by the inherent circuit delays.

Two most commonly used clocking schemes are *one-phase* and *two-phase nonoverlapping*. The one-phase clocking scheme consists of a sequence of pulses between high (logic 1 or V_{dd}) and low (logic 0 or V_{ss}) with *width W* and *period T* with an *intraclock gap g* between successive pulses. The quantity W/T is sometimes called the *duty cycle*. We will often use the phrase "during *cl*" to stand for "during the time when *cl* is high." The one-phase clock signal defines two separate *events:* the rising *cl* edge and the falling *cl* edge. These events mark the points in time when

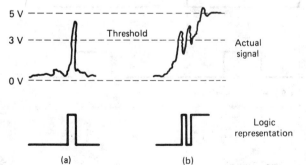

(a)   (b)

Figure 3.53 Static and dynamic hazards.

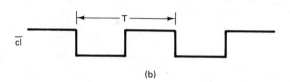

(a)

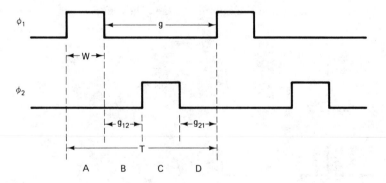

(b)

Figure 3.54 (a) Single-phase clock; (b) its complement.

the signal values at various points in the circuit will generally change to make a transition to a new state of the circuit. We will also use the complementary clock pulse $\overline{cl}$ shown in Fig. 3.54(b), particularly for CMOS circuits. The two-phase nonoverlapping clocking scheme is depicted in Fig. 3.55, and consists of two separate clock signals ϕ_1 and ϕ_2 synchronized to each other, each having a period T. Note that the signals have a nonoverlapping period and two equal *interclock gaps* $g_{12} = g_{21}$. The interclock gaps do not always have to be equal, but they are usually assumed to be so in order that the ϕ_1 and ϕ_2 clocks can be interchanged in the circuit if needed. It is sometimes convenient to refer to the four *epochs* designated A, B, C, and D in Fig. 3.55, constituting the total period T of the clock pulse. We will often use the phrases "during ϕ_1" or "during ϕ_2" with meaning as explained before. The two phases of the clock signals define four separate events: the rising ϕ_1 edge, the falling ϕ_1 edge, the rising ϕ_2 edge, and the falling ϕ_2 edge.

To proceed further with our discussion on clocked logic circuit, we need to discuss elementary storage devices. Basically, all semiconductor storage devices are either dynamic or static. In a dynamic device the storage is due to a small capacitance which holds information in the form of the presence or absence of

Figure 3.55 Two-phase nonoverlapping clock.

electric charge in it; that is, the capacitance is in a charged or discharged state. In static devices, stored information is due to the state of conduction or no conduction of a transistor. Static devices usually need more power but can hold the information indefinitely. The dynamic devices use low power but need periodic refreshment of the stored charge, which may be lost due to leakage. Additionally, dynamic devices are more sensitive to power supply noise, external magnetic/electric fields, and environmental radiation—any of which may be sufficient to add or subtract the charge at the node, thus changing the logic state. Furthermore, because of the refresh requirement, dynamic circuits must be operated at a minimum frequency.

3.6.1 Dynamic Memory Element

A memory element formed by a pass transistor and an inverter is shown in Fig. 3.56. The physical process of *latching* the input signal X consists of raising the control signal, usually a clock, to high and holding the input signal X at a *settled* or unchanged value so that the input gate capacitance of the inverter at node W can be charged to X. The output of the inverter at V reacts to X so that it can settle to its final value after a small but finite amount of delay through the inverter circuit. If we assume that the control signal is high for a finite period of time, the signal X must

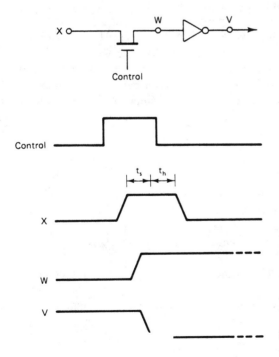

Figure 3.56 Dynamic latch in nMOS.

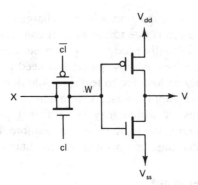

Figure 3.57 CMOS dynamic latch.

attain its final settled value for a minimum time, called the *setup time* t_s, before the falling edge of the control signal, and the signal value must not change until a minimum *hold time* t_h after the falling control signal edge, as depicted in Fig. 3.56. The hold time is necessary to allow for possible delays between the falling edge of the master clock and the falling edge of the local control clock line. This delay is also referred to as *clock skew*. In our example shown, this delay would be very small. If the control signal at the gate of the pass transistor is turned off, setting its value to zero, nodes W and V will maintain their settled value for a sufficiently long time even after the signal X is turned off. The latching operation is recognized to have taken place with the falling edge of the clock. The circuit thus acts as a device to store the information X (in the example we assume that $X = 1$; the reasoning is also valid if $X = 0$, which will make $W = 0$ and $V = 1$). The storage device is called *dynamic* since the information is held up to the time when the charge at node W dissipates beyond the threshold voltage of the inverter. A CMOS dynamic memory element using a transmission gate and an inverter can similarly be built as shown in Fig. 3.57. The operation of the circuit is very similar to the nMOS device discussed above.

3.6.2 Static Storage Element

The flip-flop is the basic static memory circuit. Of the several varieties (see Section 3.8 for further discussion), we first discuss the RS flip-flop.

The RS flip-flop is shown in Fig. 3.58(a). The "state" of the flip-flop is represented by the two complementary outputs Q and $\bar{Q}$. If the "set" input $S = 1$, and the "reset" input $R = 0$, the next state is $Q = 1$ ($\bar{Q} = 0$). If $S = 0$, $R = 1$, the state is $Q = 0$ ($\bar{Q} = 1$). When both $R = S = 0$, the device remembers which of the two inputs was the last one to be high; that is, Q does not change. The input combination $R = S = 1$ is not allowed. The operation can be summarized as

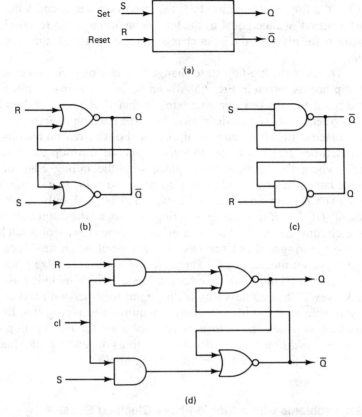

Figure 3.58 (a) Schematic of an *RS* flip-flop; (b) cross-coupled NOR gate implementation of an *RS* flip-flop; (c) cross-coupled NAND gate implementation of an *RS* flip-flop; (d) clocked *RS* flip-flop.

follows:

$S(t)$	$R(t)$	$Q(t + 1)$	$\overline{Q}(t + 1)$
1	0	1	0
0	1	0	1
0	0	$Q(t)$	$\overline{Q}(t)$
1	1	Not	allowed

The *RS* flip-flop circuit can be implemented with the cross-coupled NOR gates shown in Fig. 3.58(b) or the cross-coupled NAND gates shown in Fig. 3.58(c).

For the flip-flop to change its state, the input values to R and S must be stable for at least a period of time equal to the total delay through the feedback loop in order to propagate the effect of the input change to the output. This time is called the *latching time*.

To effect the flip-flop state change in synchrony with a clock pulse, a clocked *RS* flip-flop as shown in Fig. 3.58(d) can be used. The major difference between this circuit and the clocked dynamic latch is that if R and S values are set to 0, the flip-flop retains its state indefinitely. Also, the latching operation of the *RS* flip-flop takes place at the rising edge of the clock. For this reason the clocked *RS* flip-flop is sometimes referred to as an *edge-triggered* flip-flop as opposed to a *simple latch*, where the latching takes place with the falling edge of the clock. The simple latch can be viewed to *propagate a data value* to a storage element when *cl* $= 1$. The *RS* flip-flop *changes state* when *cl* $= 1$ only if an S or R signal is present. (If S or R are changed during the clock, the output also changes. Under these circumstances the *RS* takes on the characteristics of a latch; it is no longer clock-edge triggered but follows the data input when the clock is high.) Being edge triggered means that the inputs R and S must stabilize prior to the arrival of the clock, called the *setup time* for the inputs, which include possible delay due to clock skew. This stable value of the input together with the clock held at high must remain so for at least the period required for the flip-flop to change its state. This time is called the *hold time or latch time* for the flip-flop. Thus the terms *setup* and *hold times* have different meanings depending on what kind of latch— simple or edge triggered—we are talking about.

3.6.3 Problems with a Single-Phase Clocking Scheme

Although many digital systems have been designed using a single-phase clocking scheme, there are a number of tight constraints that have to be met with respect to circuit delays, clock period, and clock width for the correct operation of the circuit. We discuss some of these problems in this section. A better clocking scheme for MOS circuits is the two-phase nonoverlapping scheme. For CMOS circuits both two-phase clocking and single-phase clocking, together with the use of complementary clock $\overline{cl}$, are common. A critical analysis of the problems of single-phase clocking is given in Unger and Tan (1983).

To understand the problems with a single-phase clock, consider the canonic form of a sequential network with feedback loops shown in Fig. 3.59(a). Let us assume simple latches at the feedback paths, whose output defines the present state.

The clock width W must be greater than the delay time for the combinational logic, but cannot be more than the minimum delay through the network because this might cause a *multistepping* or *race* condition. It means that the present state values get changed more than once during the clock period, racing around the feedback path. If the final state of the circuit is according to the correct transition as specified in the state diagram of the sequential machine, such a race condition is called

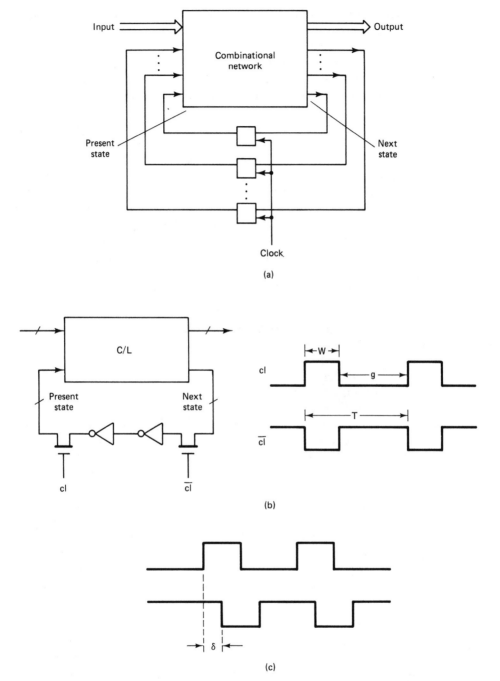

Figure 3.59 (a) Sequential machine with a single-phase clock; (b) sequential machine with a complementary single-phase clock; (c) clock skew in a complementary clock.

noncritical. During this transition, the circuit might go through some intermediate states different from the final state. If the outputs from some of these intermediate states are different from the output of the final state, spurious signals or hazards may be produced. If the output connected to the sequential machine is not sensitive to hazards, a noncritical race is acceptable. But more serious is the situation when the circuit settles in an incorrect final state because some signals traveling through the combinational network faster than others have prevented a proper state change from taking place. Such a situation is called a *critical race* and must be avoided.

There are also some constraints on the clock period T. It must be greater than the longest delay in the combinational network so that the computation of the *next state* propagates through the feedback path to be latched by the *next* clock pulse. But T cannot be indefinitely large if dynamic storage elements are used to store the next-state information, because information may be lost due to leakage. The clock period must be less than the necessary refresh time for the dynamic elements.

The interclock gap g must also satisfy some constraints in order to make the circuit insensitive to spurious hazards at the input. This can be done by making g long enough so that any transient activity due to the glitches may fade out before the arrival of the next clock and the circuit responds only to the true stable signals at the input.

Complementary single-phase clocking. Some of the problems with single-phase clocking can be avoided by using a complementary single-phase clocking that uses both cl and $\overline{cl}$. The canonic form of a sequential circuit using this kind of clocking and dynamic storage elements is shown in Fig. 3.59(b) (only one feedback line representative of a group of lines is shown). The next-state information is latched during $\overline{cl}$ into the storage elements, but is not allowed to feed back until $\overline{cl}$ turns off and cl turns on to deliver the information to the network as "present state." The clock width W must be longer than the setup time of the combinational logic (C/L) network, but it does not have to be less than the minimum delay, δ_{min}, in C/L since the signal has to pass through the $\overline{cl}$ driven pass transistors, which are turned off during this period. Thus a multistepping or race condition is completely avoided. The interclock gap g must be larger than the setup time of the latches. The clock period $T = W + g$ must be greater than the maximum delay, δ_{max}, of the C/L network, so that the "present-state" information is ready to be delivered at the next clock cl. All of the foregoing conditions can be achieved by simply making the width W and/or g larger. Furthermore, if these periods are long enough, the effect of spurious glitches or hazards in the input lines will also be eliminated since the circuit will react to the ultimate stable inputs.

The clock skew. The fundamental reason that the circuit above was able to avoid the race condition is the fact that the feedback path was never closed; the information has been staged in two steps and the stages are isolated from each other by complementary clocks. This scheme will work if the edges of cl and $\overline{cl}$ are perfectly aligned in a complementary fashion. But a major problem may arise if the clocks are misaligned due to a phenomenon called *clock skew*, which is due to the delay in a clock signal during its journey to its destination node from its source. This delay is caused by the resistance and the parasitic capacitance associated with wires that carry the clock pulse and is approximately proportional to the square of the length of the wire from the source to destination. Obviously, different amounts of skewing are experienced at different paths in the circuit, and the signal changes that are supposed to occur in coincidence with the events of the clock pulses at different parts of the circuit may never actually occur at the same time. We can simply say that the events occur almost simultaneously without being able to specify any order at these timing events marked by the clock pulse.

To understand how clock skew might pose a serious problem, consider the circuit of Fig. 3.59(b) again. Let us assume that $\overline{cl}$ is delayed with respect to its ideal timing by an amount δ and assume that $\delta > t_s$, the setup time for the feedback latches. The waveforms are sketched in Fig. 3.59(c). Under these circumstances, some of the signals produced at the output of the C/L defining the next state may actually make their way through the feedback path, which is closed momentarily, producing further changes in the output of the C/L network for the next clock period. This is equivalent to a multistepping or race condition.

Clock skew is a serious problem when a number of chips are put together to form a total system. But even within a single large and complex chip, the effects due to skew must be analyzed thoroughly to avoid faulty circuit operation.

Because of the foregoing problems, single-phase clocking is not the designer's favorite, although it leads to economical circuits. Most of the problems can be avoided by using a more expensive clocking—the two-phase nonoverlapping scheme —which we discuss next.

3.6.4 Two-Phase Clocking Scheme

The discussion of the two-phase clocking scheme is best motivated from a discussion of the operation of the D flip-flop. A D or data flip-flop is an adaptation of the RS flip-flop with provision to load external data into the device. As shown in Fig. 3.60(a), the data D are latched into the RS flip-flop if the "sense" line S is raised high. If S is set to low after this, the data are retained in the cell. In practice, a D flip-flop is never built using this circuit. In CMOS, even replacing the AND and the OR with a complex gate AND–OR–INVERT, the circuit would still require 14 transistors (see Section 3.8 for further discussion). The D flip-flop will be symbolized as shown in Fig. 3.60(b). If the data have to be latched with the falling edge of the sense line, the circuit shown in Fig. 3.60(c) can be used. By putting these two circuits in cascade as shown in Fig. 3.61, a "master-slave" flip-flop is built. Data are latched into the "master" latch when S is high and are transferred to the output "slave" latch when S goes low. Indeed, the complementary relation between the ϕ_1 and ϕ_2 clocks can be utilized to build a static shift register using D flip-flops as shown in Fig. 3.62.

The master-slave control structure discussed above leads directly to the *two-phase, nonoverlapping* clocking scheme depicted in Fig. 3.55. The key idea behind

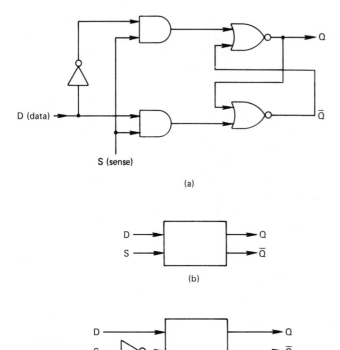

(a)

(b)

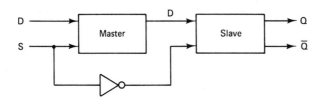

(c)

Figure 3.60 (a) D flip-flop circuit; (b) schematic of a D flip-flop; (c) D flip-flop latched at falling edge.

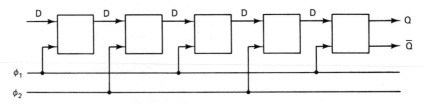

Figure 3.61 Master-slave D flip-flop configuration.

Figure 3.62 Shift register using a D flip-flop.

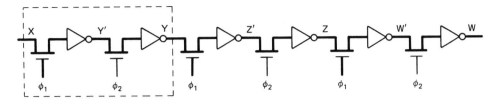

Figure 3.63 Dynamic shift register for one bit.

the two-phase clocking scheme is the complete isolation imposed by the master-slave principle. This is more clearly illustrated by the construction of a dynamic shift register using the memory element of Fig. 3.56. A shift register consists of a cascade of register stages. A register stage consists of a series connection of two memory elements clocked at alternate phases. A three-stage shift register is shown in Fig. 3.63. The part of the circuit enclosed in the dashed box represents a stage. A memory element is also sometimes referred to as a one-bit *half-register state*.

An array of *n* one-bit half-register stages, shown in Fig. 3.64, constitutes a half-register stage for an *n*-bit word. A shift register for *n*-bit words can be designed as shown in Fig. 3.65.

Arbitrary combinational logic (C/L) can be inserted between the half-registers to build a building block for a *data path*, as shown in Fig. 3.66. The logic could be a simple function or any arbitrary complex *n*-input *n*-output function preselected by the control inputs. To perform a sequence of operations on a data stream, a pipelined processing structure as shown in Fig. 3.67 can be used. Data will be processed as they stream through the various stages of the processing section. In one complete cycle of ϕ_1 and ϕ_2, two basic register-to-register operations can be performed on the data. During epoch A of the clock period, the data to the

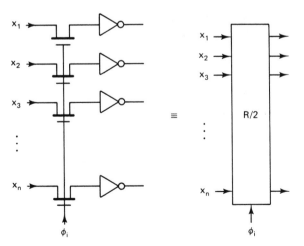

Figure 3.64 Half-register stage.

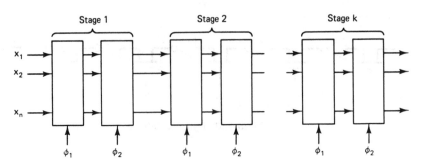

Figure 3.65 An *n*-bit shift register.

ϕ_1-controlled latches must be stable for a period at least equal to the setup time before the falling edge of ϕ_1 so that the storage nodes are charged to appropriate values. The duration of epoch A must be long enough to charge the largest capacitance of the storage nodes. During epoch B, the logic functions of the ϕ_1 stages are performed and the outputs of the combinational circuits settle. Actually, the combinational circuits may start working during epoch A as soon as the inputs to the stage have passed through the ϕ_1-clocked pass transistor. The duration of epoch B must be longer than the longest delay in the combinational logic of the ϕ_1 stage; if the logic networks start working within epoch A, the duration of B could be shorter. During epochs C and D, similar things happen with respect to the ϕ_2 stage storage nodes and combinational logic. Thus the input and the output of a stage become stable at alternate clock phases. This idea is the basis of the development of a "strict two-phase clocking" scheme (Noice et al., 1982; Noice, 1983), described later in the section. A critical discussion of the problems associated with the two-phase clocking discipline is to be found in Seitz (1980), who also formulates some of the problems connected with the concept of strict two-phase clocking discipline.

Let us illustrate the point about "stable at alternate clock phases" with respect to the circuit of Fig. 3.63 and the waveforms shown in Fig. 3.68, assuming for the moment that there is no clock skew. The binary string $X = 1010$ is applied to the

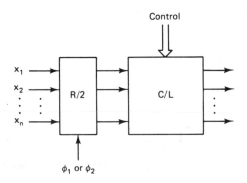

Figure 3.66 Half-stage of a data path.

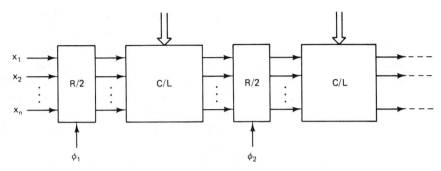

Figure 3.67 An n MOS data path.

input of the shift register. For proper latching, the value of X must settle for at least the setup time before the falling edges of consecutive ϕ_1 pulses. The waveform of X may look as shown in Fig. 3.68. Output Y' of the first inverter becomes zero after a short delay through the inverter and holds this value until the next 0 of X is latched near the rising edge of the next ϕ_1 pulse, when its value changes to 1, and so on. Notice that output Y of the first stage of the shift register changes in a similar fashion with respect to the leading edges of the ϕ_2 pulse, and that Y denotes the sequence 1010 at the end of the first and through the next four complete ϕ_1–ϕ_2 cycles. Similar waveforms can be drawn for the signals at points Z', Z, W', and W. Signals such as those depicted by Y' and Y in Fig. 3.68 are called *stable signals*. In particular, a *stable-ϕ_1* signal is defined to be a signal whose settled value extends over the period beginning and ending after a short delay δ between two consecutive rising ϕ_2 pulses. A *stable-ϕ_2* signal can be defined similarly. [The concept of stable signals has been introduced and formalized by Noice (1983); his definition is more technical in the sense that δ corresponds to the maximum skew or delay in the circuit.]

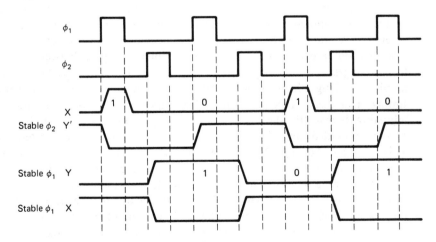

Figure 3.68 Stable signals.

The stable-ϕ_2 signal Y' and the stable-ϕ_1 signal Y in Fig. 3.68 will remain intact if X is replaced by the stable-ϕ_1 signal as shown in the bottom part of Fig. 3.68. This is because in this case a stable-ϕ_1 X signal contains the original signal X, satisfies the setup time requirements, and settles to the desired value before the rise of the ϕ_1 pulses. This leads to a "strict clocking discipline" (Noice et al., 1982), which can be informally described as follows. The discipline consists of alternating ϕ_1- and ϕ_2-clocked memory elements with combinational logic (switches and gates) between them. The inputs to the memory elements are stable; the output is stable of opposite phase; the inputs to the combinational network are stable with stable output of the same phase. Such a network structure has an additional interesting property in that its signal lines can be two-colored: one color, say red or solid lines, for ϕ_1 and stable-ϕ_1 signal lines, and another, say green or thin lines, for ϕ_2 and stable-ϕ_2 signal lines. The colors can change only at the clocked pass transistor, where a stable signal is converted to a stable signal of opposite phase. Note that the input and output of the combinational network have the same color, either red or green, because a purely combinational circuit adds a delay, which can be considered as part of δ, without altering the nature of the signal. The two-colorability property constitutes an easy check for a valid network structure obeying the strict two-phase clocking discipline.

A large number of useful circuits have been designed using the above clocking discipline. It is recommended that the reader try some examples to check the strict two-phase clocking discipline. In particular, mark the stable signal type and the color for all the signal lines for the following circuits discussed earlier: the one-bit shift register (Fig. 3.63), the shift register (Fig. 3.65), and the data path circuits (Figs. 3.66 and 3.67).

How good is the two-phase clocking scheme in eliminating problems due to race, hazards, and skew? First, we note that the complementary one-phase clocking is a special case of the two-phase nonoverlapping scheme in which epochs B and D have been reduced to zero. Obviously, therefore, two-phase clocking at worst is better than complementary single-phase clocking. Indeed, by controlling the width of epochs A and C we can satisfy the requirements with respect to setup time and clock period. Furthermore, because of complete isolation betwen ϕ_1- and ϕ_2-controlled stages, the multistepping problem is avoided. Thus the two-phase nonoverlapping clocking scheme allows us to design circuits that are guaranteed to have no timing error due to race or hazards. The importance of epochs B and D is that by adjusting their widths the circuit can also be guaranteed to have no timing error due to skew provided that all delays in the circuit are less than a maximum value of δ. This has been shown by Noice (1983) and we briefly discuss his reasoning below.

Let us return to the example of the dynamic storage device redrawn in Fig. 3.69. The timing relations in Fig. 3.69 refer to the global master clock. The specific local clock applied at the gate of the pass transistor may be skewed with respect to the master clock by a bounded delay δ. If we bound the start of the settled value for the signal X to a period δ after the master clock ϕ_1 rises, we can always guarantee the setup time by arbitrarily stretching the ϕ_1 period. This corresponds to adjusting the widths of epochs A and C. To guarantee the hold time, we cannot,

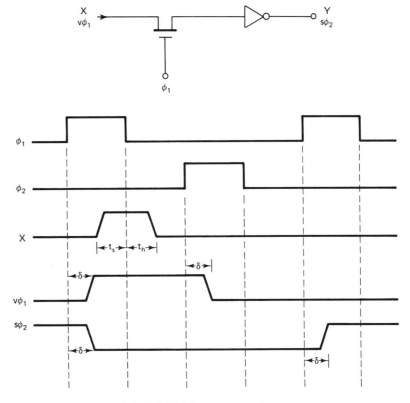

Figure 3.69 Valid signal with a latch.

however, designate the end of signal X with reference to the falling edge of ϕ_1, since these two events have to be considered concurrent and they could actually happen in any sequence because of skew. The end of signal X should be pegged to happen δ times after the rising edge of ϕ_2. Now, the hold time can be guaranteed by arbitrarily stretching the interclock gap, since it is known that all the events concurrent with the falling ϕ_1 edge must have been over before the next event, the rising ϕ_2. This corresponds to the adjustment of widths of epochs B and D. Such a signal is called a *valid ϕ_1 signal*, designated as vϕ_1. Similarly, a valid ϕ_2 signal, symbolized vϕ_2, is defined to be a signal whose settled value extends over a period δ after rising ϕ_2 to δ after rising ϕ_1.

Now, observe the output of the inverter. The storage node at the input to the inverter will be charged after a finite time from the beginning of the vϕ_1 signal, and the output of the inverter will then settle to a constant value and remain at that value until a different signal is latched, which will not happen until δ time after the next rising ϕ_1 edge. Such a signal will be called a *stable-ϕ_2 signal*, denote as sϕ_2. Similarly, a stable-ϕ_1 signal, symbolized sϕ_1, is defined to be a signal whose settled value extends over the period beginning shortly after δ between two consecutive

Figure 3.70 Generation of a gated clock.

rising ϕ_2. Note that the word "stable" and the symbols $s\phi_1$ and $s\phi_2$ denoting the stable signals now have a slightly different meaning than they had earlier, since δ now corresponds to the maximum skew.

We have just shown that using the signal types ϕ_1, ϕ_2, $s\phi_1$, and $s\phi_2$ and a strict clocking discipline, it is possible to design networks that do not have timing errors due to race, hazards, and skew. The necessity of stretching the epochs A, B, C, and D means that the price paid to achieve this is essentially to slow the clocks down or, equivalently, lower the speed of the circuit, depending on the value of the maximum skew.

In a practical circuit, besides ϕ_1 and ϕ_2, we sometimes have to use a special kind of clock called a *gated clock* or *qualified clock*, so called because it is generated by gating a control signal such as "Load" with a clock pulse, say ϕ_1, yielding a gated clock symbolized as "Load ϕ_1," as shown in Fig. 3.70. This signal is essentially a clock pulse which is turned off whenever the controlling signal is absent. The gated clock, like its parent ϕ_1 or ϕ_2, must be static hazard-free because quite often such clocks are used to latch a dynamic memory element. Any hazard present in these pulses could destroy the stored charges at the memory element. It is easy to guarantee that ϕ_1 and ϕ_2 are clean and square pulses without glitches because they are produced from separate clock generator circuits. But the gated clock is produced within the circuit and therefore might have a possible skew derived from the skew of the clock itself. Such a signal ANDed with a control signal and with a possible hazard at the beginning coinciding with the rising edge of ϕ_1 might produce a hazard at the output. However, if the Load signal is made a stable-ϕ_1 signal, it will remain at a settled value throughout the ϕ_1 period and the signal Load.ϕ_1 will be glitch-free. Thus, to work consistently with the two-phase clocking discipline, the

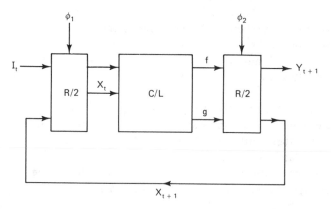

Figure 3.71 Sequential machine using two-phase clocking.

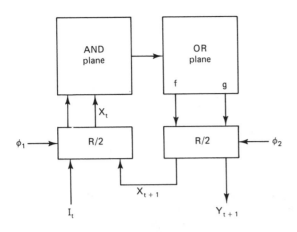

Figure 3.72 Finite-state machine with PLA and half-register.

rule to follow is always to use stable signals as the controlling signals to produce a gated clock.

A *finite-state machine* using two-phase nonoverlapping clocking can be realized by the network structure shown in Fig. 3.71. The *primary input* I_t and the "feedback" input X_{t+1}, representing *present-state* information, both of which are stable-ϕ_1, are latched into the input half-register during ϕ_1. The combinational logic computes the two functions f and g, representing the *next-state* and *output* functions, respectively, as

$$X_{t+1} = f(X_t, I_t)$$

$$Y_{t+1} = g(X_t, I_t)$$

The f and g outputs are stable-ϕ_2 and are latched into the output half-register during ϕ_2.

As we discussed in the preceding section, a convenient and flexible way to implement combinational logic is to use a PLA. Replacing the C/L part in Fig. 3.71 by a PLA, the general structure for the finite-state machine will be as shown in Fig. 3.72. Most of the PLA generator software available today has the option to generate a finite-state machine with input and output latches and clock signal lines.

3.6.5 Precharged Logic

The basic idea of *precharged logic* can be explained with respect to the inverter gate shown in Fig. 3.73. The depletion load pull-up is replaced by a clocked enhancement-mode pull-up. This allows the output node to be precharged high during ϕ_2. During ϕ_1 if the input is high, the output is discharged to zero; otherwise, it stays

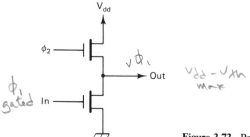

Figure 3.73 Precharged inverter.

high. The input "In" is a ϕ_1-gated clock signal. The signal waveform obtained at the output of the inverter is shown in Fig. 3.74, which also depicts a possible skew of δ. It is very similar to a valid ϕ_2 signal except that its high and low values are predetermined with respect to ϕ_2 and ϕ_1, respectively. Similarly, a ϕ_1-precharged inverter produces a valid ϕ_1-like signal at the output. The precharged inverter needs less power since there is no direct current path. Also, since the output is defined only during ϕ_1, the speed is determined by the pull-down time. The pull-down current only has to sink the charge stored in the load capacitance since the pull-up is turned off during ϕ_1. A normal pull-down in ratioed logic also sinks the current in the pull-up resistor. The precharging of the output is done during ϕ_2, when we do not care about the output. The ratio considerations for the standard inverter design are avoided for precharged logic. The output, however, cannot be raised beyond $V_{dd} - V_{th}$ for nMOS circuits. For a precharged CMOS network, such a restriction does not apply, as we will see in Section 3.7.

We discuss some examples of precharged circuits in this section (several others are discussed in Chapter 7) and point out the potential charge-sharing problems that might arise. Precharged circuits offer speed and power advantages, but one must be cautioned about the delicate timing problems, as discussed here.

Precharged bus. In the design of the OM2 machine (Mead and Conway, 1980) two standard buses were used; the registers, the function blocks, and the ALU were attached to them via two data ports. All transfers on the buses

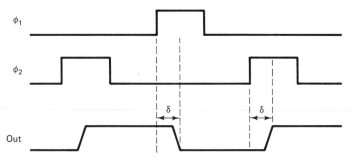

Figure 3.74 Output waveform of the precharged inverter.

occurred during ϕ_1. Du.ing ϕ_2, the computations are performed while the bus is being precharged. A schematic of the circuit is shown in Fig. 3.75(a). The data from only one source can be allowed to pull down the precharged bus during ϕ_1, but a multiple number of sources can sink the bus at different ϕ_1 periods. The output of the bus is also latched on the trailing edge of ϕ_1. The low-to-high transition at the output does not occur during ϕ_1 (evaluation phase), the levels either remain constant high or change from high to low. This accounts for higher speed for precharge networks.

Precharged multiplexer.   A circuit for a precharged multiplexer is shown in Fig. 3.75(b). The output is precharged at ϕ_2. Only one of the control signals must come high during ϕ_1 to select a source. If the source has a value 0, the output will be

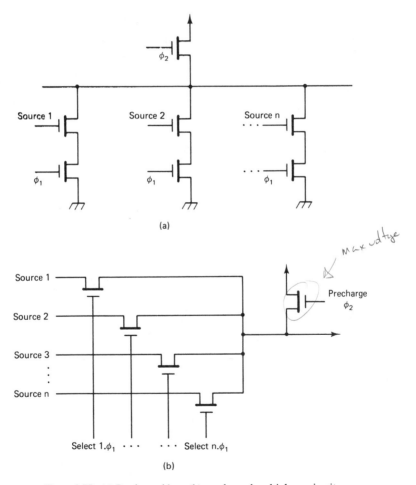

(a)

(b)

Figure 3.75 (a) Precharged bus; (b) precharged multiplexer circuit.

pulled low during ϕ_1; otherwise, it will stay high. Each source is represented by only one line, but in practice it could be an n-bit bus, in which case the logic is repeated n times.

Charge sharing.   Any glitch or static hazards present in the input to the pull-down transistors will bleed off charge from the precharged bus. Noice (1983) has investigated safe methods of controlling the pull-down path of a precharged circuit; we will briefly discuss some of his observations.

A precharged circuit might work incorrectly due to charge-sharing errors, which could occur inside the pull-down network or at the output circuit. Consider the circuit shown in Fig. 3.76(a), which has two pull-down transistors in series controlled by gated clocks. Suppose that the point Y has been charged to low in the previous cycle and the output node is precharged high during ϕ_2. Now, if during ϕ_1 a is high and b is low, the transistor controlled by a will be on and node X will share charge with node Y. This will degrade the output value and may make it indeterminate. By a similar analysis, one can see that there may be a charge-sharing problem between nodes X and Y in the circuit shown in Fig. 3.76(b). In general, it can be said that if two or more pull-down transistors have gated clock signals or if a gated clock signal is above a stable signal, there will be a potential charge-sharing problem.

The correct way to control a precharged circuit is to use a gated clock or a clock signal at the input of the bottom transistor, and all other in series with the pull-down chain must have a stable signal over the same clock phase. In other words, the signals must be stable not only over the enabling clock phase, but also for some short period of time (setup time) during the precharge, to allow the internal nodes to be charged. Two such correct circuits are shown in Fig. 3.77. The circuit in Fig. 3.77(a) is the same as that shown in Fig. 3.73. The charge sharing is avoided in Fig. 3.77(b) since a stable-ϕ_1 signal is also a valid signal on the previous ϕ_2 phase. This

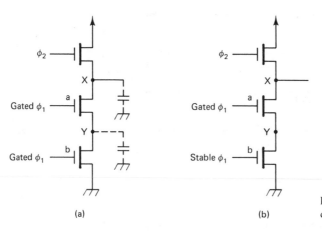

(a)   (b)

Figure 3.76 Precharged circuit with charge-sharing problem.

means that if input a is 1, point Y will be recharged high while X is charged high during ϕ_2. The stable-ϕ_1 signal settles early enough to give the internal pull-down nodes enough time to be charged high. If a is 0, there is no chance for charge sharing to occur. Thus X and Y will have the same strength during ϕ_1 if a is on and b is off, avoiding signal degradation.

The output of a precharged network is very sensitive to noise or "glitch." The circuits described below improve its noise immunity. The first scheme, shown in Fig. 3.78(a), attaches a weak (long-channel or high-resistance) device with a feedback inverter and is used for CMOS networks. Note that the precharge transistor is a p-channel device and its gate is controlled by the complement of a one-phase clock PC. The inverter provides gain to the rest of the circuits, and the feedback signal provides a path for a small dc hold current, making the precharged line less sensitive to noise. The speed of the circuit is reduced, however, since during pull-down, the large gate capacitance of the weak device has to be charged up. A faster scheme is to use the circuit of Fig. 3.78(b). Here a "bias" voltage V_{BIAS} is applied to the gate of the n-channel device to provide a constant hold current, which typically ranges between 0.2 and 1 μa. The device mimics a pull-up resistor with a very high resistance. Alternatively, a p-channel transistor with a bias voltage between V_{dd} and V_{ss} can be used.

The signal type of the output of a precharge bus is not stable; it is, as we mentioned earlier, a special kind of valid signal. To use this signal in other parts of the circuit that use only stable signals, the two circuits shown in Fig. 3.78(c) can be used. The first scheme is simply a dynamic storage element with a $8:1$ ratio inverter and is gated by a stable-ϕ_1 signal. This circuit avoids charge-sharing problems between nodes X and Y, for reasons that avoided charge sharing in Fig. 3.77; that is,

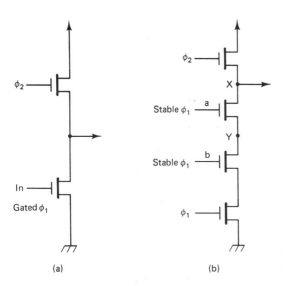

(a)

(b)

Figure 3.77 Precharged circuits with no charge-sharing problem.

the gated stable-ϕ_1 signal at the gate of the pass transistor, being on at the previous ϕ_2 phase, allows Y to get charged to 1 prior to the pull-down phase. Alternatively, the precharged output could be applied to an inverting or noninverting buffer of ratio 8 : 1 before taking it out to other parts of the circuit. The output of the buffer is a valid ϕ_2-precharged signal and can be converted to a stable-ϕ_1 signal via a ϕ_2-controlled latch.

It is, or course, possible to use the output of a precharged circuit to other parts of the network. In fact, domino logic (see Section 3.7 and the Exercises) uses precharged output directly.

Summarizing, the precharged circuits should obey the following rules to avoid charge-sharing problems: The pull-down chain of transistors can have only one

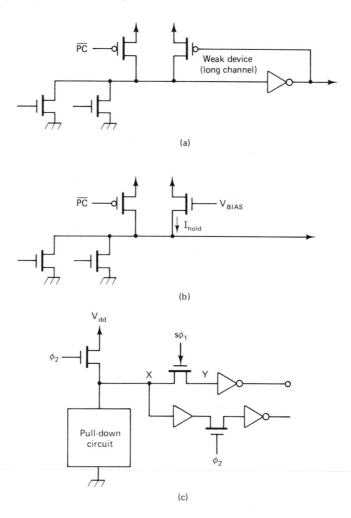

Figure 3.78 Safe methods of taking outputs from a precharged circuit.

gated clock or ϕ_1 clock input for the bottom transistor. The other inputs should be stable-ϕ_1. The precharged output can drive other gate inputs or could go through a stable-ϕ_1 latch to other gate inputs.

3.7 Clocked CMOS Logic

Clocked CMOS circuits are dynamic circuits that generally work based on the principle of precharging the output node to a particular level when the clock has a value 0. During the precharged phase the inputs to the circuits change. When the clock has a value 1, the output may be pulled to a complementary value depending on the input conditions. Clocked CMOS circuits can be designed using both two-phase and single-phase clocks. A more commonly used clock is the complementary single-phase clock. In this section we present several clocked combinational networks, dynamic storage devices, precharged domino logic, and precharged CMOS PLAs.

3.7.1 Clocked Inverter

The inverter circuit is shown in Fig. 3.79(a). Whenever $in = 1$ and $cl = 1$, the output becomes 0 because the pull-down path is closed connecting output to V_{ss}. If $in = 0$, and $cl = 1$, the p-transistors will conduct (since $\overline{cl} = 0$), sending a 1 at the output.

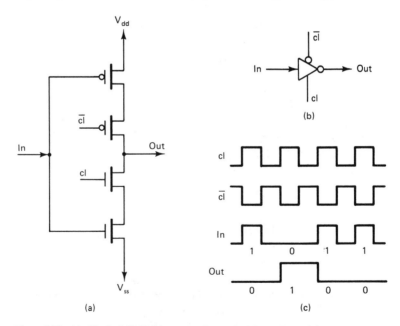

Figure 3.79 (a) Clocked CMOS inverter; (b) symbol for a clocked CMOS inverter; (c) output waveforms from the clocked inverter.

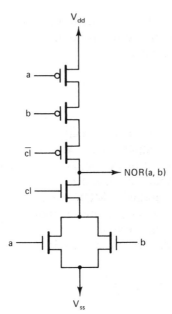

Figure 3.80 Clocked CMOS NOR network.

For all other combinations, the output holds the previous value. A symbol for the inverter is shown in Fig. 3.79(b), and Fig. 3.79(c) gives an example of a typical pulse sequence depicting the operation of the inverter. Note that the circuit would function the same way if the $\overline{cl}$ and the input connections were permuted for the p-transistors, and similarly if cl and the input were permuted for the n-transistors, yielding a total of four inverter configurations. It is possible to build more complex gates by extending the basic principle of this circuit. For example, a clocked NOR circuit is shown in Fig. 3.80. The reader should verify its function as a NOR gate. An alternative method of designing CMOS circuits is to use the precharged principle described in the next section.

3.7.2 Precharged Domino CMOS Logic

A precharged CMOS inverter circuit is shown in Fig. 3.81(a). The output is precharged to 1 during the time when $cl = 0$. If $in = 1$ during $cl = 1$, the output will be pulled to ground; otherwise, it will stay at 1. By a similar argument, one can verify that the circuits in Fig. 3.81(b) and (c) perform NAND and NOR functions, respectively. A generalization is obvious, as shown in Fig. 3.82(a); the box marked f is the pull-down network, which realizes an arbitrary combinational function $f(X)$ of n variables $X = (x_1 \cdots x_n)$ as a series–parallel interconnection of n-transistors, as discussed in Section 3.5.1. The single p-transistor precharges the output to 1 during $cl = 0$; this is the usual *precharged phase*. During this phase the inputs X to the f-network must change. During the next phase, called the *evaluate phase*, the network function is produced at the output of the f-network. If $f = 1$ during $cl = 1$, the output is pulled down to 0, realizing $\bar{f}(X)$ as the output. To ensure that it occurs in coincidence with the clock, a single n-transistor is attached in series with the f-network.

The precharge technique has the advantage that it avoids the use of the pull-up network, at the expense of one p-transistor and one additional n-transistor. This leads to almost a 50% savings in silicon area for larger circuits and reduction of the output capacitance, resulting in higher speed. Furthermore, the pull-down circuit does not have to sink the pull-up current (as in the case of the pseudo-nMOS circuit), which also results in higher speed. Also, there is no static power dissipation except for the small amount of power needed to precharge the output high every cycle (in case the output was pulled low in the previous cycle).

There is a major disadvantage of this circuit if no feedback is provided to hold the precharged line high (as discussed in Fig. 3.78). The precharge operation will probably fail due to noise and charge coupling. Consider what happens if output was supposed to remain high but another line (on another layer) crossed over and was changing from 1 to 0, as shown in Fig. 3.82(b). Where the lines cross, there will be a very small coupling capacitance; this may be large enough to cause a 0.1-V to 0.2-V shift down in precharged output. If the output crosses over a few more lines,

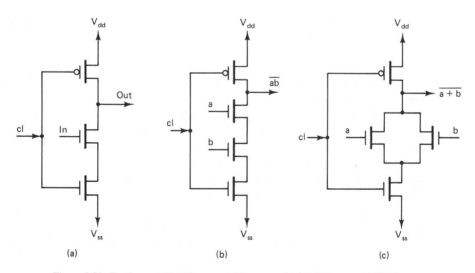

Figure 3.81 Precharged CMOS gates: (a) inverter; (b) NAND gate; (c) NOR gate.

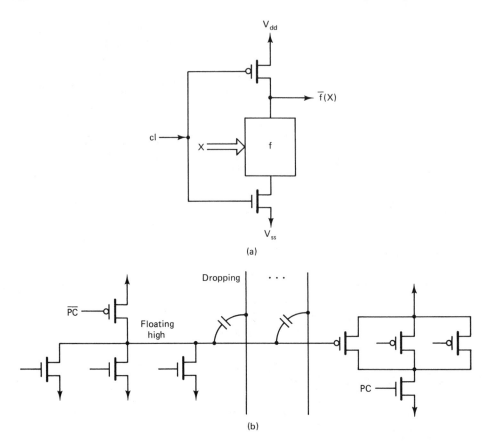

Figure 3.82 (a) Precharged network for arbitrary combinational function; (b) the capacitative coupling with a precharge line.

this might lead to serious problems, particularly if the next stage of the circuit is also precharged and only needs an input to drop an amount equal to the threshold voltage V_{thp} to turn on.

There is also a serious problem with the circuit if multiple stages have to be used together to realize a function. Consider the two-stage precharged circuit shown in Fig. 3.83(a). Ignore the dashed inverter for the moment. Suppose that the output y of the first stage is supposed to go to 0 during the evaluate phase. Since there is a finite delay involved in this transition, its effect may not be sensed by the f_2-network until the end of this delay into its evaluate phase. Since $y = 1$ during this time, the transistors controlled by y will turn on. Depending on the structure of the network and the input values assigned to X, this might momentarily produce a 0 output at z when it is supposed to be 1. This is equivalent to generating a glitch or a hazard at the output.

The problem can be remedied by passing the output of the first stage through an inverting buffer stage, as indicated by the dashed inverter in Fig. 3.83(a). This

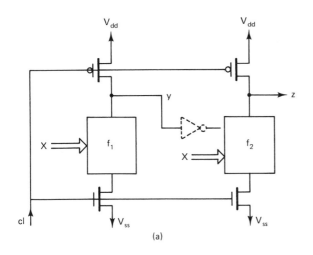

(a)

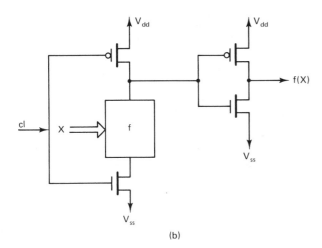

(b)

Figure 3.83 (a) The glitch problem in a domino logic network; (b) domino logic stage.

might necessitate making changes in the f_1-network and f_2-network. The output of the buffer is now 0 after the precharge period and will not turn on any transistor of the f_2-network during the initial delay associated with the evaluate phase of the f_1-network, thereby avoiding the generation of any glitches.

A precharged logic circuit with an output inverter buffer stage is shown in Fig. 3.83(b). It consists of a precharged logic stage of Fig. 3.82(a) whose output is connected to an inverting buffer stage. During precharge, the output of the logic stage is 1, so the buffer output is 0. This means that all transistors of the following stages driven by the output are turned off during the precharge phase, which guarantees that there can be no glitches in any nodes in the circuit. Furthermore, there can only be one possible transition for each node during the evaluate

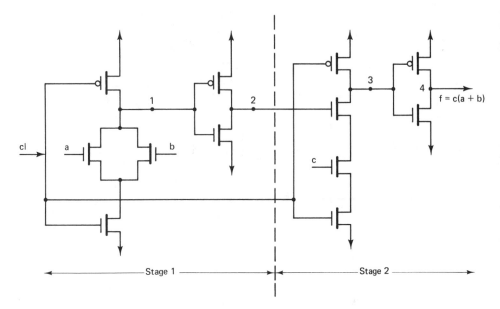

Figure 3.84 Two-stage domino network for $f = c(a + b)$.

phase—the logic stage output possibly changing from 1 to 0 and the buffer output changing from 0 to 1. These values can again change at the next precharge phase.

An example of a two-stage domino CMOS circuit is shown in Fig. 3.84 for the function $f = c(a + b)$. During precharge ($cl = 0$), nodes 1 and 3 are high, and node 2 and the output node 4 are low. Assume the input values $a = b = 1$ and $c = 0$. During the evaluate phase ($cl = 1$), node 1 goes low, causing node 2 to go high. Since $c = 0$, nodes 3 and 4 maintain their previous values. However, if $c = 1$, node 3 will be pulled low, causing node 4 to go high. The chain of actions propagating in such a fashion is comparable to a row of falling dominos, hence the name *domino logic*. The extra delay due to the inverter stage in domino logic can be avoided if alternating precharge logic is used: precharge high with pull-up followed by pre-charge low with pull-down. Not only does this eliminate the gate delay of the inverter, but precharge logic requires only an input slightly over the threshold voltage to turn on. This leads to faster circuit operation. However, this technique is useful when each gate is composed of relatively few transistors and the gates are built close together.

3.7.3 Dynamic Shift Register

Dynamic shift registers can be built in CMOS using simple latches and single- or two-phase clocking. The circuit for a single stage of a dynamic shift register using a single-phase clock is shown in Fig. 3.85. When $cl = 0$, the input is passed through the transmission gate T_1 to charge up the input capacitance C_1 of the first inverter.

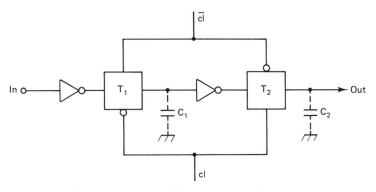

Figure 3.85 CMOS dynamic shift register stage using a one-phase clock.

When $cl = 1$, T_1 is turned off, isolating the charge in C_1, which is then inverted and delivered at the output via transmission gate T_2. At the end of one clock period, the output becomes equal to the input. Note if T_1 and T_2 were placed before the two inverters, the circuit will still be a shift register. A cascade of n such cels will form an n-bit register.

An obvious adaptation of the circuit using two-phase, nonoverlapping clocking is shown in Fig. 3.86. Again, n such stages will form an n-bit shift register.

One possible way to avoid the use of the transmission gate is to use the clocked inverter circuit of Fig. 3.79 to constitute a half-register stage. The circuit is shown in Fig. 3.87. It consists simply of two clocked inverters in cascade. The input is complemented once during ϕ_1 and again once more during ϕ_2 so that at the end of a complete clock cycle the output equals the input. The circuit also has the advantage that its leads to regular and compact layout.

It is now possible to build a CMOS data path similar to the nMOS circuit shown in Fig. 3.66, by inserting arbitrary combinational logic between the half-register stages. To perform a sequence of operations on a data stream, a pipelined processing structure similar to the nMOS data path shown in Fig. 3.67 can be used. The notion of strict two-phase nonoverlapping clocking discipline is also applicable to this class of circuits. An alternative approach to the data path is to use a precharged domino logic stage, as explained in the next section.

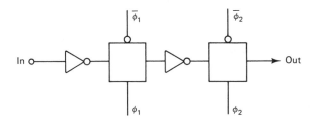

Figure 3.86 CMOS dynamic shift register stage with a two-phase clock.

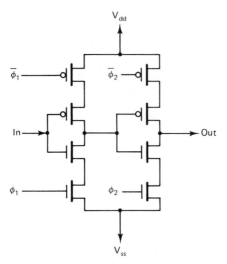

Figure 3.87 Shift register stage using clocked inverter.

3.7.4 Data Path Using Domino Logic

The circuit shown in Fig. 3.88 is a dynamic shift register stage using precharge logic and a single-phase clock. When $cl = 0$, the output transmission gate is open and capacitors C_1 and C_2 at the output of the inverters are precharged to 1 and 0, respectively. When $cl = 1$, the transmission gate is closed and the input *in* is transmitted to the output after double inversion. Thus the circuit produces a one-stage delay equal to the clock period. A cascade of n such cells will form an n-bit shift register.

If we replace the single n-transistor attached to *in* (enclosed in the dashed box) by the arbitrary n-transistor logic network f of Fig. 3.82, and use several such

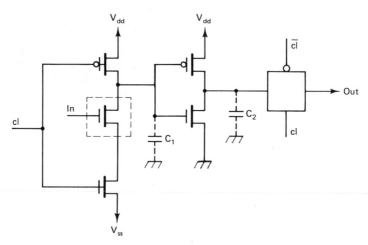

Figure 3.88 Dynamic shift register stage using domino logic.

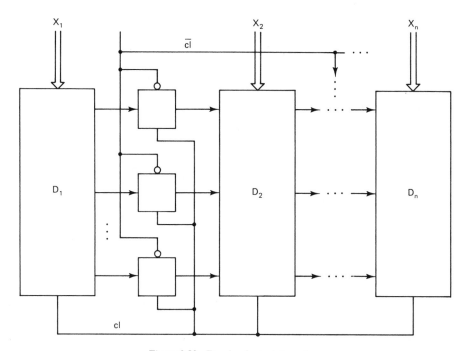

Figure 3.89 Domino logic data path.

circuits together, we get the domino logic data path represented schematically in Fig. 3.89. As in the case of an nMOS network, the pipelined structure performs a sequence of operations on the data stream $X_1, X_2, \ldots, X_n$.

3.7.5 Precharged CMOS PLA

The general structure of the PLA is similar to the nMOS PLA described in Section 3.4.2. It consists of an AND plane which generates the terms to be used as input to an OR plane, which computes the function. The actual implementation is done using either NOR or NAND planes.

An example of a precharged CMOS PLA to realize the functions z_1, z_2, and z_3 as given in Section 3.4.2 is shown in Fig. 3.90. The circuit is essentially a two-stage precharged domino network and uses a single clock. When $cl = 0$, the outputs z_1, z_2, and z_3, as well as the outputs from the NAND stages, are precharged to 0. When the clock comes high, some of the output may change to 0, depending on the input values and the arrangement of the n-transistors in the AND and OR planes that realize the desired functions.

Many variations of the basic PLA structure are possible to suit the needs of a particular design. The input and output could be held in clocked half-register stages which will require two clock cycles for the PLA. A two-phase clocking scheme as

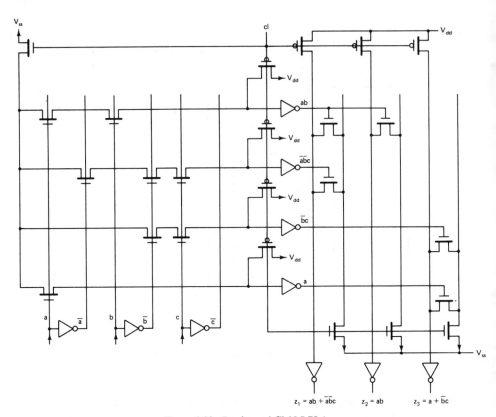

$z_1 = ab + \overline{abc}$ $z_2 = ab$ $z_3 = a + \overline{bc}$

Figure 3.90 Precharged CMOS PLA.

used for nMOS PLA can also be used. Finite-state machines can be synthesized by using appropriate feedback connections from the output to some of the input. Most of the optimization techniques discussed earlier for nMOS PLA are also applicable to CMOS PLA (Weinberger, 1979; Wood, 1975; Glaser, 1982; Hachtel et al., 1980).

3.8 Register Stage

In our discussion so far we have implicitly assumed that the charge leakage will not cause the storage nodes to degrade too much and that all storage nodes are refreshed often enough. If this is not true, the storage nodes have to be refreshed. A *semistatic register* is a dynamic memory element in which the charge is refreshed every clock cycle. We will discuss several such nMOS and CMOS circuits in this section.

A cross-coupled CMOS (or nMOS) inverter, as shown in Fig. 3.91, can be used to form the basic storage element of a *static* register cell. Whenever the input is equal to the output, the network will make a transition to a stable state in which the input and output are complementary. Note that the circuit does not need refreshing

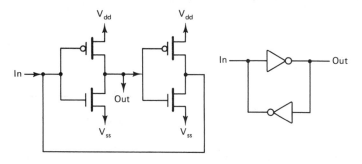

Figure 3.91 Register stage with two inverters.

since it will hold its state information indefinitely. It is very much like the *RS* flip-flop circuit except that it has only one input.

To be able to control the state of the flip-flop, a data value has to be loaded into the latch by using one of the clocked schemes shown in Fig. 3.92 for CMOS and in Fig. 3.93 for *n*MOS. In the schemes in Fig. 3.92(a) and in Fig. 3.93(a), the input data are forced into the latch during $cl = 1$. There is a potential problem if point *P* was holding a value opposite to the value of *X*. This might drive the final state of the latch to an undefined state. A better scheme for CMOS is shown in Fig. 3.92(b), which uses a gated clock LD.ϕ_1, the symbol LD standing for "load." During LD = 0, the transmission gate T_2 is closed and T_1 is open, which allows recirculation of the bit information to refresh point *P*. During LD = 1, the transmission gate T_2 is open, isolating point *P* from the feedback inverter and exposing it to the input driver, which charges *P* to the value of input signal *X* via gate T_1. One disadvantage of this scheme is that the line *X* has a dynamic capacitance load since the input transistor gate (point *P*) is connected to *X* if the transmission gate is closed. If the two transmission gates are on for a short time during the transition of the clock signals, the signal on the input (x) may be driven by the output inverters. This coupling from output to input may cause real problems or may cause some simulators to be unable to deal with this circuit. This circuit is rarely used in practice. A circuit that avoids this problem is shown in Fig. 3.92(c). This circuit uses 10 transistors and is frequently used. However, $\overline{X}$ may be weak depending on the design and cannot be heavily loaded without slowing the latch. A circuit in which both *X* and $\overline{X}$ are strongly driven is shown in Fig. 3.92(d).

A semistatic *n*MOS register stage is shown in Fig. 3.93(b). If the load signal LD is high, the signal value *X* will be latched into the ϕ_1-clocked memory element and will emerge as a stable signal at the output. If the LD signal is turned off, the information can be retained via the recirculating path controlled by the complement signal $\overline{LD}$ until new information is fed in by turning on LD. Note that the circuit provides both *X* and $\overline{X}$ as stable-ϕ_2 signals (the input is stable-ϕ_1) and should be latched into a ϕ_2-clocked latch or network at the output.

The register stages of Figs. 3.92 and 3.93 do not obey the strict two-phase clocking discipline, as one can easily verify. A semistatic *n*MOS register cell that obeys the strict clocking discipline is shown in Fig. 3.94. If the load signal LD is

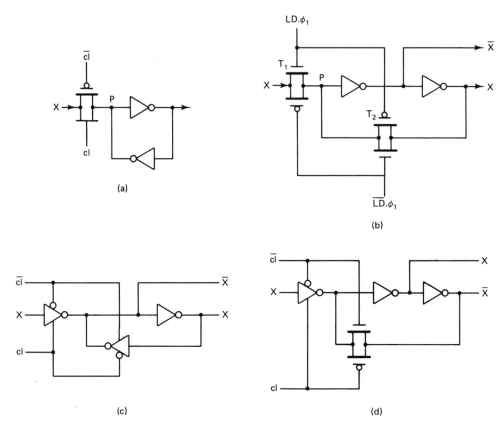

Figure 3.92 Several designs of a register stage in CMOS.

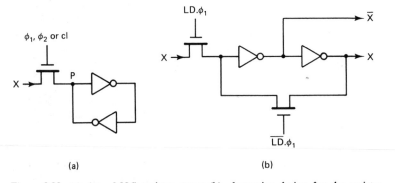

Figure 3.93 (a) An nMOS register stage; (b) alternative design for the register

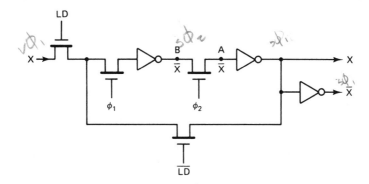

Figure 3.94 Safe semistatic register stage.

high, the signal value X will be latched into the ϕ_1-clocked memory element and will emerge as a stable-ϕ_1 signal via the ϕ_2-clocked memory. If the LD signal is turned off, the information can be retained through successive clock periods via the recirculating path controlled by the complement signal $\overline{LD}$ until new information is fed by turning on LD. The signal LD could be level, but a more practical circuit uses gated clock signals by replacing LD and $\overline{LD}$ by LD.ϕ_1 and $\overline{LD}.\phi_1$.

One other advantage of this circuit is that it provides both true and complemented outputs X and $\overline{X}$, as shown. The complemented output $\overline{X}$ is safe and is stable at ϕ_1. The output at point A, although stable at ϕ_1, is not very safe because if it is connected to a large load capacitance, the storage node might lose its charge and destroy the stored information. The complemented output at point B is safe, but it is stable at ϕ_2. The CMOS equivalent of the circuit of Fig. 3.94 is shown in Fig. 3.95. The circuit uses two-phase clocks and their complements. If the input is valid or stable Φ_1, the output of the circuit is stable ϕ_1.

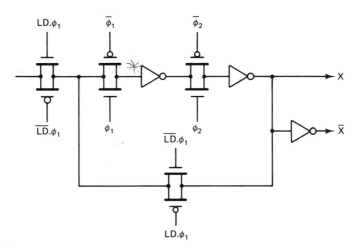

Figure 3.95 Safe semistatic register stage in CMOS.

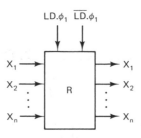

Figure 3.96 Selectively loadable register stage.

An array of recirculating register cells of Figs. 3.92 through 3.95 can be arranged to form a *selectively loadable* register stage as symbolized in Fig. 3.96 and can be used as a subsystem to hold, for an indefinite period of time, information whose content can be altered at the control of the load signal. The input and output of the register stage are stable at ϕ_1 if we use the cells of Fig. 3.94 or Fig. 3.95. If we use the cells of Fig. 3.92(b) or Fig. 3.93(b), the input and output will be stable at opposite phases. A number of such registers connected in series form a push-down stack. The information can be put into it from the left and can be held in it for an indefinite period of time. There is, however, a more elegant circuit to do stack operations (see Chapter 7).

3.9 Summary

This chapter has presented the basic logic design methods for synthesizing arbitrary combinational and sequential circuits in both nMOS and CMOS technology. These tools will be used to design building blocks for subsystems in Chapter 7. At this point an understanding of the basic semiconductor processing technology, design rules, simple layout techniques, and electrical parameters is essential to appreciate the abstractions used in the hierarchical design methodology. The following three chapters are concerned with these topics.

REFERENCES

Ashenhurst, R. L., "The Decomposition of Switching Functions," *Ann. Comput. Lab. Harvard University*, Vol. 29, No. 30. Cambridge, Mass.: Harvard University Press, 1959.

Caldwell, S. H., *Switching Circuits and Logical Design*. New York: Wiley, 1958.

Glaser, A. B., "A Feedback Reduced PLA," *MIT Tech. Rep.*, 1982.

Hachtel, G. D., L. A. Sangiovanni-Vincentelli, and A. R. Newton, "Some Results in Optimum PLA Folding," *Proc. IEEE International Conference on Circuits and Computers*, p. 1040. New York, Oct. 1–3, 1980.

Hennie, F. C., *Iterative Arrays of Logic Circuits*. Cambridge, Mass.: MIT Press, 1961.

Krambeck, R. H., C. M. Lee, and H. S. Law, "High-Speed Compact Circuits with CMOS," *IEEE J. Solid-State Circuits*, Vol. SC-17, No. 3, 1982, pp. 614–619.

McCluskey, E. J., *Introduction to Switching Circuits*. New York: McGraw-Hill, 1965.

Mead, C., and L. Conway, *Introduction to VLSI Systems*. Reading, Mass.: Addison-Wesley, 1980, Chaps. 1 and 3.

Mukhopadhyay, A., and H. Stone, "Cellular Logic," Chap. VII in *Recent Developments in Switching Theory* (Ed. A. Mukhopadhyay). New York: Academic Press, 1971.

Muroga, S., *VLSI System Design*. New York: Wiley, 1983.

Noice, D. C., "A Clocking Discipline for Two-Phase Digital Integrated Circuits." Ph.D. dissertation, Stanford University, Stanford, Calif., Jan. 1983.

Noice, D., R. Mathews, and J. Newkirk, "A Clocking Discipline for Two-Phase Digital Systems," *Proc. IEEE International Conference on Circuits and Computers*, Sept.–Oct. 1982.

Seitz, C. L., "Systems Timing," Chap. 7 in *Introduction to VLSI Systems* (Eds. C. Mead and L. Conway). Reading, Mass.: Addison-Wesley, 1980.

Shannon, C. E., "A Symbolic Analysis of Relay and Switching Circuits," *Trans. Am. Inst. Electr. Eng.*, Vol. 57, 1938, pp. 713–723.

Shannon, C. E., "The Synthesis of Two-Terminal Switching Circuits," *Bell Syst. Tech. J.*, Vol. 28, No. 1, 1949.

Sutton, J. A., II, private communication, Harris Corp., Melbourne, Fla., 1985.

Unger, S. H., and C. J. Tan, "Optimal Clocking Schemes for High Speed Digital Systems," *Proc. IEEE International Conference on Computer Design: VLSI in Computers*, Oct.–Nov. 1983.

Weinberger, A., "Large Scale Integration of MOS Complex Logic: A Layout Method," *IEEE J. Solid-State Circuits*, Dec. 1967, p. 182.

Weinberger, A., "High-Speed Programmable Logic Array Adders," *IBM J. Res. Dev.*, Mar. 1979, p. 163.

Wood, R. A., "A High Density Programmable Logic Array Chip," *IBM J. Res. Dev.*, July 1975.

Ullman, J. D., *Computational Models in VLSI*. Rockville, Md.: Computer Science Press, 1983.

EXERCISES

3.1. Prove that the network proposed in Fig. 3.11 can realize an arbitrary switching function of three variables with a switching matrix size not exceeding 4×4. Obtain a specific realization for

$$f = \overline{x}\overline{z} + \overline{x}y + \overline{xy}z$$

3.2. Complete Table 3.2; that is, obtain the functions g_0 and g_1 for $i = 0, 2, 3, 4, 5, 10, 11, 12, 13, 15$.

3.3. If the input and output terminals of a 4-to-1 multiplexer circuit is reversed, can we use the circuit as a two-variable decoder? Justify.

3.4. What is the output of the circuit shown in Fig. E3.4? What problems do you see?

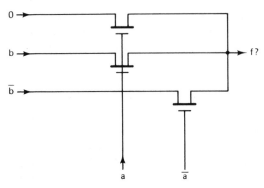

3.5. Draw a stick diagram for an efficient modified tally circuit with three inputs (x_1, x_2, x_3) and two outputs (z_3, z) where $z = z_0 + z_1 + z_2$. Note that tying outputs z_0, z_1, and z_2 of Fig. 3.22 will not constitute an efficient solution.

3.6. Design iterative networks to detect each of the following patterns in an input binary string:

(a) The ith output is 1 if the binary sequence preceding this bit consists of groups of 1's and 0's such that each group contains an even number of bits.

(b) The ith output is 1 and all other outputs are 0 if and only if this bit is the first 1 bit in the group of bits preceding the ith bit.

(c) There is only a single output from the network, which is 1 if and only if the input sequence contains two consecutive groups of three 1's separated by at least one 0.

3.7. Of the two possible inverter circuits shown in Fig. E3.7, why is the depletion load pull-up configuration usually preferred? What is the ratio of the pull-ups if the inverters are used in cascade as shown?

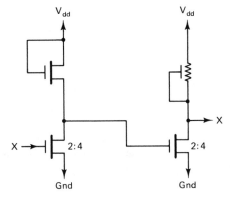

3.8. Construct the stick diagram for an nMOS PLA which implements the adder stage of two 2-bit numbers $N_1 = (a_1 a_0)$ and $N_2 = (b_1 b_0)$ with input carry C_i producing two sum bits $S = (s_1 s_0)$ and a carry output of bit C_o. Then explore all the optimization methods on this problem: logic minimization, encoded PLA, folding, Weinberger array, and FRPLA.

3.9. A *JK* flip-flop is very similar to the *RS* flip-flop except that the input combination $(1,1)$ is allowed and acts like a toggle input, changing the state of the flip-flop. It is described by the following table:

J	K	Q_{n+1}
0	0	Q_n
1	0	1
0	1	1
1	1	$\overline{Q}_n$

Design an *n*MOS circuit for the *JK* flip-flop using two-phase clocking. Then design a precharged CMOS network for this flip-flop.

3.10. Consider a canonic form of a sequential circuit similar to the one shown in Fig. 3.59(a) with the simple latches being replaced by *D* flip-flops. Analyze the circuit to determine what restrictions have to be put on the width and period of the clock in order that the circuit will work free of errors due to hazards or races.

3.11. A shift register circuit using a complementary single-phase clock is shown in Fig. E3.11. Describe how the circuit might malfunction if there is a clock skew between *cl* and $\overline{cl}$.

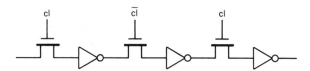

3.12. A variation of the semistatic register is shown in Fig. E3.12. During ϕ_1 information is latched into the cell; during ϕ_2 the load capacitance C is driven simultaneously, refreshing the memory element via the feedback pass transistor. There is a serious charge-sharing problem with this circuit. Find this problem and suggest methods to remedy it.

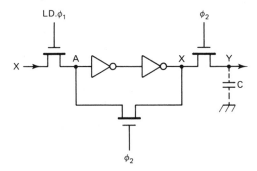

3.13. (a) Design a two-port semistatic register stage, each port being connected to a distinct bus to which it can write and read from during ϕ_1. The register stage is refreshed during ϕ_2.

(b) Design a register stage that can read and write data on a single bus using precharged logic in both *n*MOS and CMOS.

3.14. The circuit shown in Fig. E3.14 is a shift register stage in which the depletion load pull-up has been replaced by a clocked enhancement pull-up. Analyze the circuit to verify its function over a complete ϕ_1-ϕ_2 cycle and determine whether or not any constraints are imposed on the sizes of the transistors for proper functioning. What will be the general structure of a data path using this type of register cell for dynamic storage?

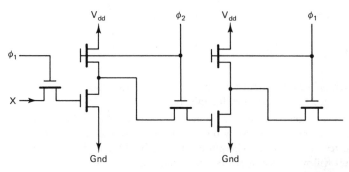

3.15. (a) The circuits shown in Fig. E3.15(a) are called "ratioless" circuits. Note that the circuit here differs from the circuit in Fig. E3.14 in that the connection to the clock line to the pull-up transistor is altered from the output pass transistor to the input pass transistor. Analyze the circuits to verify that it is acting as a shift register stage. Are there any constraints on the sizes of the transistors for proper functioning? Is there a potential charge-sharing problem? Show how the principle can be used to realize arbitrary combinational circuits.

(b) A "ratioless" dynamic register stage that does not use a power supply is shown in Fig. E3.15(b). It consists essentially of two inverter stages connected in cascade, each of which works based on the precharge principle. Explain the operation of the circuit. Is there a potential charge-sharing problem in the circuit? Show how arbitrary combinational circuits can be synthesized using the basic principle of this circuit. What practical difficulties do you see in implementing such circuits?

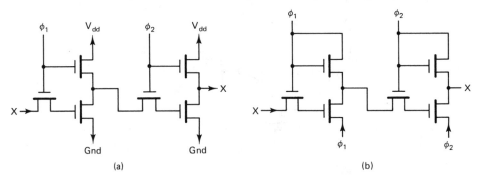

(a)   (b)

3.16. The circuit in Fig. E3.16 is a precharged circuit whose output may be pulled down during ϕ_1. Do you see any charge-sharing problem? Explain.

The principle of operation of this circuit is very similar to the principle of domino logic for CMOS. Formalize this concept to develop general techniques to realize arbitrary combinational and sequential circuits in domino nMOS logic.

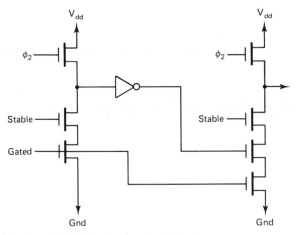

3.17. Design CMOS circuits for the following:
(a) A *D* flip-flop with a set line.
(b) A *D* flip-flop with a reset line.
(c) A *D* flip-flop with a set and a reset line.

3.18. Design an *n*MOS and a CMOS circuit for a clocked "toggle" flip-flop. A toggle flip-flop is a storage device with a control input *T*. It $T = 1$ in coincidence with a clock input, the state of the flip-flop changes; otherwise, it remains the same as the previous state. Try to use a semistatic register as discussed in the chapter.

3.19. The state diagram of a 0101 sequence detector is shown in Fig. E3.19. Assume that the detector starts in state *A* and that *C* is the accepting state. The labels on the arrows indicate the input/output values associated with the indicated transitions. Construct an encoded state transition table and the Boolean functions denoting the next-state functions. Then design a CMOS PLA implementing the state table using complementary single-phase and two-phase nonoverlapping clocking.

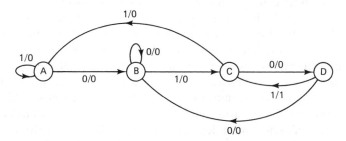

3.20. The precharged output of a domino logic stage may not stay charged properly due to the leakage current in the pull-down network, particularly if the pull-down network is large. Suggest a solution to remedy this problem.

3.21. An alternative approach to fight the glitch problem in a domino circuit is to use "N-P" logic, in which the pull-down circuits of two consecutive stages are made of *n*-channel switches and *p*-channel switches, respectively. What other changes are needed in the circuit? Sketch an N-P circuit for the function $f = c(a + b) + d(a + b)$. What will be the advantages and disadvantages of such circuits?

4

<hr>

The Technology of Semiconductors

4.1 Introduction

This chapter describes basic concepts and processes used in the production of semiconductor devices. The modern microelectronics industry is one of the most sophisticated and precise engineering disciplines that has ever been developed. This chapter provides a glimpse at it, leaving the engineering details to books specifically written to describe these processes. A designer does not have to deal with these processes, but a brief description of the background will help the reader understand the physical basis of the interface between the design and fabrication processes. As we will see later, this interface takes the form of a set of design rules and values of electrical parameters predetermined by the fabrication process.

The implementation of an integrated-circuit chip involves three major steps: design, mask making, and fabrication. For each step, a whole world of technology and tools has evolved. The architecture of the chip originates at the "design house." Here a system-level specification undergoes several levels of translations via the function, logic, and gate/switch levels to produce a layout description of the chip. The tools applicable to carry on this process are discussed in Chapter 9. The layout description gives a detailed and exact specification of all the geometrical shapes that must be mapped onto the silicon wafers in layers of conducting (metal), semiconducting (polysilicon, diffusion, p-well, p^+ regions, etc.), and insulating material (silicon dioxide, etc.) corresponding to the circuit to be designed. This description is usually expressed in a standard interchange language called CIF (Caltech Intermediate Form: see Chapter 9 for details), which is capable of describing planar geometrical structures using simple and clear notations for individual layers. The basic task of the fabrication process is to map these structures onto the silicon surface in specified layers. The first step in this process is *photolithography*, which converts the CIF files into a set of *masking plates* containing exact images of the structures in either opaque or transparent shades. This job is carried out at "mask houses." The actual fabrication of the chip is done in a "fab house," which is a modern, sophisticated chemical factory that makes "prints" of the masks onto the silicon wafers. The wafers, which are then tested and packaged, themselves form two complex technologies outside the scope of this book. We first describe the basic fabrication techniques that are common in several of the processes.

4.2 Basic Fabrication Techniques

4.2.1 Wafer Fabrication

The basic raw material is sand or silicon dioxide, which is plentiful on the earth. Sand (SiO_2) is purified to polycrystallize silicon by reacting it with carbon (C) and is then crystallized by a special growth process. Silicon is melted at 1500°C on a crucible and a seed crystal is brought in contact with it and gradually withdrawn from the molten silicon, as shown in Fig. 4.1(a). As the seed is withdrawn, silicon atoms attached to the seed cool and assume the crystalline structure of the seed. Controlled quantities of doping material are inserted into the crucible to produce a

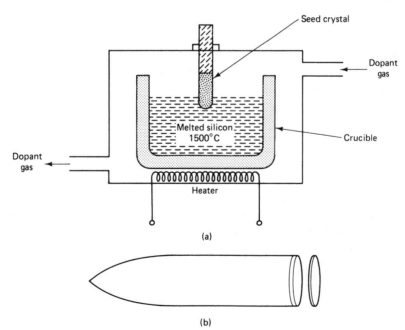

Figure 4.1 (a) Epitaxial growth apparatus; (b) the ingot and the wafer.

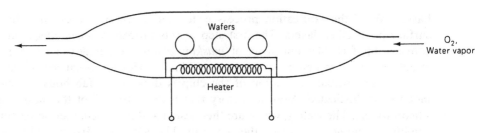

Figure 4.2   Oxidation process.

uniform light concentration of *p*- or *n*-type impurity. The result of the process is a cylindrical silicon bar called an "ingot," several centimeters long and about 8 to 10 cm in diameter. Wafers about 0.3 mm thick are then cut from this crystal [Fig. 4.1(b)].

4.2.2 Oxidation

The oxidation step is performed to grow a thin layer of protective silicon dioxide on the silicon oxide. The wafer is heated in a 1000°C furnace in an oxygen-rich (dry oxidation) or water vapor (wet oxidation) environment, as sketched in Fig. 4.2. Dry oxidation is used to produce thin but robust oxide layers, but it takes longer than the wet process, which produces a thicker and slightly porous layer.

4.2.3 Patterning

Patterning is the process of "printing" the geometrical shapes of the design onto the various layers of the silicon. The first step is to produce a *masking plate* for a given layer. The masking plate has opaque geometrical shapes, corresponding to areas on the wafer surface where certain photochemical reactions have to be prevented (or take place). To produce this plate, a *reticle*, which is simply a photographic plate 10 times the size of the original shape, is exposed by light flashes in areas corresponding to the 10 × image of the shape of the design. The optical apparatus that performs this task is called a *pattern generator*. A typical example of such a machine is the Mann 3000. The exposed plate is then developed, forming the reticle for one layer. The reticle is then photoreduced to the actual size and a photorepeater projects the image on another photographic plate of the size of the wafer by a "step and repeat" procedure to create an array of geometrical shapes in one layer over the entire plate. During this process the position and the angle of the reticle are precisely aligned with the help of two *fiducial marks* incorporated in the pattern generation files of all layers in the same relative position with respect to the entire chip.

Modern mask houses use electron-beam (E-beam) mask generation equipment, which generates the masking plate in one step. The geometric data are converted into a bit map of 1's and 0's on a raster image. The electron beam sweeps the rows in a

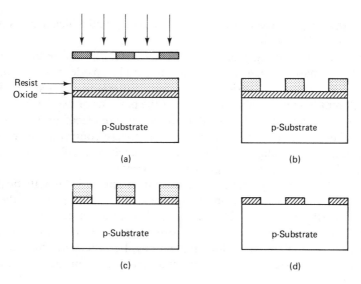

Figure 4.3 Patterning steps.

repeating S pattern, *blanking* or *unblanking* the beam according to the input bit value, 0 or 1. Using the E-beam technique several different chip types can be imprinted on the same set of masks. Multiproject chips (MPCs) are produced by E-beam methods. For further details, the reader is referred to Hon and Sequin (1980). The main disadvantage of the E-beam process is that it is sequential. A better method is to use soft x-ray lithographic techniques that can irradiate the entire chip simultaneously. X-ray techniques also provide higher resolution, which is essential for scaling the device dimensions. For further details, the reader is referred to Glaser and Subak-Sharpe (1979).

The actual patterning process called *photolithography* consists of selectively removing material (silicon dioxide, polysilicon, silicon nitride, or metal) from areas corresponding to the transparent windows of the masking plate. Removal of the oxide layer will now be described as an illustration of the typical process and is sketched in Fig. 4.3.

The entire surface with a grown oxide layer on top is coated with a photosensitive emulsion called *photoresist*. The masking plate is placed in contact with the wafer in a precise position. Exposure of the surface by ultraviolet rays via the masking plate will cause polymerization of the emulsion under the transparent windows [Fig. 4.3(a)]. The polymerized emulsion is then chemically dissolved, leaving islands of unpolymerized resist corresponding to the opaque regions of the mask [Fig. 4.3(b)]. The wafer is then baked in an oven to improve the hardness of the resist.

The actual removal of the material is done simply by immersing the wafer in a suitable etching solution (typically hydrofluoric acid for SiO_2 and polysilicon and phosphoric acid for nitride and metal). The solution eats out the oxide but leaves the resist intact [Fig. 4.3(c)]. The final step is to remove the photoresist material by another step of selective chemical reaction by acid which attacks only the resist, not the oxide [Fig. 4.3(d)].

The patterning process is now complete. The end result is an engraving of the original mask in islands of the base materials—oxide, polysilicon, metal, or nitride.

In a typical fabrication process, patterning is used at different stages on different base materials. In order that the patterns of two layers are properly aligned, special alignment marks called *parity marks* are placed on the masking plate to help overlay several patterns. In spite of the precise control of the patterning process, a tolerance or misalignment has to be sustained. We will see later that this leads to a set of design rules for the particular process.

It is also important to know that in spite of the statistical variation of the process parameters, the fabrication steps actually do produce the intended devices on the silicon surface. The masking plate contains certain *critical dimensions*. To test the fabrication process, test structures are placed on each wafer that can be used to determine line widths, separations, FET characteristics, and other parameters necessary to assess the reliability and performance of the fabricated circuits.

4.2.4 Diffusion

The diffusion process causes simultaneous creation of p or n doped regions where the silicon surface is not protected by oxide. It consists of two steps: *predeposition* and *drive-in*. In the predeposition step, dopant atoms such as phosphorus or boron mixed with an inert gas such as nitrogen are introduced into the furnace at 1000°C. The atoms diffuse on to a thin layer on the surface of the silicon, forming a saturated solution of solid and gas for the given temperature. The impurity concentration of the layer goes up with temperature until about 1300°C and then it drops. The depth of penetration depends on the amount of time the process is carried on. In the next, drive-in step, the wafer is heated in nitrogen gas (without dopant). This distributes the surface diffusion deeply in the body, depth of penetration depending on the temperature and the total time of the process. The net result is as sketched in Fig. 4.4. It is possible to grow an oxide layer right after the drive-in process, thereby providing a thin protective oxide layer. In most modern fabrication houses, diffusion is usually done by ion-implantation techniques, which provide more accurate control of the diffusion process.

4.2.5 Ion Implantation

Dopant gas is passed through an ionizer, which ionizes the gas. The ions are then accelerated between two electrodes, maintaining a voltage difference of about 150 kV. The gas is passed through a strong magnetic field, which separates the impurity

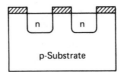

Figure 4.4 Diffusion step.

dopant ion stream on the basis of the molecular weight of the dopant ions. The stream is deflected to the wafer held in the path of impact of the ions. The principle is the same as that used in mass spectroscopy. The ions strike the silicon surface at a high velocity and are lodged in the silicon to a depth determined by the accelerating field and the concentration of ions in the dopant gas. This is followed by a drive-in step, which redistributes the ions and increases the depth of penetration.

The proper mask must be used to select the regions where implantation is needed. Thick oxide, photoresist, or metal can serve as masks. If the oxide layer is thin, implantation can be done through the layer. This is typically used to adjust the MOS transistor threshold. Implantation through thin polysilicon is used to create diffusion under the polysilicon for creating "buried contact," as we will discuss later. Ion implantation is also used to create a p-well in an n-type substrate or an n-well in a p-type substrate in a typical CMOS process.

4.2.6 Deposition

Layers of silicon dioxide, silicon nitride, or polysilicon can be deposited on the wafer surface using a chemical vapor deposition technique in a high-temperature chamber.

To deposit silicon dioxide, a mixture of nitrogen, silane, and oxygen is introduced at 300° to 500°C. Silane reacts with oxygen to produce silicon dioxide, which is deposited on the wafer. Polysilicon is deposited simply by heating silane at 1000°C, which releases hydrogen gas from silane and deposits silicon. This is used in any silicon gate process. To deposit silicon nitride, silane and ammonia are heated at about 700°C to produce nitride and hydrogen. Aluminum is deposited by vaporizing aluminum from a heated filament in a high vacuum. Note that the silicon dioxide is deposited at a considerably lower temperature than the boiling point of aluminum (650°C), so that this step can be performed even after metallization. The step is therefore usually used for overglassing, followed by a mask that exposes the bonding pads and puts protective layers over the rest of the wafer.

4.3 Fabrication Processes

There are a large number and variety of basic fabrication steps used in the production of modern semiconductor devices. The process could be designed for MOS (nMOS or pMOS or CMOS) devices. The gate could use metal or polysilicon. The substrate could be bulk silicon or silicon-on-sapphire. Finally, there are variations in techniques to isolate the devices in the wafer to avoid parasitic transistors.

Metal-gate (sometimes called aluminum-gate) processes are the simplest to describe. We present them first to serve as an introduction to other processes. Then we describe the silicon-gate processes that will concern us most in this book.

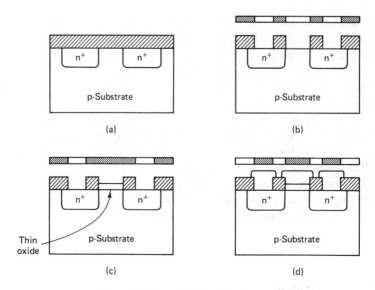

Figure 4.5   The nMOS metal-gate process.

4.3.1 Metal-Gate nMOS Process

The first five steps are exactly as explained in Figs. 4.3 and 4.4. So far, only one mask has been used to define the diffusion area. Now a thick (15,000-Å) oxide is grown on the surface [Fig. 4.5(a)]. A second mask is used to pattern areas for gate, source, and drain [(Fig. 4.5(b)]. A thin oxide is now grown over these three regions and then etched out in all areas except the gate region, using a third mask [Fig. 4.5(c)]. Aluminum is now deposited over the entire surface. A fourth mask is then used to remove aluminum from regions where it is not wanted, leaving metallic connections to terminals of the transistors [Fig. 4.5(d)]. Finally, a fifth mask is used for overglassing the entire wafer, leaving exposed areas only for the bonding pads.

4.3.2 Metal-Gate CMOS Process

The sequence of steps for the metal-gate CMOS process is conceptually very similar to the process described above except for the fact that the diffusion step to create the drain and source regions is preceded by a p-well implantation step. The p-well defines the region for the n-channel transistors, whereas the p-channel transistors are built in the n-type substrate [Fig. 4.6(a)]. The implantation is done through a thin oxide with p-ions (boron) using a photoresist mask. A drive-in step follows in an oxygen environment. Photoresist is then etched out. The next step is to create the source and drain regions for all p-channel transistors together with p-channel stops in the p-well. Channel stops or plugs are heavily doped regions of the same type as that of the substrate and are usually embedded near the boundary of the regions

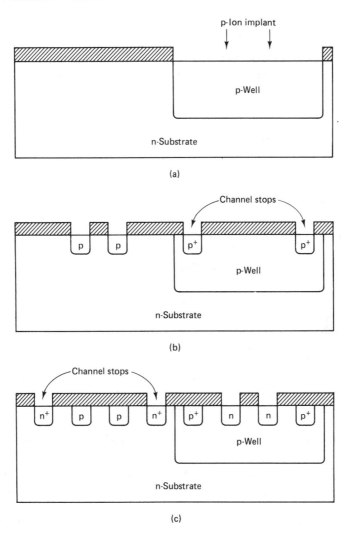

Figure 4.6 Formation of transistors and channel stops in a metal-gate CMOS process.

[Fig. 4.6(b)]. They are also called *guard rings* because they prevent the formation of undesirable parasitic transistors under metal lines. A similar diffusion step is necessary to create the drain, source, and channel stops for the *n*-channel transistors [Fig. 4.6(c)]. The next few steps are very similar to the steps illustrated in Fig. 4.5(a) through (d) for the metal-gate *n*MOS process: thick oxide growth; selection of drain, source, and gate regions; thin oxide growth over gates; and metal interconnect to terminals. A final overglassing step is followed at the end.

One major difficulty with the metal-gate process is that a separate mask is used to delineate the gate region. To allow for tolerances, the gate area is deliberately overlapped with the drain and source areas. This not only uses more area, but also increases the gate capacitance, making the device slow. The silicon-gate process described in the next section is *self-aligned*; that is, a single mask region can be used to define the gate as well as the drain and source regions. In addition, the metal-gate process provides only one layer, namely metal, for all terminal interconnects. This increases the total area of the chip.

4.3.3 Silicon-Gate *n*MOS Process

The important distinguishing characteristics of the silicon-gate *n*MOS process are that it is self-aligning, creates depletion loads, has two layers of interconnect and a special structure called *buried contact* to connect between polysilicon and diffusion wire, produces high-functional-density chips, and has good power and speed performance. It has been one of the most widely used processes and is a step toward the evolution of bulk CMOS processes which might dominate future processes. The steps in the silicon-gate *n*MOS process can be described as follows.

Definition of active regions.   A sandwich of thin oxide (by oxidation), silicon nitride (by chemical vaporization), and another thin oxide (by vaporization) is prepared on the surface of a clean wafer. The first oxide acts as a buffer to relieve mechanical stress due to unequal thermal expansion of the substrate and nitride. The nitride also acts as a shield to prevent oxidation of the active region during field oxide growth. A layer of photoresist is deposited on the entire surface [Fig. 4.7(a)]. Using a *diffusion layer* mask D, the thin oxide and nitride are etched out. The open area is called the *field-oxide region*. A p^+ region is created by ion implantation which act as a channel stop [Fig. 4.7(b)]. A thick oxide (about 1 μm) is now grown on top of the implanted region, which together with channel stops serves to isolate the active region from neighboring active regions. Note that the field oxide region is self-aligned with respect to the active region by the same mask D and that the silicon surface area for the field oxide is slightly etched so that the field oxide sinks into it from the surface line. After this, the remaining oxide and nitride are removed, resulting in the state as shown in Fig. 4.7(c). The active area will now be used for source, drain, and channel regions for all transistors, plus any diffusion wire needed for interconnects. Figure 4.7(d) shows the top view of a possible active area for an inverter circuit.

Definition of depletion Loads.   An *implantation mask I* is used to pattern areas where depletion-mode transistors are going to be located. The areas are then implanted with arsenic or antimony. A thin layer of oxide (about 700 Å) is then grown over the exposed silicon surface.

Polysilicon – diffusion interconnect.   Using a *buried contact mask B*, holes are opened on the thin oxide on areas where polysilicon and diffusion wires have to interconnect. The holes are then etched to remove the thin oxide in these regions [Fig. 4.7(e)].

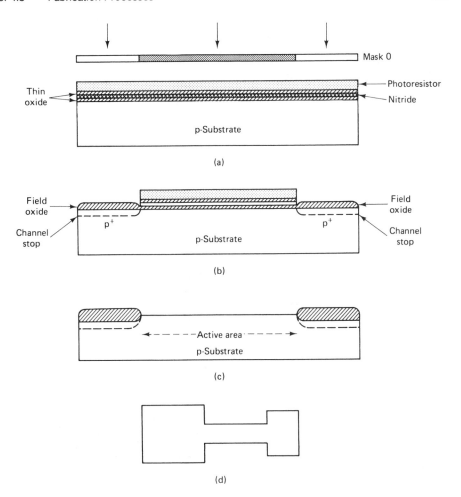

Figure 4.7   Silicon-gate nMOS process.

Definition of transistors and polysilicon – diffusion contacts.   A layer of polysilicon (500 Å) is deposited by chemical vaporization over the entire surface and then, using the *polysilicon mask P*, polysilicon gate regions and connecting wires are created.

Diffusion step.   Using mask $D \wedge (\sim P)$, a thin oxide layer is etched out from regions under the active area not covered by polysilicon. An implantation step with phosphorus or arsenic will dope the source and drain regions and diffusion wire regions. A drive-in step follows. Since the polysilicon is n-type doped, during the drive-in step, n-type impurities diffuse from polysilicon directly to silicon in regions where polysilicon is in direct contact with silicon, that is, in regions defined by buried mask. The drive-in step also reduces the resistivity of the polysilicon wires. There is also some amount of outdiffusion from the source or drain region which

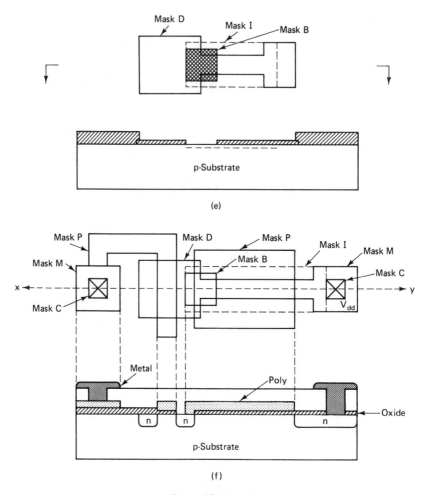

Figure 4.7 (*Cont.*)

spreads into the areas within the buried mask as shown in Fig. 4.7(f), which gives a top and a cross-sectional view along the *xy*-line of an inverter circuit. The layout does not show the Gnd connection explicitly. Note that the edges of the polysilicon gate define the beginning of the source–drain regions and the process is, therefore, *self-aligned*. Also, notice that the channel region *does not* have any diffusion, although in all layout artwork the channel regions are colored green or are shown indistinguishable from the source or drain regions.

Polysilicon – metal and diffusion – metal interconnects.   A thick layer of oxide mixed with phosphor (P_2O_5) is deposited over the surface and reflowed to round the corners. A *contact cut mask C* is used to cut holes and the oxide is etched away. At a later step aluminum is deposited to cover these holes, which establish the appropriate interconnect. Sometimes, the aluminum simply diffuses right into the shallow drain or source regions, essentially shorting the reverse-biased junctions corresponding to the source–drain diffusions and the substrate. This is called the *spike-through* problem. This can be prevented if the source–drain regions can be doped again, essentially forming a deeper junction into the bulk of the substrate. Another solution is to provide a barrier of palladium silicide locally before the metallization step, which prevents the aluminum from penetrating the diffusion area.

Metallization.   The whole surface is now covered with aluminum by vacuum vaporization. A *metal layer mask M* is now used to pattern conducting regions over the entire circuit, which necessarily includes the contact cut regions. The final layout and cross section are shown in Fig. 4.7(f).

Annealing and passivation.   The wafer is then annealed (heated at about 400°C for about an hour) to remove radiation damage caused by electron-beam bombardment. This improves the robustness of the metal layer. After a thick layer of oxide is deposited on the entire surface, an *overglass mask G* is used to open windows corresponding to the bonding areas, which are connected by gold wire to the outside world. The remaining thick oxide acts as a protective layer for the wafer.

The whole process needs seven masks: D, I, B, P, C, M, and G. These symbols are also used to denote the respective layers. In earlier processes, the mask layer B was not used since polysilicon–metal connection was achieved by a special structure called a butting contact—a concoction of polysilicon, diffusion, metal, and contact cut, which is now obsolete.

A substantial amount of research is under way to incorporate a second metallization layer for interconnect. Conceptually, the process should be very simple: Select cut areas for the second metal layer through lower levels, deposit metal over the entire wafer, and use a second metal mask to pattern the desired interconnection. The problem is if both these layers use the same metal: During vacuum deposition of the second layer, the first layer may evaporate out. The second layer must be applied at a temperature lower than 500°C, the melting point of aluminum. A suggested solution is to use molybdenum (melting point 2600°C) or tungsten, which has a much higher melting temperature, as the first layer of metallization followed by a second level of aluminum at a lower temperature. This approach has created some problems: Bonding of gold wire with molybdenum or tungsten is not very smooth because of the electrochemistry of the contact, which causes corrosion at the contact point. Also, the etching process to cut holes through appropriate layers is more complex.

4.3.4 Silicon-Gate CMOS Process

During the last few years silicon-gate CMOS has emerged as a dominant process technology for VLSI. CMOS provides lower delay and lower dc power consumption; however, CMOS requires greater process complexity and increased chip area (about

10 to 30% compared to *n*MOS). As we will see later, CMOS also provides better scalability and "ratioless" logic circuits. But unlike *n*MOS, CMOS process technology is undergoing extensive experimentation and there are large variations in processing steps and parameters. It seems that this state of affairs will continue and it will be a while before a "standard" CMOS process will evolve.

The basic problem in CMOS design is to have both *p*-channel and *n*-channel transistors on the same wafer. This leads immediately to two different approaches to CMOS: the traditional *p*-well approach with proven reliability and the newer *n*-well approach. With the *p*-well approach, the starting substrate is *n*-type, which contains the *p*-channel transistors and deep *p*-type doped areas called the "*p*-well," where *n*-channel transistors are created. The *p*-well approach produces balanced performance of the *p*- and *n*-channel transistors because of two opposing factors. The *n*-channel transistors have higher conductivity, about twice the conductivity of *p*-channel transistors; and transistors of any type produced in a well have less speed than those built into a clean substrate by doping. It is also easier to produce a *p*-well in an *n*-substrate rather than an *n*-well in a *p*-substrate. The *n*-well approach, on the other hand, produces the *n*-channel transistors in *p*-type substrate and the *p*-channel transistors in an *n*-well, which actually increases the performance difference between the *p*- and *n*-channel transistors. Proponents of the *n*-well approach claim that this process is compatible with the *n*MOS process using about 20% more processing steps. The *n*MOS and CMOS processes can share processing steps and substrate material, which is an advantage for high-volume production. The *n*-well process also provides reduced latch-up sensitivity, improved *n*-channel performance, and low cost.

We first describe a typical *p*-well process. The steps are sketched in Fig. 4.8(a) through (d).

p-well formation.   The field oxide and *p*-well are produced in the same manner as described for the metal-gate CMOS process [Fig. 4.8(a)].

Active area definition.   This step is the same as used for the *n*MOS silicon-gate process [see Fig. 4.7(a) through (c)] except that the active areas are chosen in both the *p*-well and *n*-substrate regions. The next step in the *n*MOS process—implantation to produce a depletion load— is not necessary. A thin oxide is grown over the exposed surface to form the gate insulator [Fig. 4.8(b)].

Definition of transistors.   A layer of polysilicon is deposited by chemical vaporization and then, as in the *n*MOS process, polysilicon gate regions and wires are created by selectively etching the thin oxide [Fig. 4.8(c)].

Source – drain diffusion.   Two separate masks are used—one to define the *p*-channel transistors and the other for *n*-channel transistors. These two masks are usually complementary. Next, a coating of *n*-type dopant (phosphorus) is deposited by vaporization on the wafer and then removed from regions containing *p*-type

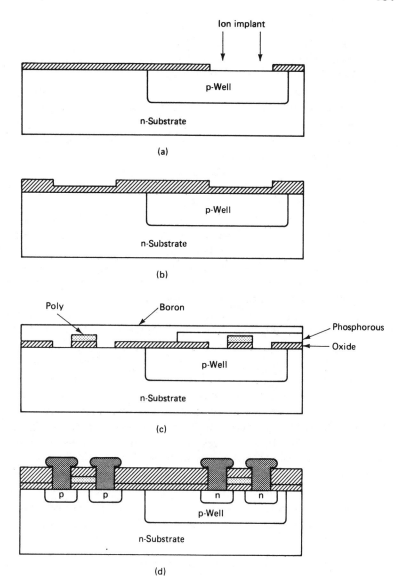

Figure 4.8 Silicon-gate CMOS process.

transistors. Next, a similar layer with *p*-type dopant (boron) is deposited on the entire wafer surface as shown in Fig. 4.8(c). Then in a single high-temperature drive-in step, the source and drain diffusion areas are created simultaneously. This step also dopes the polysilicon gates slightly to reduce resistivity. In most modern processes the diffusion process is replaced by ion implantation, in the previously defined diffusion step, which aligns the areas with respect to gates. So the process is also self-aligned. The polysilicon mask also prevents doping underneath, thereby reducing gate–drain and gate–source overlap. The next few steps are exactly the same as those used for the *n*MOS process: contact cuts for polysilicon/diffusion-to-metal interconnect, metallization, annealing, and passivation (overglassing). The final structure is shown in Fig. 4.8(d).

Note that the process did not allow buried contact. To connect polysilicon directly with diffusion, we need to dope polysilicon in two ways, depending on whether the polysilicon wires are connecting to *p*- or *n*-type underlying diffusion. When two such polysilicon wires are directly connected (i.e., for a CMOS inverter, the two gates of a pair of *p*-channel and *n*-channel transistors are connected together to form the input), a diode will be formed. So additional contact cuts and metallic connections will be needed to avoid this problem. For this reason, buried contacts are not used for CMOS processes.

The sketches in Fig. 4.8 do not show the additional details that are needed to properly bias the *p*-well region for correct transistor operation. This is done by "plugs," a localized heavily doped region. The p^+ plugs and additional metallization are used to connect the well to ground potential. Similarly, n^+ plugs are used outside the active region to connect to the *n*-substrate for proper biasing. These plugs also reduce latch-up sensitivity (see Section 4.4).

A typical *n*-well process follows exactly the steps described above, except that the substrate is *p*-type, in which an initial *n*-well is started first. There will be a lot of experimentation before the *n*-well process matures. For example, the CMOS process at the University of Berkeley uses two different masks to define two active regions—one for *n*-channel and the other for *p*-channel devices. Separate implantation steps are used for the two regions to control the threshold voltages separately. A separate p^+ implantation is also used for the *p*-channel region similar to the ion implantation for the *n*MOS process, in order to adjust the threshold voltage of the *p*-transistors. In the processes described here, diffusion of *p*- and *n*-type islands is done in one drive-in step after depositing *p*-doped and *n*-doped oxide in the respective active regions. An alternative approach that is very similar to *p*-well is to use both *n*- and *p*-wells. This is called the *twin-well or twin-tub* process. The substrate is a highly resistive *n*-type. Two separate wells are formed for the *p*- and *n*-type transistors. A typical two-well process takes about six ion implants and nine masking levels.

4.3.5 Oxide-Isolation CMOS Process

The *oxide-isolation* CMOS process is a self-aligned silicon-gate process with high density, speed, and yield. The process is also known as the *isoplanar or selectively oxidized* CMOS process. The steps of the process are sketched in Fig. 4.9(a) through (h) and are very similar to those of the process described by Lipman (1981) [see also Sequin (1982)].

First, a thin oxide is grown on a lightly doped *n*-substrate; then on top of the oxide a nitride layer is deposited [Fig. 4.9(a)]. As in the *n*MOS process described in Section 4.3.3, the oxide acts as a buffer to relieve mechanical stress due to unequal thermal expansion of the substrate and nitride. Using the *first* mask, the active areas

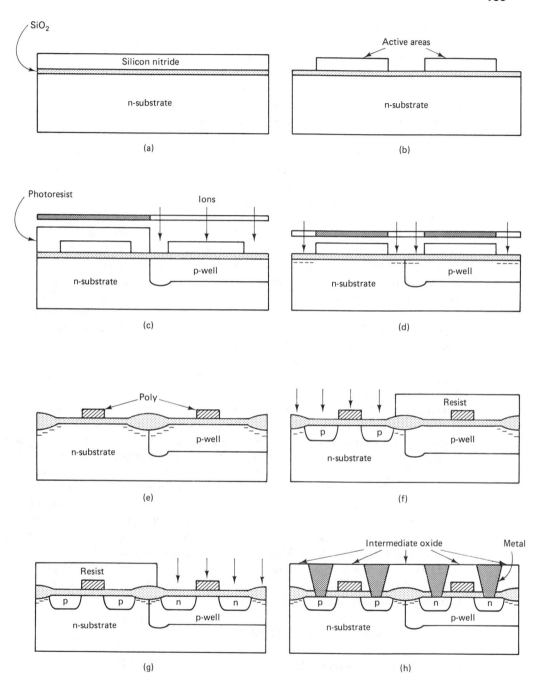

Figure 4.9 Oxide-isolation CMOS process.

in the form of nitride islands are then defined as shown in Fig. 4.9(b). The nitride layer serves as a shield to prevent oxidation of the layer under it during the field oxidation process. Next, the *p*-well is formed by using a *second* mask. A thick photoresist is used to cover up the parts of the wafer that will be *n*-type bulk. High-energy *p*-type ions are implanted through the nitride and oxide layer [Fig. 4.9(c)]. Note that the implantation that travels through the oxide penetrates deeper into the substrate compared to regions under the nitride islands. Now a low-energy high-dosage *p*-type implant [Fig. 4.9(c)] is applied to form channel stops in the *p*-well. This implant is shielded by the photoresist as well as the nitride, which self-aligns the implantation in the future thick oxide area. The photoresist is now removed and a low-energy *n*-type implant is applied which produces the channel stops in the *n*-type bulk regions. The nitride shields this implant and makes it self-aligned. The dose of this implant is much less than the dose of the *p*-type implant, so that the *p*-channel stops are not inverted [Fig. 4.9(d)]. This is now followed by a field oxidation, which produces a thick field oxide over the implanted region. The nitride layer prevents oxidation underneath but allows the field oxide to grow. Again, note that as in the *n*MOS process, the field oxide region is self-aligned and the field oxide sinks into the substrate from the surface line, undercutting the thin oxide. After the formation of the field oxide, the nitride islands are etched out. At this stage a special p^+ ion implantation step can be performed to adjust the threshold voltage of the *p*-transistors. The remainder of the process is very similar to the silicon-gate CMOS process. A layer of polysilicon is deposited and the transistor gates and polysilicon wires are carved out [Fig. 4.9(e)] using a *third* mask. Next, the drain and source regions of the *p*-channel and *n*-channel transistors are formed. Implantation techniques are more commonly used now rather than the older methods of dopant deposition followed by drive-in [as explained in Fig. 4.8(c)]. The wafer is covered with photoresist and a *fourth* mask opens the source and drain regions for the *p*-channel transistors. A *p*-type implantation now produces the active *p*-type islands. The process is now repeated to form the *n*-channel transistors using *n*-type implantation [Fig. 4.9(g)]. The wafer is covered with photoresist and a *fifth* mask opens the source and drain regions for the *n*-channel transistors. This is now followed by a diffusion or implantation step to produce the actual *n*-type regions. The remainder of the steps include deposition of an intermediate oxide layer which is flowed to give a smooth surface. A *sixth* mask is used to open contact windows through the oxide layers to get to the *p*- and *n*-regions, the *p*-well, and the polysilicon layer [Fig. 4.9(h)]. A *seventh* mask defines the metallization pattern, and an *eighth* mask completes the passivation with a protective nitride layer.

4.3.6 Silicon-on-Sapphire CMOS Process

We conclude this chapter with a brief discussion of SOS CMOS. Conceptually, this is the simplest process since the designer does not have to be concerned with wells, guard rings, and thin oxide. The designer deals only with structures such as transistors and wiring which have real meaning with respect to the circuit. It has

therefore been suggested (Sequin, 1982) that SOS CMOS is best suited for instruc-
tional purposes.

In this technology, the starting wafer is sapphire (aluminum oxide, Al_2O_5) on
which silicon, whose crystal lattice is compatible with that of sapphire, is grown [Fig.
4.10(a)]. Rather than diffusing wells and p- or n-type islands in the substrate, the
active islands are built on top of the substrate by patterning and are then properly
doped for transistor action. The steps in the process are sketched in Fig. 4.10. The
active regions are isolated by chemically etching the silicon between the areas rather
than growing thick field oxide [Fig. 4.10(b)]. The first mask defines the active
regions. The sapphire base acts as a good insulator and is also responsible for
reduced junction capacitance and therefore the resulting circuits are faster. The
active islands are then doped to be n-type or p-type by implantation steps [Fig.
4.10(b)]. This can be done by using two separate masks for the n- and p-type
implantation or by a uniform implantation followed by a counterdoping step using a

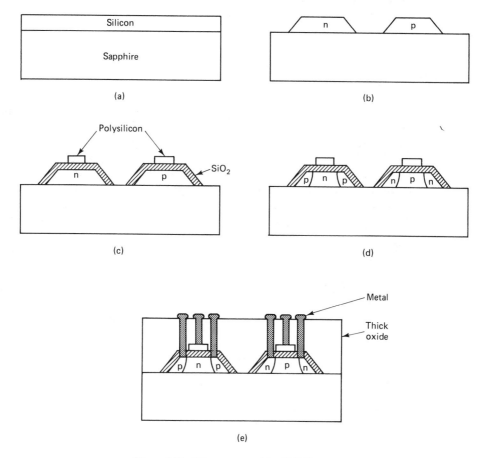

Figure 4.10   Silicon-on-sapphire CMOS process.

single mask. Thin oxide is now grown by oxidation. If buried contact is to be made, a fourth mask is used to expose the silicon area where contact is to be made, as in the case of the *n*MOS silicon-gate process described earlier. The polysilicon is deposited on top of it to form the gate terminals [Fig. 4.10(c)] and is patterned with a fifth mask. The source and drain regions are now formed by implantation [Fig. 4.10(d)] using two separate masks or a pair of complementary masks. This implantation step is made rather deep so that doping penetrates all the way through the silicon. A thick layer of oxide is now grown over the entire wafer to protect the structure. An eighth mask is used to cut contact holes. Metal is now deposited and patterned by a ninth mask. The last mask is used to create a protecting layer that covers everything but the bonding pads. The final structure is sketched in Fig. 4.10(e). One disadvantage of the SOS process is that the dielectric constant of sapphire is high compared with that of silicon. This results in a higher coupling capacitance in the adjacent wires, which gets worse with scaling, offsetting the reduced junction capacitance. This affects the speed adversely. Furthermore, sapphire is an expensive raw material which may be more suitable for jewelry than for transistors.

4.4 The Latch-Up Problem

Bulk CMOS processes produce certain parasitic *npn*-transistors. If these transistors are improperly biased, a phenomenon called *latch-up* can cause large currents to flow, destroying the MOS devices. To understand latch-up, we need to understand the basic operation of a *silicon-controlled rectifier* (SCR), which is a *pnpn*-device with three terminals called *anode*, *cathode*, and *gate*, as shown in Fig. 4.11. The two-transistor equivalent circuit for the SCR is shown in Fig. 4.12. The arrows indicate the direction of flow of negative currents. If the gate current I_g is increased, this causes an increase in the collector current I_{c_2} of the *npn*-transistor; but I_{c_2} is also the base current I_{b_1} of the *pnp*-transistor, and therefore its collector current I_{c_1} is also increased, resulting in a further increase of the base current I_{b_2} of the *npn*-transistor. The current increases here refer to increases in magnitude. If the "gain" of the two transistors α_1 and α_2 are such that $\alpha_1\alpha_2 \geq 1$, the feedback action will turn the device on permanently, and the increased current will self-destruct the

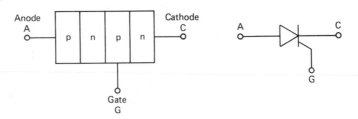

Figure 4.11   Silicon-controlled rectifier (SCR).

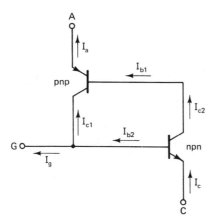

Figure 4.12 Equivalent transistor circuit of SCR.

device. The SCR is said to be *triggered* under these conditions. The formation of the parasitic transistors in the silicon is illustrated in Fig. 4.13(a) together with the equivalent electrical circuit shown in Fig. 4.13(b), where R_{sub} (substrate resistance) is equal to the parallel combination of R_{sub1} and R_{sub2} and R_{well} denotes the well resistance. If a positive impulse is applied at node G, the voltage across the well resistance R_{well} will induce a positive feedback current. Similarly, if point A surges above V_{dd}, a current will flow in R_{sub}, inducing a positive feedback current. This type of situation may arise when power is turned on; a race situation may exist between the input and the power voltage, and point G, being an input node, may receive such a pulse. Also when inputs are switched, point A may go beyond V_{dd}, due to a phenomenon called "inductive kick." Finally, during dynamic switching in CMOS—which demands a large current for a very short duration—the power supply may drop, inducing a latch-up. In all these situations, if the induced current is I and the voltage IR_{sub} or IR_{well} is enough to forward-bias the base-emitter junction of one of the parasitic transistors, latch-up takes place provided that $\alpha_1 \alpha_2 \geqslant 1$.

Two methods have been used to reduce the chances of latch-up. One is to reduce the values of R_{sub} and R_{well} as much as possible (thereby lowering the base-emitter bias voltage) by using what is called *guard rings*. These are low-resistivity connections to supply voltages built around the CMOS *p*-channel and *n*-channel transistors. This, of course, has the effect of reducing the gate density by requiring more space between the *n*- and *p*-channel transistors. Placing *p*-plug or *n*-plug near the supply nodes also has the equivalent effect of reducing the values of R_{well} or R_{sub}, respectively. The other alternative is to control the values of gains α_1 and α_2 of the parasitic transistors. The value of α_2 is set by the process parameters; α_1 can be reduced by increasing the distance between *p*-well and *p*-diffusion, effectively increasing the width of the base region for *pnp*-transistors. Here, again, some area is wasted, reducing the gate density. In bulk CMOS, there is no absolute protection against latch-up; the chance of latch-up can be reduced only at the expense of silicon areas. The latch-up phenomenon is most prevalent near the pads which are the most susceptible part of the circuit because of connection to output.

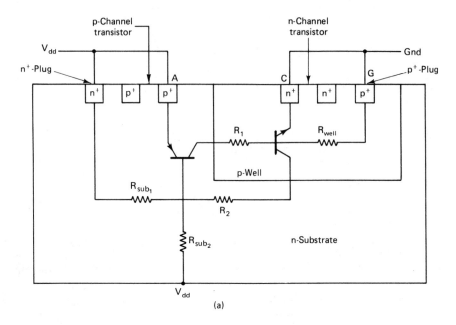

(a)

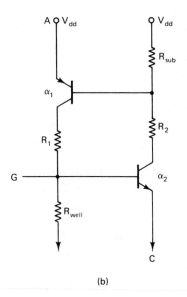

(b)

Figure 4.13 Latch-up phenomenon.

4.5 Summary

In this chapter we have provided brief descriptions of some of the popular semiconductor processes used to fabricate nMOS and CMOS devices. We have left out many processes that are currently being developed and used by the industry: the twin-well process, dielectric isolation process, high-temperature and radiation-hard process, and two-layer metal interconnect process, among others. Our purpose was to give the basic ideas of the engineering and concentrate on only two processes, silicon-gate nMOS and bulk CMOS, which are available to many students and universities via silicon foundries. The present trends indicate that CMOS will replace the nMOS process. Of the CMOS processes, the p-well process has several advantages over the n-well in terms of reliability and reduced sensitivity with respect to radiation, and it balances out the performances of nMOS and pMOS in the same substrate. The n-well process has reduced latch-up sensitivity and is highly compatible with the existing nMOS processes. The twin-well process combines the advantages of the p-well and n-well processes.

REFERENCES

References cited at the end of the Appendix A are also relevant for this chapter.

Chwang, R., and K. Yu, "C-IIMOS—An n-Well Bulk CMOS Technology for VLSI," *Lambda*, 4th quarter 1981, pp. 42–47.

Glaser, A. B., and G. E. Subak-Sharpe, *Integrated Circuit Engineering*. Reading, Mass.: Addison-Wesley, 1979.

Gulett, M. R., "The CMOS Controversy," *VLSI Design*, 4th quarter 1981.

Hon, R. W., and C. H. Sequin, "A Guide to LSI Implantation System," Xerox Palo Alto Research Center, Palo Alto, Calif., Jan. 1980.

Lipman, J., "A CMOS Implementation of an Introductory VLSI Design Course," *Lambda*, Vol. 2, No. 4, 1981, pp. 56–58.

Sequin, C. H., "Generalized IC Layout. Part II: Application to CMOS Processes," Computer Science Division, University of California at Berkeley, Berkeley, Calif., Jan. 7, 1982.

Wollesen, D. L., "CMOS-LSI: The Computer Component Process of the 80's," *Computer*, vol. 13, p. 59, Feb., 1980.

Yu, K., R. Chwang, M. Bohr, P. Warkentin, S. Stern, and C. N. Berglund, "HMOS-CMOS: A Low Power High Performance Technology," *IEEE J. Solid-State Circuits*, Vol. SC-16, Oct. 1981, pp. 454–459.

EXERCISES

4.1. The cross-sectional view of the nMOS device shown in Fig. 4.7(f) corresponds to the center horizontal line xy in the layout. Show the corresponding cross-sectional view if the line were chosen in the center of a standard $4:1$ ratio nMOS inverter layout (see Chapter 5 for a layout).

4.2. The layout of a CMOS inverter circuit is shown in the next chapter. Illustrate the silicon-gate CMOS p-well fabrication of this inverter by giving both the top and cross-sectional views (running across the center line) at each step of the process.

4.3. Based on the descriptions of the oxide-isolation p-well process in this chapter, describe the steps in an oxide-isolation n-well process and an oxide-isolation twin-well process.

5

Design-Fabrication Interface

5.1 The Clean Interface

One of the fascinating developments in recent years in LSI/VLSI technology is the emergence of a set of powerful design tools that enable designers to deal with designs at a level that does not involve the gory details of process- and fabrication-dependent parameters. The situation is very similar to that of an artist with a camera and film—his or her design tools—who creates his or her own "artwork" and then delivers the films to a processing house for development. In order that the developed film is what the artist intended it to be, he or she must follow certain "design rules" regarding the setting of the film speed, shutter speed, and lens aperture. The custom VLSI designer is also an artist; his or her artwork is the set of geometrical shapes that define the mask layers and his or her "design rules" correspond to a set of permissible geometries of the mask layers that allow for certain process variations and tolerance and still guarantee that the circuit will function properly and electrically. This enables the designer to work behind the walls of a clean interface, concentrating on design innovation and not getting embroiled in the complex process-dependent activities.

5.2 Why Design Rules?

There are several error mechanisms that contribute to deviations in feature shapes from the shapes defined in a designer's artwork. These include mask misalignment, variations in the photoresist edges due to variations in exposure, undercutting of thin oxide under the corners of the photoresist, overetching, spreading of the diffusion and implantation under the gate or near the source–drain end, and tolerance of the field oxide windows. A feature's size may alter during the operation of the circuit. For example, if the current density exceeds certain limits in metal wire, a phenomenon called *metal migration* might occur, which causes the metal atoms to move toward the direction of current flow. A relatively thinner part of the metal wire will get thinner since the current density here will exceed the average current density in the wire and may ultimately blow out as a fuse. To prevent this, a minimum width of metal wire, corresponding to the current density requirement, has to be guaranteed.

The purpose of the design rules is to guarantee that under the worst cumulative variations of the processes noted above, the circuit does not fail catastrophically; that is, separate features do not merge and small features do not split, thereby preserving the original topology of the intended circuit. The rules must also guarantee that the electrical parameters, such as the resistance and capacitance of the wires, which are determined by the physical dimensions of the features (such as length and width), are not altered by the process variations to a point that may cause a serious degradation in performance.

One of the fundamental difficulties in specifying design rules is that the fabrication processes are undergoing rapid evolutionary changes, and there is a tradeoff between complexity and circuit (area) efficiency. As a result, the industrial design rules are complex and undergo constant change. The single most important parameter that characterizes any process is the feature size. As long as the processing steps are not altered drastically, a set of design rules expressed in terms of feature size will have the best chance of survivability. With improvements in process technology, the feature size will scale down without changing the design rules. Mead and Conway (1980) formulated a set of such design rules for the relatively stable silicon-gate *n*MOS process which has become a sort of university standard. Similar attempts are under way for design rules for CMOS and other generic processes. We describe these in the following sections.

5.3 Mead–Conway Design Rules for the Silicon-Gate *n*MOS Process

Mead and Conway (1980) chose an elementary *length unit*, λ, to formulate the design rules. As of 1983, the value of λ was approximately 2 μm (1 μm equals one-thousandth of a millimeter; 1 mil, which is one-thousandth of an inch, equals 25.4 μm). The quantity λ may be thought of as the maximum deviation of a feature on the wafer from its intended position on the artwork. Two features on different mask layers can then be misaligned by as much as 2λ on the final wafer. If the crossing of these two features is catastrophic for the design, they must be separated by at least 2λ in the original artwork. If the crossing of the edges is undesirable but not catastrophic, the maximum movement of an edge is 0.5λ on one side. In this case, the edges must be separated by at least 1λ. These two "meta" rules (Lyon, 1981) form most of the Mead–Conway rules, but stricter exceptions are needed to

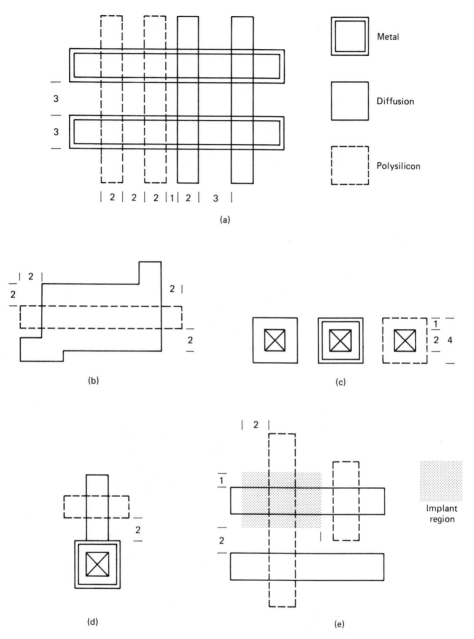

Figure 5.1 (a) Width and spacing rules for metal, diffusion, and polysilicon; (b) extension rule for transistor; (c) overlap rule for contact cuts; (d) contact cut clearance rule; (e) depletion-mode implant rules.

deal with specific features such as metal width/spacing or diffusion spacing, for reasons related to physical processes.

Mead–Conway design rules specify constraints on *width, separation, extensions,* and *overlaps* of features in three mask levels of conductors (metal, polysilicon, and diffusion), contact cut windows, and the implantation layer for the depletion load. These rules are summarized in Fig. 5.1(a) through (e). The numbers in these figures are lengths in units of λ.

The minimum feature size on one layer is 2λ, except for metal, whose minimum width is 3λ. The minimum polysilicon–diffusion spacing is 1λ, polysilicon–polysilicon spacing is 2λ, and for both metal–metal and diffusion–diffusion the minimum spacing is 3λ. Diffusion must be spaced 3λ apart since the depletion regions have a tendency to spread outward and make accidental connection. Metal is patterned at the end and runs through rough terrain. To be conservative, the minimum width and spacing are set at 3λ. We will see in a later chapter how the minimum width has to be increased to carry adequate amount of currents so that metal migration does not take place. The 1λ separation between polysilicon and diffusion is adequate since even if these edges move closer than 1λ, the net effect is to increase the coupling capacitance, which might slow down the speed but does not cause the circuit to fail; it also has the effect to narrow the diffusion region which is not catastrophic. The width and spacing rules are shown in Fig. 5.1(a).

A transistor is formed whenever a polysilicon wire of width $\geq 2\lambda$ crosses over a diffusion wire of width $\geq 2\lambda$. To avoid a short-circuit path between source and drain by diffusion, the polysilicon gate must extend on both sides by at least 2λ. Similarly, to guarantee the existence of diffusion regions to carry the channel current, the source–drain diffusion must extend over the polysilicon by at least 2λ. This is shown in Fig. 5.1(b).

The contact cut window has a minimum of a 2λ square size and must be surrounded by 1λ overhang of the material (polysilicon, diffusion, or metal) to ensure sufficient contact area [Fig. 5.1(c)]. The aluminum cover of the entire contact area is to ensure low resistance and high current density at the contact point. A polysilicon wire must be at least 2λ away from any contact cut in diffusion [Fig. 5.1(d)] to avoid accidental contact.

Finally, there is the implantation overlap and clearance rule [Fig. 5.1(e)]. For a depletion-mode transistor, the implantation region must extend by 2λ into the source and drain region and by 1λ into the gate overhang region. Also, if there is any enhancement-mode transistor nearby, a possible extension of the implantation device toward the channel of the enhancement device must be avoided. If the gate of the depletion and enhancement devices is the same, this might make the resultant threshold of both devices positive, thereby destroying the depletion device. A clearance of 2λ between the implant region and the diffusion region of the enhancement device is necessary. If the source or drain of the enhancement device is shared with the depletion device, this coupling is not that serious and only 1λ clearance is adequate. A simpler but more conservative statement of the rule is "2λ

everywhere," meaning that the extension of implantation must be 2λ in all directions and the clearance from any enhancement transistor is also 2λ.

5.4 Lyon's Buried Contact Rules

Buried contacts have replaced the use of butting contacts for connecting polysilicon with diffusion. Lyon (1981) defined a set of design rules applicable to buried contact, as shown in Fig. 5.2. As with a contact cut, a 2λ square of polysilicon–diffusion–buried is the minimum area of overlap. This structure guarantees an overlap area of 2λ square independent of misalignment less than 1λ. The "polysilicon–surround" structure in Fig. 5.2(b) could reduce the overlap area below 2λ square, even with a misalignment of less than 1λ for polysilicon or diffusion. Such is also the case with

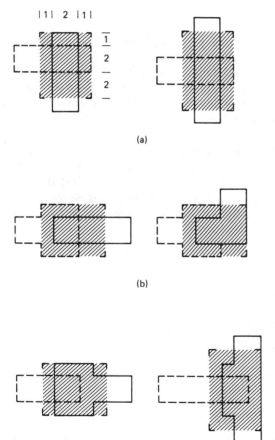

(a)

(b)

(c)

Figure 5.2 Buried contact rules (shaded region denotes the buried mask):
(a) crossing contacts;
(b) polysilicon–surround contact;
(c) diffusion–surround contact.

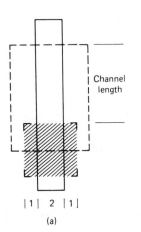

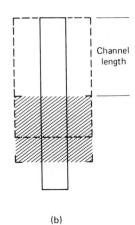

|1| 2 |1|

(a)

(b)

Figure 5.3 Buried contact for pull-up transistor: (a) Lyon's gate–source contact on channel end; (b) MOSIS gate–source contact on channel end.

the "diffusion–surround" structure of Fig. 5.2(c), but this structure has a larger "pit" area—the area from which thin oxide has been removed and polysilicon has been etched out. The buried mask should extend by 1λ beyond all sides of the overlap region to make sure that the thin oxide is completely etched out from the overlap region. To prevent the possibility of a parasitic enhancement device in series with the diffusion line, the buried mask is extended an additional 1λ toward the diffusion line.

Another variation of the buried contact structure is shown in Fig. 5.3, used to connect the gate and source in a pull-up transistor. Here the overlap region is defined in one direction. Lyon's structure shown in Fig. 5.3(a) does not need a buried extension at the edge and the channel length should be measured from the buried edge. Chips fabricated by MOSIS (MOS Implementation Facility, supported

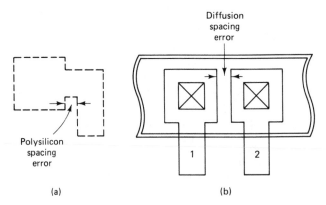

(a)

(b)

Figure 5.4 (a) False polysilicon spacing error; (b) false diffusion spacing error.

by ARPA and NSF primarily for U.S. university researchers and teachers and located at the Institute of Science of the University of Southern California) using Lyon's structure worked. But MOSIS recommends the more conservative structure shown in Fig. 5.3(b). Here the overlap region is $2 \times 3\lambda$, the buried area extends on the sides by 2λ, and the channel is measured starting 1λ beyond the buried edge.

A single conservative but simple "2λ everywhere" rule is used to define the clearance rule for buried contact. This rule is used to prevent breakdown of the 1λ-wide gate-oxide layer near the buried mask, producing accidental shorts.

Design rules usually do not incorporate any knowledge of the underlying electrical network and the connectivity relations of its nodes. This sometimes leads to "false" design rule violation, as illustrated in Fig. 5.4. In part (a), the 1λ spacing error for polysilicon can be ignored since the entire polysilicon area is electrically

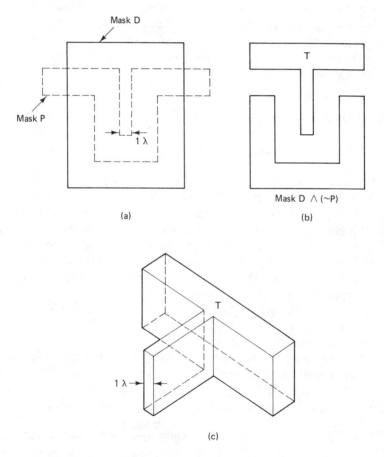

Figure 5.5 Lithographic errors: (a) transistor layout; (b) the mask $D \wedge (\sim P)$; (c) the 1λ thin wall.

connected. In part (b), the 1λ diffusion spacing error is again false, since lines 1 and 2 are connected to each other via the contact cuts by the overlying metal line. Most design-rule-checking software (see Chapter 9) incorporate electrical connectivity information to avoid such false error reporting.

But there is a special situation where such 1λ spacings might really cause a problem, leading to what is known as *lithographic error*. This error has no electrical significance. Consider the transistor layout shown in Fig. 5.5(a). The 1λ polysilicon spacing is not a design rule error, as we have just discussed. But recall that the diffusion step (Section 4.3.3) in the silicon-gate nMOS process uses a mask D $\wedge$ ($\sim$ P) to etch out thin oxide from the active T of Fig. 5.5(b). This will translate into a photoresist of the shape of Fig. 5.5(c). The thin 1λ wall shown in this figure may sometimes collapse for the simple reason that it is too thin and tall and fails to hold by itself. These fallen walls actually act as debris floating over the surface of the wafer and produce unpredictable errors in the etching process. Since both the positive and negative of the masks can be used for patterning, a lithographic error is avoided by making sure that there is no 1λ-width error in any part of the D $\wedge$ ($\sim$ P) mask. A good, clean rule to avoid lithographic error is to make sure that all width and spacing are at least 2λ wide.

5.5 CMOS Design Rules*

5.5.1 CMOS-pw Design Rules

Since the formulation of the Mead–Conway design rules for nMOS, several attempts have been under way for the development of similar portable design rules for the CMOS process. One of the major problems in formulating such design rules for CMOS was the complexity and variations of the CMOS process. Recently, MOSIS is offering foundry services for SOS CMOS. The groundwork for the design rules for the bulk CMOS p-well process was laid at the Jet Propulsion Laboratory (Griswald, 1982). The Berkeley VLSI software (Ousterhout, 1981) tools (see Chapter 9) have incorporated this technology together with its design-rule-checking software, which are currently being used by most U.S. universities with VLSI programs. A second twin-well process has been supported by Stanford University. We now present the design rules for the CMOS-pw process.

The design rules for the CMOS-pw technology are similar to the Mead–Conway rules. The following layers are the same as in nMOS technology: polysilicon (red or dashed region), contact cut (crossed region), and metal (blue or region enclosed by double-stranded lines), with their accompanying width and spacing rules being the same as for nMOS. The diffusion layer is called thin oxide (represented by green or the region enclosed by single-stranded lines). The same set of design rules that apply to diffusion for nMOS with respect to polysilicon, contact cut, or metal apply here also.

* For MOSIS Scalable and Generic CMOS Design Rules, see Appendix B, p. 349.

A transistor is formed whenever a polysilicon wire of width $\geq 2\lambda$ crosses a thin oxide layer of width $\geq 2\lambda$, but the transistor must be totally embedded in the p-well (brown or the region enclosed by cross-point lines) or p^+ (orange or the region enclosed by alternating dots and dashes). All n-channel transistors are laid out in the p-well region and all p-channel transistors are laid inside the p^+ region. There are some special rules that apply to p-well, p^+, and thin oxide layers.

p-well and thin oxide rules [Fig. 5.6(a)].   An enclosed thin oxide must have a 3λ spacing from its p-well boundary. An outside thin oxide must be spaced at least 5λ away from a p-well. The minimum width of a p-well is 4λ; the minimum separation is 2λ.

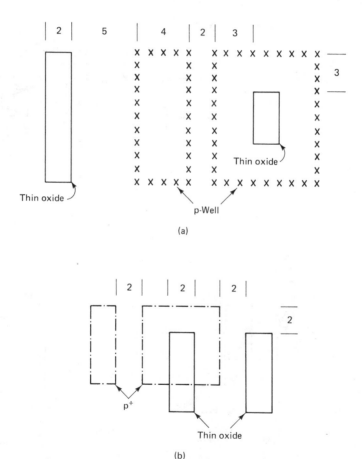

(a)

(b)

Figure 5.6 CMOS-pw design rules: (a) p-well and thin oxide rules; (b) p^+ and thin oxide rules; (c) contact cut rules; (d)–(e) split contacts; (f) p^+ and gate rules.

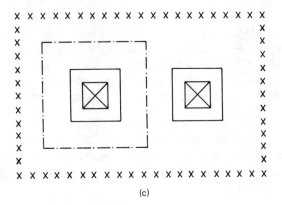

(c)

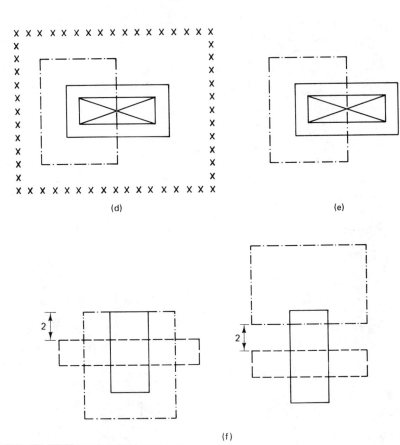

(d)   (e)

(f)

Figure 5.6 (*Cont.*)

p^+ and thin oxide rules [Fig. 5.6(b)].   The minimum separation between thin oxide and p^+ is 2λ. The thin oxide can cross the p^+ region boundary, but whenever it does so the part that remains within the p^+ region must have a clearance of at least 2λ from the p^+ boundary. The minimum separation and width of the p^+ region are 2λ. The minimum separation and width of thin oxide are the same as for diffusion in nMOS.

Contact cut rules [Fig. 6.6(c)].   The regular contact cut structures for metal–polysilicon and metal–diffusion in nMOS are also used for metal–polysilicon and metal–thin oxide in CMOS, respectively. There are no butting or buried contact equivalents. A special contact cut structure [Fig. 5.6(c), left part] is used for connecting the p-well to V_{ss} or ground. It consists of a metal–thin oxide cut surrounded by 2λ p^+. The whole structure is then embedded in a p-well region, keeping a clearance of at least 3λ between the thin oxide and the well boundary. The thin oxide and metal can be connected together by a contact cut within a p-well region without bonding it to the p-well [Fig. 5.6(c), right part]. This is usually used to get a signal out in metal from within a p-well region, as we will see for an inverter circuit layout (Fig. 5.14). If the thin oxide needs to protrude out of the p^+ region, a *split contact* [Fig. 5.6(d) and (e)] is used. It consists of a $4 \times 8\lambda$ metal–thin oxide contact, half of which protrudes out of the p^+ region. The split contacts are used to connect transistor sources to supply or ground via the substrates.

p^+ and gate rules [Fig. 5.6(f)].   A p-transistor's channel area must be surrounded by p^+ implantation 2λ all around, much like the ion implantation for the depletion-mode transistor. If the transistor is of n-channel type (i.e., laid within a p-well), its gate cannot come closer than 2λ to a neighboring p^+ region.

p^+ and p-well separation rule.   There should be a minimum spacing between the p^+ and p-well region. This rule has no electrical significance and follows from the general precaution taken to avoid a lithographic error, which says that no photoresist should be allowed to have less than 2λ width on the wafer during the photolithographic steps.

5.5.2 Magic Design Rules for CMOS

The latest (circa spring 1985) distribution of the Berkeley VLSI design tools (Ousterhout et al., 1983) called *Magic* specifies a new set of design rules for CMOS for both 3-μm and 1.25-μm processes (Katz and Scott, 1958), and a continuous incremental design rule checker (see Chapter 9 for further details). This means that when the layout is done and is not flagged for any error, the design rule checking is also automatically done. Two levels of interconnect by metal are provided by the technology. The layout rules are defined for both p-well and n-well. Contacts have to be placed for both n-diffusion and p-diffusion; depending on what type of technology is actually fabricated, one type of well will be ignored. The layout system of Magic uses only "abstract" layers, which do not correspond exactly to mask layers. The actual dimensions of the contacts or diffusion may also be changed during fabrication. The general rule is that the two levels of metal and the polysilicon and diffusion layers correspond to actual layers, but the designer does not have to specify implants or wells, which are generated automatically. The designer is thus relieved of the burden of well placement and the associated design rules, but could request the system to display the well layer when the layout is done. The abstract layers and the rules for their minimum width and spacing are given in Table 5.1.

A p-channel transistor is formed when polysilicon overlaps with p-diffusion with pfet defining the channel region. Similarly, an n-channel transistor is formed when polysilicon overlaps with n-diffusion with nfet in the channel region. The appropriate wells and implants for the transistors are generated by magic automatically. The transistor "extension" and "clearance" rules are similar to those for the nMOS process: a 2λ overhang in both the polysilicon and diffusion direction is required and the channel region must maintain a clearance of 1λ from any nearby contact. These rules are illustrated in Fig. 5.7(a) and (b). The spacing between polysilicon and ndiffusion/pdiffusion must be at least 1λ, except where a transistor is formed. The polysilicon and diffusion contacts could be adjacent to each other and must be recognized as an exception to the rule.

When polysilicon connects to metal via a pcontact, the width of the contact must be at least 5λ in the direction the metal enters or leaves the contact. Large pcontact must be an even multiple of 4λ plus 1λ in the metal direction and only a

TABLE 5.1

Layer	Color	Symbol	Minimum width, λ	Minimum spacing, λ
First-level metal	Blue	m1, metal1	3	3
Second-level metal	Purple	m2, metal2	3	4
Polysilicon	Red	p, poly	2	2
n-diffusion	Green	ndiff	2	3
p-diffusion	Brown	pdif	2	3
p-transistor	Brown stripes	pfet	2	2
n-transistor	Green stripes	nfet	2	2
Polysilicon–metal1 contact	Cross	pcontact	4(5)	4
n-diffusion–metal1 contact	Cross	ndcontact	4	4
p-diffusion–metal1 contact	Cross	pdcontact	4	4
n-diffusion–p-well contact	Gridded squares	pwcontact	4	4
p-diffusion–n-well or n-substrate contact	Gridded squares	nwcontact	4	4

multiple of 4λ in the other direction. Large ndcontact or pdcontact similarly should be an even multiple of 4λ in both directions. These rules are illustrated in Fig. 5.7(c) and (d).

The connections between metal1 and metal2 layers are made by "metal2 contact" or "m2 contact," or "via." The m2 contact must have width and length equal to an even multiple of 4λ. The m2 contact is shown in Fig. 5.7(e). There are two additional rules: There cannot be any polysilicon, ndiffusion/pdiffusion underneath the contact area, but polysilicon or diffusion could be outside at the edge of the m2 contact. If the contact area is completely contained within the polysilicon or diffusion area with 1λ clearance on all sides, the structure is acceptable. These rules are illustrated in Fig. 5.7(f), (g), and (h). A general rule for contacts is that all contacts have m1 and exactly one other material, and contacts cannot be stacked. This means, for example, that to connect from poly to n-diff a short metal connection is required.

The continuous area formed by pdiffusion and pfet (or ndiffusion and nfet) must contain a well contact. The "pwcontact" is used for p-wells and must be tied to the ground; the "nwcontact" is for n-wells and must be tied to V_{dd}. A common structure in Magic layout is the juxtaposition of a well contact with a diffusion contact, as illustrated in Fig. 5.7(i). The polysilicon edge must be at least 2λ apart from the contact regions. It is recommended to place one nwcontact for each p-transistor and one pwcontact for each n-transistor. This rule is derived from the observation that placing too few contacts might increase the chance of latch-up.

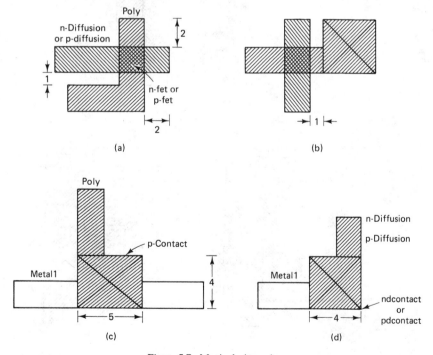

Figure 5.7   Magic design rules.

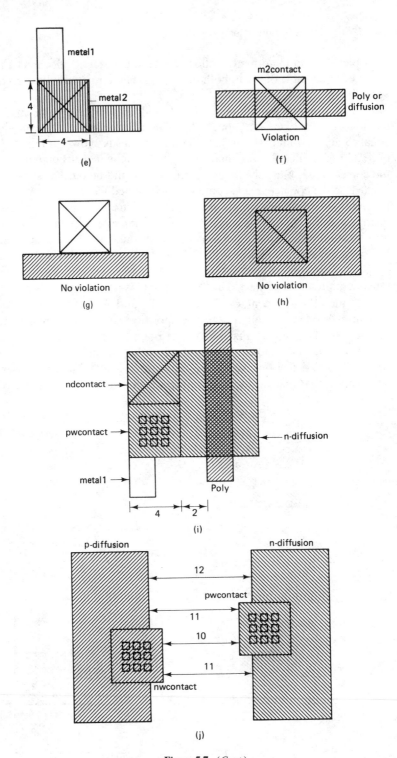

Figure 5.7 (*Cont.*)

Diffusions and well contacts of the n-type must satisfy minimum spacing constraints with respect to pdiffusion and p-well contacts, as shown in Fig. 5.7(j). The minimum well-to-well spacing is 10, well-to-diffusion is 11, and diffusion-to-diffusion is 12.

Hierarchical design sometimes needs overlapping subcells. The overlap may be used to establish a connection as long as the overlap is consistent with the electrical integrity of the circuit. For example, a polysilicon layer of a cell could overlap with a polysilicon–metal contact of another to establish an electrical connection, but polysilicon of one cell cannot overlap with the diffusion of another cell because it will create a new circuit structure (i.e., a transistor). A contact in one cell may not overlap the contact of another cell unless the contacts are of the same type and size.

There are some special layers that are used only for special types of design, such as pads. Two such layers are called ''p-ring'' and ''n-ring.'' As the names imply, these layers are used to create guard rings for latch-up prevention: nring for guard rings around n-wells and pring for guard rings around p-wells. These layers create appropriate types of diffusion areas (n or p) and will also generate the wells over the area.

5.6 Electrical Parameters

Adherence to design rules guarantees that the circuit as specified by the layout artwork will be produced correctly if the process parameters vary with a total tolerance of λ. In order that the designer can make approximate estimates about the performance parameters (i.e., speed, power, and current), it is necessary to know the values of the electrical parameters—the resistance and capacitance of the features. A large number of circuit simulation programs have recently been developed for the analysis and prediction of the electrical behavior of VLSI circuits. We discuss some of these tools in later chapters. The complexities of the underlying models and algorithms determine how accurate these calculations are. In this section we discuss some fundamental ideas regarding the circuit parameters and simple procedures to compute them. In later chapters we will see how to use these values to estimate circuit performance.

5.6.1 Resistance

The electrical resistance R of a sample of semiconductor material of length l, width w, and thickness t (Fig. 5.8) is given by

$$R = \rho \frac{l}{wt} = \rho \frac{l}{A}$$

where $A = wt$ is the area of the sample perpendicular to the direction of current flow and ρ denotes the resistivity of the material, which is a function of hole and electron

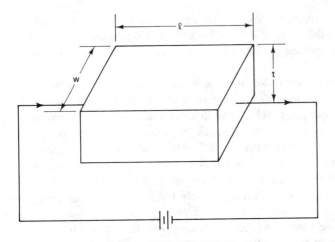

Figure 5.8 Sample semiconductor resistor.

concentration and their mobilities. If $l = w$, that is, for a square material, the resistance R, denoted $R_\square$, is

$$R_\square = \frac{\rho}{t}$$

$R_\square$ is independent of the length of the square and is called the *sheet resistivity*. The quantity t is fixed for a given sample, and ρ is expressed as ohm · cm. Thus the unit of $R_\square$ is ohm and is more commonly expressed as ohms per square ($\Omega/\square$).

For diffused layers, ρ varies with the depth of the diffused layer and the surface and substrate impurity concentrations. The average value of ρ can then be taken to compute $R_\square$. Typical values of $R_\square$ are given below for the silicon-gate *n*-channel process with 4-μm line width ($\lambda = 2 \; \mu$m). These values are very approximate and are average values used for delay calculations.

Channel: $15 \times 10^3 \; \Omega/\square$
Polysilicon: 20–$150 \; \Omega/\square$
Diffusion: $15 \; \Omega/\square$
Metal: $0.02 \Omega/\square$

Similar figures for the bulk CMOS process are given below. These figures are again very approximate and are used for delay calculators.

p-Channel: $15 \times 10^3 \; \Omega/\square$
n-Channel in p-well: $6 \times 10^3 \; \Omega/\square$
Polysilicon: 20–$50 \; \Omega/\square$
n-Diffusion: 15–$60 \; \Omega/\square$
p-Diffusion: 150–$200 \; \Omega/\square$
Metal: 0.02–$0.06 \; \Omega/\square$

These figures show that the value of $R_\square$ varies considerably; metal offers the least resistance and is therefore the most commonly used material to carry heavy currents. The channel offers the most resistance. Depletion-mode transistors are used primarily to build a high resistance. The polysilicon resistance can be controlled by external doping and a drive-in step and could be made as low as that of diffusion.

It is useful to be able to make rough estimates of the resistance of rectangular-shaped regions or wires. We can write the resistance formula as

$$R = R_\square \times \left(\frac{l}{w}\right)$$

This formula can be understood by considering the rectangle as a parallel connection of w rows, each row consisting of a series connection of l-unit resistors $R_\square$. The formula is not very exact because it does not take into account possible changes in the values of l and w due to lateral diffusion of dopant as well as the possible variation of ρ with depth. For an arbitrary area bounded by horizontal and vertical lines, R can be computed by laying the area out as an end-to-end placement of rectangles of length l_i and width w_i, which are connected in series. Thus

$$R = R_\square \times \sum \left(\frac{l_i}{w_i}\right)$$

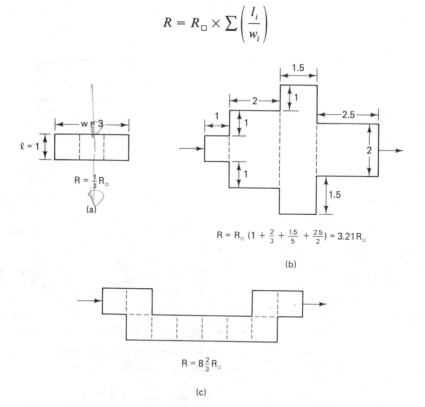

$$R = \frac{1}{3} R_\square$$

(a)

$$R = R_\square \left(1 + \frac{2}{3} + \frac{1.5}{5} + \frac{2.5}{2}\right) = 3.21 R_\square$$

(b)

$$R = 8\frac{2}{3} R_\square$$

(c)

Figure 5.9 Resistance computation: (a) a simple rectangle; (b) arbitrarily shaped rectangular region (corners ignored); (c) a sample with corners.

Figure 5.9 illustrates a few such computations using the formula above. In part (a), the current flows along with the side with $l = 1$; since $w = 3$, the structure can be considered as three $R_\square$ resistors connected in parallel. Hence the effective resistance $R = \frac{1}{3}R_\square$. In part (b), the current flows in the horizontal direction and the effective resistance equals the sum of resistances of four rectangles as indicated by the dashed lines, in series, giving $R = 3.21R_\square$. In part (c), the resistance computation is complicated by the fact that the current flows in the corner squares are not uniform. As a rough estimate, the value of a corner square resistance is taken as $0.66R_\square$. The effective resistance is then equal to the resistance of six squares and four corner squares all connected in series, yielding $R = 8\frac{2}{3}R_\square$. Three interesting transistor structures and the computation of their channel resistances are shown in Fig. 5.10. The quantity $R_\square$ here refers to the sheet resistance of the channel. In part (a), the effective resistance equals the resistance of three $R_\square$'s in series plus a resistance which is a parallel combination of two corner squares of dimension 1×1 and a central piece of dimension 1×2, yielding a total resistance of $3.4R_\square$. In part (b), the effective resistance equals the parallel resistance of three squares and one

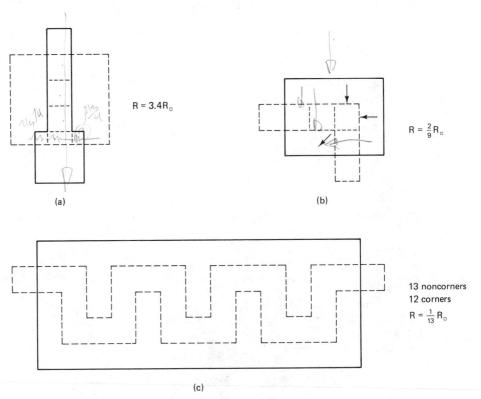

$R = 3.4R_\square$

(a)

$R = \frac{2}{9}R_\square$

(b)

13 noncorners
12 corners

$R = \frac{1}{13}R_\square$

(c)

Figure 5.10 Computation of the channel resistance: (a) a pull-up transistor; (b) a bent transistor; (c) a serpentine transistor.

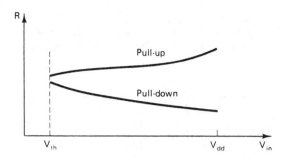

Figure 5.11 Variation of pull-up and pull-down resistance with input voltage of the inverter.

corner, yielding $R = \frac{2}{9}R_\square$. In part (c), the resistance of the long serpentine transistor equals the parallel combination of 13 squares and 12 corners, yielding $R = \frac{1}{13}R_\square + \frac{0.66}{12} R_\square$. [Resistors formed out of a zigzag-shaped diffused path have been called *meander* resistors (Glaser and Subak-Sharpe, 1979) in the literature. The terminal connectors to the resistor (usually, contact cuts) also contribute about 20 to 150 Ω to its value, which we have ignored.]

The procedure used to compute the effective resistance of wires and transistors as discussed above is very approximate. It ignores nonlinear effects due to the distributed nature of the parameters and nonuniform flow of currents in different regions of the wires. Furthermore, the resistance of a MOS transistor depends on its terminal voltages, the type of transistor (p-, n-, enhancement or depletion mode), and the context of its use (pull-up, pull-down, or pass transistors) (Terman, 1983). For example, the variation of the pull-up and pull-down resistance of an nMOS inverter with the input voltage V_{in} at the gate can be depicted as shown in Fig. 5.11. The dependence of the channel resistance on its type is due to the fact that the threshold voltages are different, which affects the terminal characteristics. A transistor used as a pass transistor works as a linear resistance since V_{ds} is small, but when used in a pull-down or pull-up network, its resistance varies significantly, since it switches from linear to saturated mode, and vice versa. Despite these variations, an average value for the channel resistance, as given above, can be taken to be the "effective" channel resistance for the purpose of estimating the delay of circuits. (In his simulation model Terman defines three effective resistances for each transistor: the static resistance, which is the effective resistance; and two dynamic resistances, corresponding to high-to-low and low-to-high current transitions in the transistors, to take into account the basic asymmetry of their characteristics.)

5.6.2 Capacitance

The importance of capacitance is that it determines the delay or the speed of the circuit. In this section we present some basic facts about capacitance in a MOS device. The computation of delay or switching speed is presented in Chapter 6.

The capacitance of a MOS network depends on a number of factors, such as the physical structure of the MOS devices, the terminal voltages that control the formation and the depth of depletion regions, and the network topology. We discuss these factors here briefly.

Gate capacitance.   The structure of a MOS transistor is that of a parallel-plate capacitor, as shown in Fig. 5.12. The oxide layer acts as the insulator between the two conducting plates: the gate (in polysilicon or metal) and the substrate. The formula for parallel-plate capacitance per unit area is

$$C_{ox} = \frac{\varepsilon \vartheta}{T_{ox}} \qquad \text{farad/cm}^2$$

where $\varepsilon = 3.9$ is the dielectric constant of the oxide, $\vartheta = 8.85 \times 10^{-14}$ F/cm^2 is the free-space permittivity, and T_{ox} is the oxide thickness in cm. The value of C_{ox} remains relatively constant but shows a slight drop with increase in substrate voltage. For semiconductors, the capacitance values are very small and are usually expressed in units of picofarad (10^{-12} farad, denoted pF) or femtofarad (10^{-15} farad, denoted fF). Similar parallel-plate capacitances are also formed if there are any overlap regions between the gate and source and the gate and drain. Even if the process is self-aligned, a certain amount of channel capacitance between gate-to-source and gate-to-drain will exist. These capacitances are denoted C_{gs} and C_{gd}, respectively, and must be added to C_{ox} to compute the total gate capacitance C_g of the MOS transistor since they are all connected in parallel. The values of these capacitances are specified by the manufacturer with gate voltage V_{gs} negative or zero, that is, without any depletion region under its gate.

As the gate voltage is increased, the phenomena of accumulation, depletion, and inversion start to happen (see the Appendix for details). A depletion capacitance

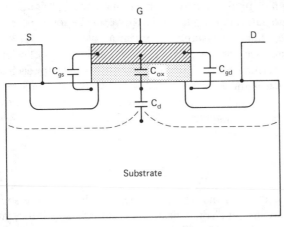

$$C_g = C_{gs} + C_{gd} + C_{ox}$$

Figure 5.12 MOS capacitor.

C_d is formed between the gate and the boundary of the depletion region, which is connected in series with C_{ox} and lowers the effective gate capacitance C_g. This is depicted in Fig. 5.13. The value of C_d depends on the depth of the depletion region. The initial depth of the depletion region depends on the built-in contact or barrier potential (typical value 0.7 V) and then increases with increasing V_{gs}. As V_{gs} is increased so as to exceed the threshold voltage V_{th} (for p-transistors these quantities have opposite signs), inverison takes place and the channel forms the conducting plate instead of the substrate, the depletion capacitances no longer exist, and the total capacitance shows a marked increase to its original oxide capacitance value.

In the discussion above we have assumed that the gate voltage is static or varies very slowly so that the phenomena of accumulation, depletion, and inversion take place in proper sequence. If, however, V_{gs} varies very rapidly (i.e., when the signal frequency of the gate voltage is rather high), the channel may not be formed and disappear that quickly, and on the average the device will appear to be in the depleted state all the time, bringing in the effects of the depletion capacitances C_d to reduce the total capacitance (dotted horizontal line in Fig. 5.13). This dynamic behavior of the capacitance does not really concern us since we will have to operate at considerably lower frequency to guarantee proper switching (and hence the formation of the channels) of each transistor.

We mentioned earlier that the capacitances C_{gs} and C_{gd} are due to the gate overlaps at the source and drain sides. Even if there is no overlap, it is useful to visualize C_g as consisting of a parallel connection of C_{gs} and C_{gd} since the channel can be viewed as a physical extension of the source and the drain diffusion islands (McCarthy, 1983). When the transistor is operating in the linear region, the two component capacitances have almost equal value; that is, $C_{gs} = C_{gd} = 0.5C_g$. When the transistor is saturated, the channel vanishes near the drain end and we can assume that $C_{gs} = C_g$.

The fact that C_{gd} is very small or practically vanishes near saturation has some beneficial effect on the circuit operation. If the transistor is used in a pull-down

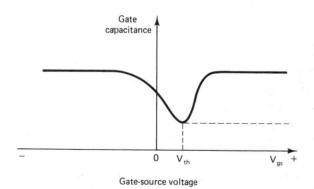

Figure 5.13 Variation of the gate capacitance with gate voltage.

network, the voltage across C_{gd} is of opposite polarity to that of the input voltage V_{gs}. The net effect of this doubles the effect of C_{gd} (see the Exercises). This is known as *Miller capacitance* and has an adverse effect on the performance of the circuit. This emphasizes the importance of self-aligned processes, which attempt to avoid the overlap capacitance completely.

For most of the operating regions of MOS transistor for digital circuits, the channel capacitance Cg can be approximated by C_{ox}, which ranges between 0.4 and 0.7 fF/μm^2. Typical values of C_{gs} and C_{gd} range between 0.2 and 0.4 fF/μm. Note these capacitances are along the source and drain edge of the FET. Thus, the average gate capacitance of a $2\lambda \times 2\lambda$ transistor with $\lambda = 2.00$ μm is about 12.00 fF. (2.4 f due to Cgs and Cgd and 9.6fF due to the channel). Typical values of wiring capacitance for polysilicon is about 0.05fF/μm^2, and that of metal is about 0.03fF/μm^2.

Interconnect capacitance.   The so-called "stray" or "wiring" capacitances of the paths connecting the active channels could become significantly high, depending on the length of the wire, so as to account for a major part of the circuit delays. This is because the wiring capacitances need to be charged up together with the gate capacitances for signals to be detected by the transistors. It is therefore convenient to be able to compute the wiring capacitances in terms of the gate capacitance of a transistor of some standard size. It is also important to recognize that the diffusion wires run buried in the substrate, and their capacitance depends on two factors: the total area of the wire and the perimeter of the wire. The perimeter determines the so-called "side-wall" capacitance, which is at least 20% of the capacitance determined by the total area.

We will now present an estimate of the interconnect capacitance (Sutton, 1985). We will use the following terminology (the subscripts p and n denote p- and n-channels, respectively): W_p, W_n, gate widths; L_p, L_n, gate lengths; L_{sdp}, L_{sdn}, source–drain lengths; W_{sdp}, W_{sdn}, source–drain widths; L_{mtl}, L_{poly}, length of runs of metal and polysilicon wires, respectively; W_{mtl}, W_{poly}, the corresponding widths of runs. Since the diffusion wire is highly capacitive, a run of wires made of diffusion should always be avoided. The relevant capacitances will be denoted as:

C_g: gate capacitance (pF/μm^2)

C_{pln}: p-channel source–drain capacitance

C_{nln}: n-channel source–drain capacitance

C_{mf}: metal-to-field oxide capacitance

C_{mp}: metal-to-polysilicon capacitance

C_{mt}: metal-to-oxide capacitance

C_{pf}: polysilicon-to-field oxide capacitance

The values of the parameters above are extremely process-dependent and are constantly changing with the evolution of processes. Typical values of C_{mf} and C_{pf}

come near 0.05 fF$/\mu$m^2 and source–drain capacitance ranges between 0.4 and 0.5 fF$/\mu$m^2. We have seen earlier that C_g is on the order of 1 fF$/\mu$m^2.

To calculate the effective metal and polysilicon capacitance load on a typical gate, it is first necessary to estimate the extent of the runs of different wires on a typical chip. Sutton proposes the following table for such an estimate, selected by visual examination of a large number of sample layouts.

	Over			Under metal
	Field	Polysilicon	T_{ox}	
Metal	60%	35%	5%	—
Polysilicon	100%	—	—	40%

This yields the following two expressions for metal and polysilicon capacitance per unit area, C_{mtl} and C_{poly}, respectively:

$$C_{mtl} = 0.60C_{mf} + 0.35C_{mp} + 0.05C_{mt}$$

$$C_{poly} = 1.00C_{pf} + 0.40C_{mp}$$

The basic capacitance equation of a driver can now be written as the sum of four load capacitances: gate capacitances of the inputs driven by the driver, the source–drain capacitances to which the output of the driver is connected, the wiring capacitance due to metal, and the wiring capacitance of the polysilicon at the output of the driver.

$$C = \left(W_p L_p + W_n L_n\right)C_g + W_{sdp}L_{sdp}C_{pln} + W_{sdn}L_{sdn}C_{nln}$$
$$+ L_{mtl}W_{mtl}C_{mtl} + L_{poly}W_{poly}C_{poly}$$

Note that the lengths and widths used in the equation above denote the sum of all effective lengths and widths that are attached as load to the driver. The expression above can be simplified somewhat by making the load appear to look like one large inverter. Let us make the simplifying assumption that $L_p = L_n = L$, since the channel length is usually fixed. Then the equivalent gate width W_{eq} of the "load inverter" can be expressed as

$$W_{eq} = \frac{C}{LC_g} = W_p + W_n + k_1 W_{sdp} + k_2 W_{sdn} + k_3 L_{mtl} + k_4 L_{poly}$$

where

$$k_1 = \frac{L_{sdp}C_{pln}}{LC_g} \qquad k_2 = \frac{L_{sdn}C_{nln}}{LC_g} \qquad k_3 = \frac{W_{mtl}C_{mtl}}{LC_g} \qquad k_4 = \frac{W_{poly}C_{poly}}{LC_g}$$

Frequently, $k_1 = k_2$ and $k_3 = k_4$ or will be close enough to be averaged. Typical values of these quantities are $k_1 = k_2 = 0.8$ and $k_3 = k_4 = 0.15$. Thus

$$W_{eq} = W_p + W_n + 0.8(W_{sdp} + W_{sdn}) + 0.15(L_{mtl} + L_{poly})$$
$$= W_p + W_n + 0.8W_{sd} + 0.15L_{interconnect}$$

where W_{sd} denotes the average source–drain width and $L_{interconnect}$ denotes the average interconnect wiring length. As an example, consider a complex gate driving a metal bus of 1200 μm and a polysilicon wire of 200 μm to three other gates whose effective (W_p, L_n) values are $(21, 7)$, $(30, 20)$, and $(10, 10)$. Then W_{eq} is given by $W_{eq} = (21 + 30 + 10) + (7 + 20 + 10) + 0.8(0) + 0.15(1200 + 200) = 308$ μm. We will see in Chapter 6 how driver sizes can be computed using this formula.

We have so far explained the procedures for estimating the resistance and capacitance of different circuit elements. A knowledge of these circuit parameters is essential for estimating the delay and performance of the circuit, estimating the power consumption of the circuit, and developing models for simulation programs to analyze the logical behavior of the circuit. We will study these problems in Chapter 6. We conclude this chapter with a discussion of some layout styles for simple and illustrative circuits.

5.7 Simple Layout Examples

The art of translating a circuit into a layout can be learned only by hands-on experience. One attempts to minimize the area of the layout subject to design rules, but there are several systems-level factors that determine the layout in a complex manner. These factors are flow of data and control information, communication and routing, power and ground routing, currents, power, and delay. The subject matter is better treated in the context of subsystem design discussed in Chapters 7 and 8. Here we introduce a couple of sample layouts to give the reader an opportunity to apply his or her knowledge of design rules, and to introduce a few of the layout strategies.

First, let us do the inverter circuit in nMOS and in CMOS. For the nMOS inverter, we have to realize a given ratio, say $8 : 1$. We could start with the pull-up transistor shown in Fig. 5.14(a) with $L/W = 4/1$. We then put the buried contact at the bottom of the channel using Lyon's definition of the channel layout [Fig. 5.14(b)]. Then the 2λ implantation is placed to make this a depletion-mode transistor. The pull-down transistor is then placed, keeping in mind the "2λ everywhere" separation rule for the buried contact and also the rule that the implantation must be at least 2λ away from the transistor gate. Since the overall size of the inverter is $8 : 1$, the L/W for the pull-down is taken to be $2/4$. At this time, we have to decide how we want the output of the inverter. Since the gate and source of the pull-up are connected, we can read the output in diffusion or in polysilicon to the right or to the left. Some of the possibilities are shown in Fig. 5.14(c). Finally, the V_{dd} and the Gnd connections are placed using contact cuts, which completes the layout. Completely different layouts are possible for this inverter by choosing a different L/W ratio for the pull-up, by laying the transistor sideways, and so on, to

(a)   (b)   (c)

Figure 5.14 Layout of an nMOS inverter.

suit a specific design style. It is precisely this flexibility in layout that makes development of general-purpose automated layout tools an extremely difficult task.

To lay out a CMOS inverter, we could start with the minimum-size p- and n-transistors as shown in Fig. 5.15(a). But the source and drain of these transistors could not be connected to output via thin oxide since thin oxide cannot cross the p-well boundary. We need a concoction of three contact cuts [Fig. 5.15(a)], two in

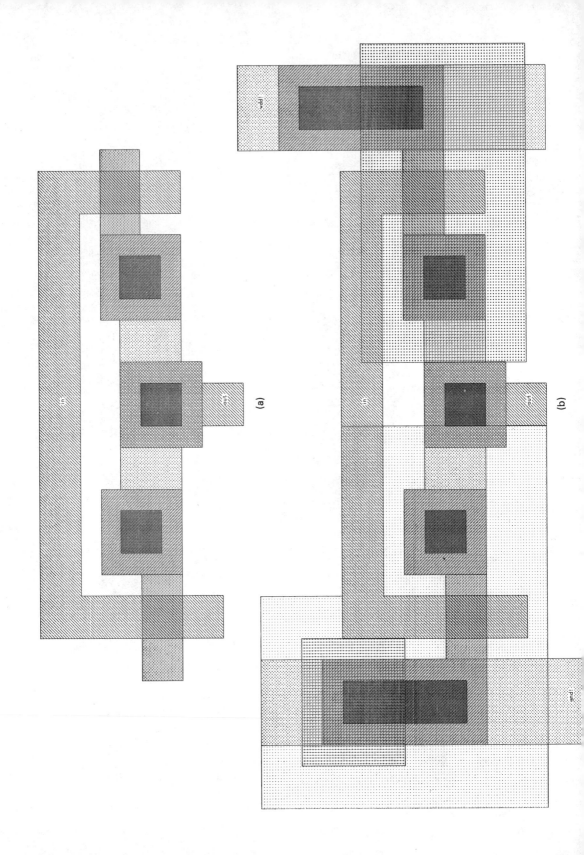

thin oxide (one in 3λ within the p-well and the other 2λ within the p^+) connected together by metal to the output polysilicon contact cut. The inputs of the two transistors can now be connected by polysilicon. The layout is now completed by grounding the p-well substrate with a split contact and making the V_{dd} and Gnd connection [Fig. 5.15(b)].

The layout of a CMOS NAND gate is shown in Fig. 5.16, and that of an AND–OR–INVERT gate realizing $f = ab + cd$ is shown in Fig. 5.17. The latter layout illustrates a general approach to the layout of a CMOS network based on the fully complementary structure shown in Fig. 3.37. It consists of a p^+ and a p-well region placed sideways, containing the dual networks F and $\bar{F}$, respectively. The input variables run in metal over both the regions.

Switch networks in nMOS yield themselves to most compact layouts since they use minimum-size pass transistors and have no p-well or p^+ regions to worry about. Two such layouts are shown next. The layout in Fig. 5.18 is a realization of the two-variable function block of Fig. 3.13 with the difference that it uses polysilicon lines to carry the control signals and yellow transistors at points where the diffusion will have to carry signals straight through. An exact translation of the stick diagram of Fig. 3.13 would have used more area. The layout in Fig. 5.19 is a realization based on the stick diagram of the basic cell of the sequence detector circuit given in Fig. 3.18. These two layouts illustrate an organizing principle of VLSI circuits that defines a good layout strategy. The principle states that the flow of data and control signals in the circuit must be organized in a way that minimizes the communication overhead between circuits. In these examples, the data flow in the horizontal directions and the control flows in the vertical directions, allowing interaction between control and data at regularly defined points in the circuit.

When a cell is used many times in a design, certain global considerations are important. Some of these strategies have been discussed in Section 1.3 and it is recommended that the reader read that section again. Recall that a good global routing strategy is defined to be the one that improves the regularization factor by allowing easy iteration, sharing, abutment, mirroring, and rotation of the basic or the "leaf" cell and of the composition cells, which consist of an interconnection of some of the basic cell types. Also, a good layout starts with a clearly defined strategy of flow of global information in horizontal or vertical directions such that mutually noncontacting layers (i.e., metal and polysilicon or metal and diffusion) are used to carry these signals. We will illustrate these principles with some examples.

Consider the sequence detector cell of Fig. 5.19. It has been designed for easy abutment in the horizontal directions. An array of two such cells is shown in Fig. 5.20. Note the flow of information in the circuit: Control is carried in metal in the vertical direction and data in diffusion in the horizontal direction. The layout of the function block of Fig. 5.18 is such that it can easily be stacked up in the vertical direction, sharing the common control or variable lines. An array of two such blocks is shown in Fig. 5.21. Such a network can be used to generate an arbitrary set of two-variable combinational functions. The layout of a clocked inverter stage is

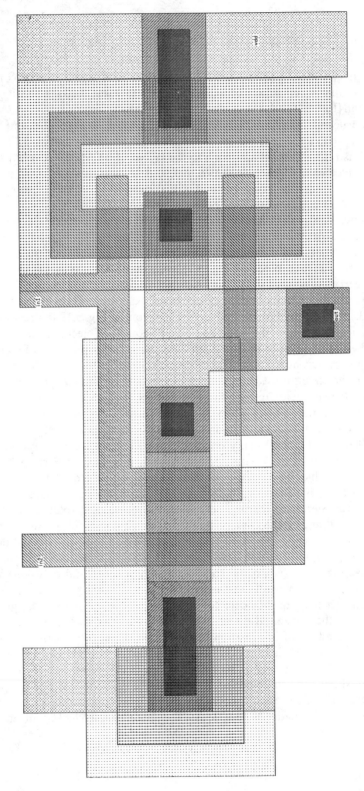

Figure 5.16 The NAND gate in CMOS.

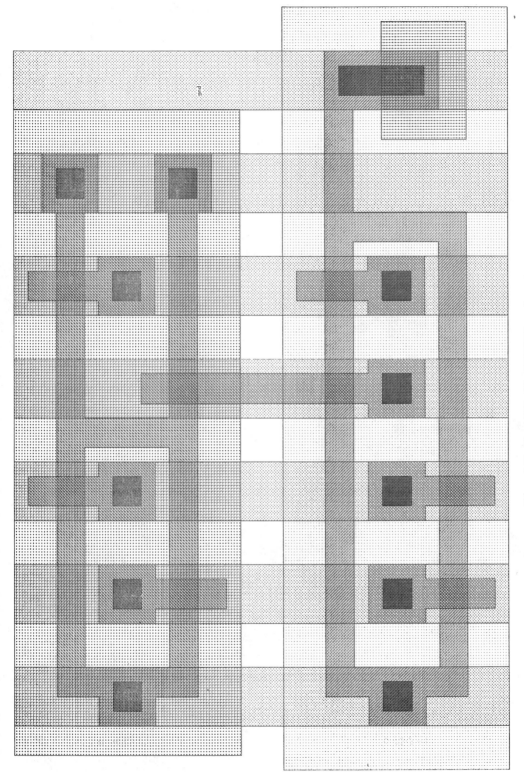

Figure 5.17 The AND–OR–INVERT gate layout.

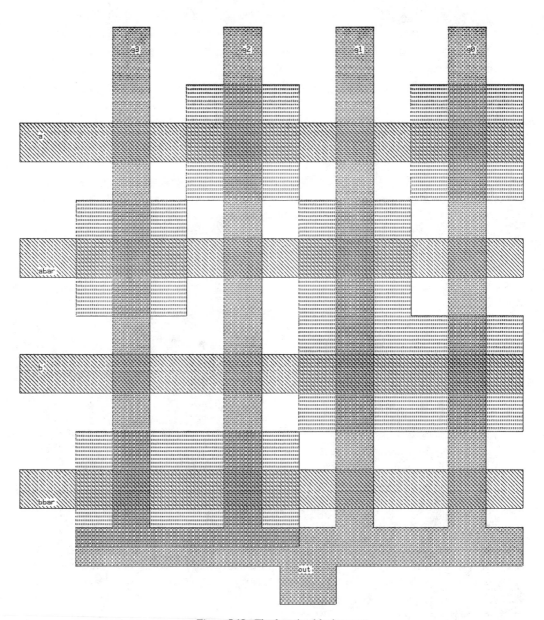

Figure 5.18   The function block.

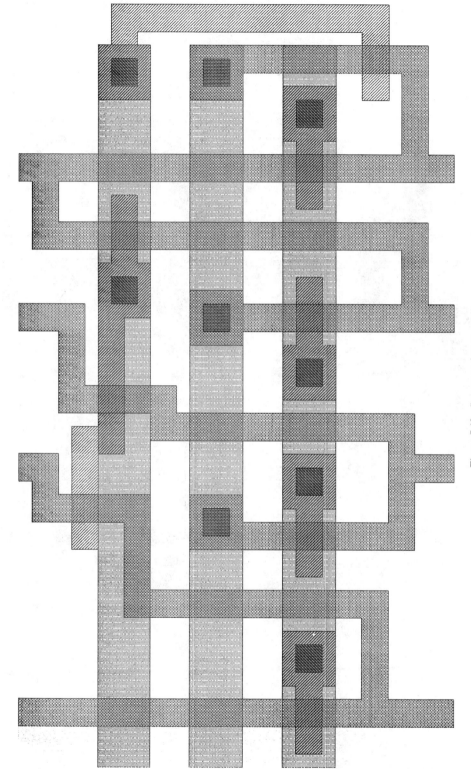

Figure 5.19 Sequence detector cell.

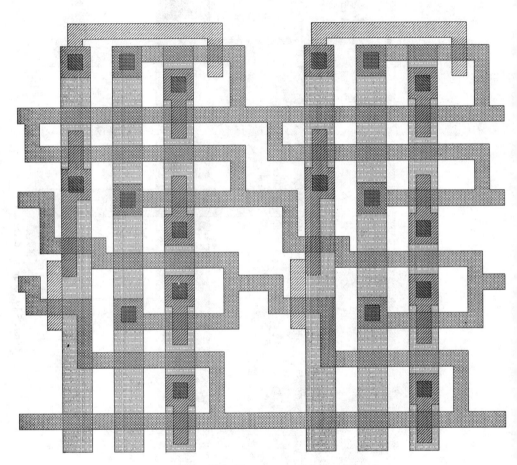

Figure 5.20 Two sequence detector cells.

Figure 5.21 Two function blocks stacked up.

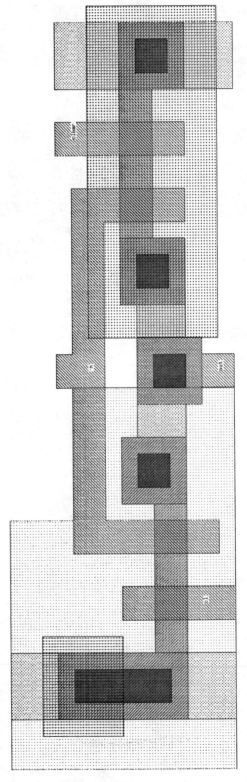

Figure 5.22 Clocked inverter stage.

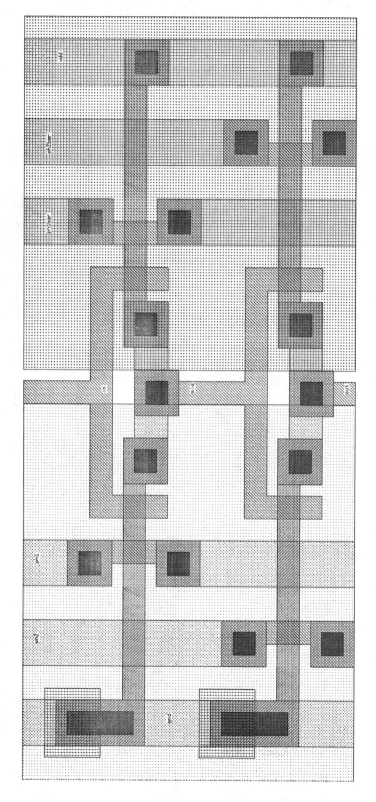

Figure 5.23 Dynamic register stage.

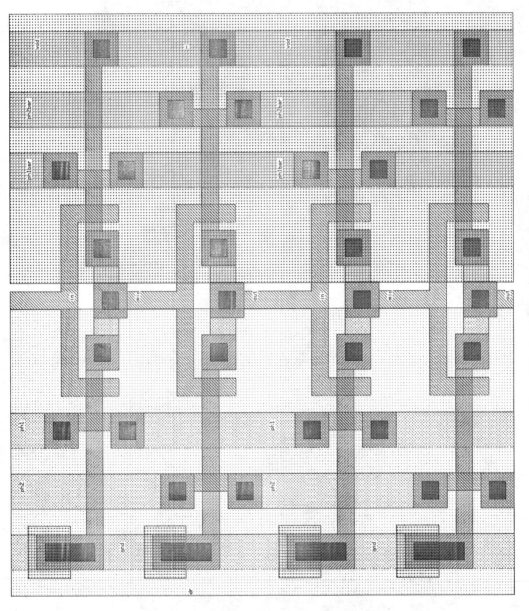

Figure 5.24 Array of two dynamic register cells.

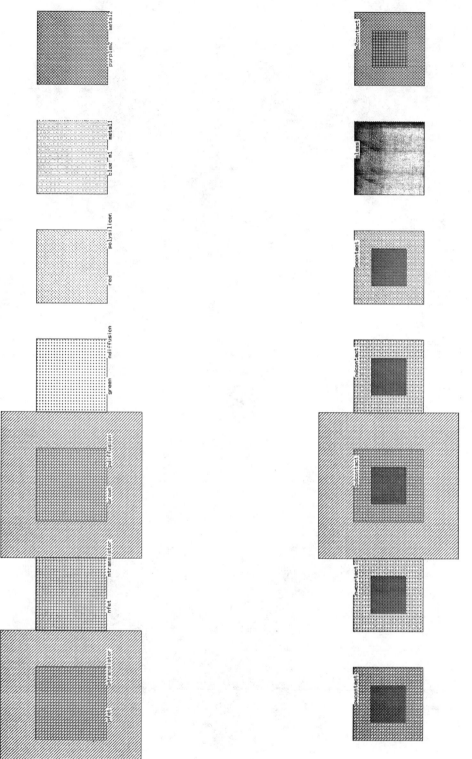

Figure 5.25 Stipple patterns for Magic layers and contacts.

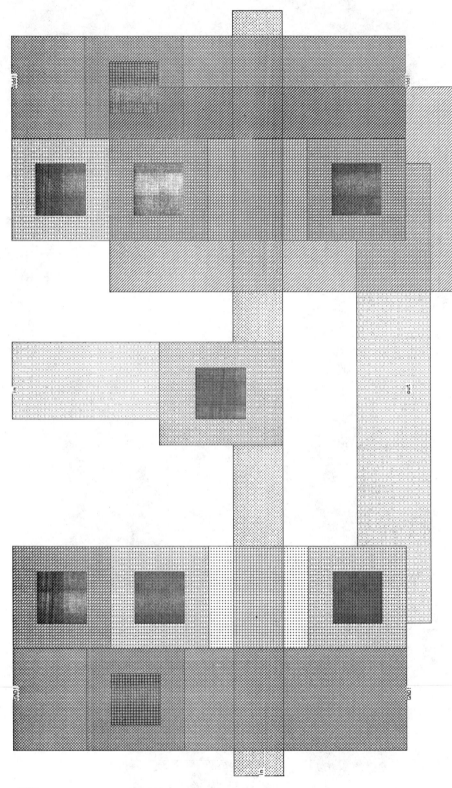

Figure 5.26 Layout of a CMOS inverter in Magic.

shown in Fig. 5.22. A dynamic register stage in CMOS using two clocked inverter stages is shown in Fig. 5.23, corresponding to the circuit of Fig. 3.87. The layout of this circuit is such that it can easily be abutted in both the horizontal and vertical directions and at the same time can be mirrored to share the V_{dd} and V_{ss} lines. An array of two such stages is shown in Fig. 5.24. Note that the φ_1 and φ_2 clocks and the power lines run in the horizontal direction in the layout. Finally, consider the layout shown in Fig. 1.1, which is an *n*MOS implementation of the circuit in Fig. 1.3. The special constraint imposed on this layout was that all the signal lines (data and control) and the power lines have to be run in metal in the horizontal direction and that the cell must be good for abutment in both horizontal and vertical directions. This will allow stacking up eight cells in the vertical direction to form a character comparator. The symbols p and t denote corresponding bits of 8-bit characters P and T, respectively. The output line, which runs in polysilicon in the vertical direction (a pull-up is attached to the top of this line, not shown in the layout) is pulled down to 0 if $P = T$. By putting n such comparator cells in the horizontal direction, a pattern P of n characters (held in the circulating shift register) can be compared against a text character T (bused and applied in parallel) simultaneously.

The stipple patterns for the different CIF layers of Magic are shown in Fig. 5.25. A layout of an inverter in Magic is also shown in Fig. 5.26. Note that the input and output of the inverter are in the metal1 layer, and the V_{dd} and Gnd are in the metal2 layer. Note the use of "pwcontact" to ground the *p*-well (*p*-well not shown in Fig. 5.25). The designer does not have to be concerned about the wells that are automatically generated in the CIF files. Similarly, an "nwcontact" has been used to connect the *n*-well to V_{dd}. If the process is not a twin-well process, one of these contacts will be discarded at the fabrication stage. Note also the use of the juxtaposition of a well contact with a diffusion contact near the V_{dd} and Gnd connection similar to the structure illustrated in Fig. 5.7(i).

The layout examples discussed above cover only a partial set of heuristics. The routing of power lines requires consideration of current density and power. We have also given a formula for computing the R and C values, but so far have not attempted to analyze the layout to compute the circuit delay. We also need to find ways of reducing the delay if it becomes excessive. The following chapter deals with these topics.

5.8 Summary

This chapter has presented the design rules for the silicon-gate *n*MOS and the bulk *p*-well CMOS processes. The design rules have been stated in terms of a fundamental resolution-length unit λ whose value is approximately 2 μm. The design rules have been formulated by abstracting several processes rather conservatively with respect to spacing and width of the basic feature. As a result, it is expected that this set of design rules will have a reasonable longevity in the rapidly evolving world of semiconductor technology.

The scaling down of the linear dimensions of the devices should not affect the design rules, but in practice it does because of the physical phenomena that occur at the submicron level. The most obvious effects of scaling down the linear dimensions of a process will be to increase the chip complexity because it will enable the designer to pack circuits more densely. However, the designer will have to be aware of the changes in the values of the electrical parameters of the devices and circuits. Here we will simply make a few observations assuming that the linear dimensions of all the features have been scaled down by a factor α. The supply voltage is also assumed to be scaled down in order to avoid large electric fields, and the currents are necessarily scaled down. Then the resistance $R_\square$ scales up by α and the capacitance per unit area scales down by α. The gate delay due to electron transit time goes down by a factor α, but the circuit delay (equivalently the RC product) remains constant.

REFERENCES

Glaser, A. B., and G. E. Subak-Sharpe, *Integrated Circuit Engineering*. Reading, Mass.: Addison-Wesley, 1979.

Griswold, T. W., "Portable Design Rules for Bulk CMOS," *VLSI Design*, Sept.–Oct. 1982.

Katz, R., and W. Scott, private communication, University of California at Berkeley, Berkeley, Calif., 1985.

Lyon, R. F., "Simplified Design Rules for VLSI Layouts," *Lambda*, 1st quarter 1981.

McCarthy, O. J., "MOS Device and Circuit Design." New York: Wiley, 1982.

Mead, C., and L. Conway (Eds.), *Introduction to VLSI Systems*. Reading, Mass.: Addison-Wesley, 1980.

Ousterhout, J. K., "Caesar: An Interactive Editor for VLSI Layouts," *VLSI Design*, 4th quarter 1981, p. 34.

Ousterhout, J. K., G. T. Hamachi, R. N. Mayo, W. S. Scott, and G. S. Taylor, "A Collection of Papers on Magic." *Rept. UCB / CSD 83 / 154*, University of California at Berkeley, Berkeley, Calif., Dec. 1983.

Sutton, J. A., II, private communication, Harris Corp., Melbourne, Fla., 1985.

Terman, C. J., "Simulation Tools for Digital LSI Design." Ph.D. dissertation, Massachusetts Institute of Technology, Cambridge, Mass. (MIT/LCS/TR-30-304), 1983.

EXERCISES

5.1. A signal is applied to the input of the nMOS and CMOS inverters of Figs. 5.14 and 5.15. The signal is carried by a 4λ-wide and 25λ-long ($\lambda = 2$ μm) metal line which changes to polysilicon via a contact cut. Calculate the resistance and capacitance of the input line in each case. Express the capacitances in units of $C_{g\square}$, the capacitance of a $2\lambda \times 2\lambda$ transistor. Assume values of sheet resistance and capacitance per unit area as given in this chapter.

5.2. Consider the layout of the two-sequence detector cells shown in Fig. 5.20. Assume that $x = 1$ for the first cell and $x = 0$ for the second cell. Calculate the capacitive loading for each of the four input lines under these conditions.

5.3. The stray capacitance of an inverter input is C_s, the gate capacitance is C_g, and the gate-to-drain overlap capacitance is C_{gd}. Evaluate the total change in charge Q at the input terminal if the input voltage is changed from 0 to V_{dd}. Then calculate the total effective gate capacitance.

6

Delay and Power

6.1 Introduction

This chapter is concerned with speed and power estimates of logic gates and circuits. The proposed estimates are based on simplified models and are intended to provide the designer with an understanding of the fundamental parameters that affect the performance of the circuit. The analysis will provide the designer with rough figures. The utility of such an analysis lies in providing insight into the physical parameters affecting the performance, which can then be incorporated into the models of simulation programs written to handle large and complex circuits.

We begin the chapter with an explanation of the basic notions of *circuit delay* and *propagation delay*. We then develop simplified formulas to estimate these delays for inverters, pass transistors, transmission gates, and steering logic and several other logic circuits. We then discuss driver circuits. The later part of the chapter is concerned with power estimates of logic circuits.

The circuit delays are produced because the voltage across the capacitance does not rise or fall instantaneously. This is illustrated in Fig. 6.1(a), which shows a simple circuit with a resistance R in series with a capacitance C. If the switch S is put in position 1, an initial current $I_0 = V/R$ flows into the circuit and charge begins to accumulate around C, increasing its voltage, which opposes the flow of current until C is charged to V, when current flow stops. At any time t, the current in the circuit $I(t)$ and voltages across the resistance R and the capacitor C, denoted

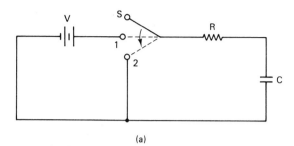

(a)

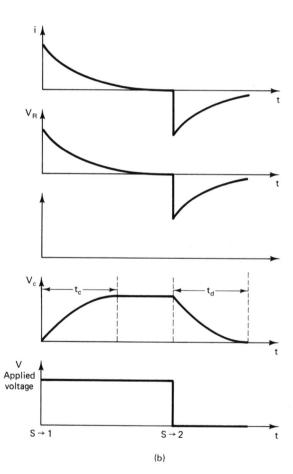

(b)

Figure 6.1 (a) Simple RC circuit; (b) current–voltage waveform with step input.

by $V_R(t)$ and $V_C(t)$, respectively, are given by

$$I(t) = I_o e^{-t/RC} \tag{6.1}$$

$$V_R(t) = I(t)(R) = Ve^{-t/RC} \tag{6.2}$$

$$V_C(t) = V - V_R(t) = V(1 - e^{-t/RC}) \tag{6.3}$$

When the switch S is thrown to position 2, the capacitor discharges, an initial current I_0 flows in the opposite direction, the voltage across C drops until it is 0, and the flow of current stops. The quantity RC, which has the dimension of time, is called the *time constant* of the circuit and is equal to the time for the capacitor to charge to 63% of its maximum value. It takes about $5RC$ for the voltage across C to rise or fall to the applied voltage, which we will call *charge delay*, t_c, or *discharge delay*, t_d, respectively, as shown in Fig. 6.1(b).

It is also useful to define a *propagation delay* in terms of the length of time required for a switching transition to occur at the output of a circuit in response to a transition of the input of the circuit. We will follow Terman's (1983) definition of propagation delay t_p to be

$$t_p = t_{out} - t_{in} \tag{6.4}$$

where t_{out} is the time when the output voltage V_{out} of the circuit crosses a predefined output threshold voltage, and similarly t_{in} is the time when the input voltage crosses an input threshold voltage. For MOS logic circuits, we will take this threshold to be V_{inv}, the logic threshold or switching point voltage. We will assume that the threshold voltages are such that $t_{out} \geq t_{in}$ and that the input or output voltages do not go into an undefined state due to charge sharing.

Equations (6.1) through (6.3) cannot be applied directly to analyze the delay of MOS networks, for several reasons. First, as we noted in Chapter 5, the resistance of a MOS transistor is a nonlinear function of its terminal voltages, the type of the transistor (p-, n-, enhancement or depletion), and the context of its use (pull-up, pull-down, or pass transistor). Second, to be precise, we need to distinguish between the static (steady state) and dynamic (during fast transition) values of the resistances. Simulation programs take this into account by computing sets of values for the resistances under different operating conditions. We will partially account for this in our analysis presented later. Third, capacitances in MOS circuits also depend nonlinearly on the terminal voltages. Our approximation of the capacitance values will be based on the procedure presented in Chapter 5. Finally, the input waveforms are not step voltage; they are in practice, ramp waveforms.

6.2 Simple Model for Estimating nMOS Inverter Delay

The model discussed in this section is based on Terman's work (Terman, 1983).

6.2.1 Low-to-High Output Transition

Let us first analyze the delay for the nMOS inverter circuit shown in Fig. 6.2 with a ratio k and driving a load capacitance C_l. We will first estimate the propagation delay with a ramp input voltage which starts at V_{dd} at $t = 0$, crosses the logic threshold V_{inv} at $t = t_{in}$, drops to threshold voltage V_{th} at $t = t_{off}$ when the pull-down transistor is turned off, and becomes zero eventually. The output waveform starts at low output voltage V_{lo} at $t = 0$, rises to V_{inv} at $t = t_{out}$, and becomes approximately equal to V_{dd} at $t = t_c$. Obviously, t_c equals the charging delay and $t_p = t_{out} - t_{in}$ is the propagation delay.

To estimate t_c and t_p, we have to estimate the load current I_l into the capacitance C_l. Then we can determine t_p and t_c by using the equations

$$C_l(V_{dd} - V_{lo}) = \int_0^{t_c} I_l \, dt \tag{6.5}$$

$$C_l(V_{inv} - V_{lo}) = \int_0^{t_{out}} I_l \, dt \tag{6.6}$$

We note that I_l is simply the difference betwen the pull-up current I_{pu} and the pull-down current I_{pd}. For most of the output transition, we can make the simplifying assumption that the pull-up is in saturation and its current is given by

$$I_{pu} = \frac{\beta_u}{2}(|V_{dep}|)^2 = I_{max} \tag{6.7}$$

The pull-down current starts at the linear part of the characteristic curve (since V_{ds} is small and equals V_{lo}), where the current remains more or less constant (see Fig. 6.3). Let us assume that this constant value of the current is maintained up to time $t = t_{in}$, when V_{in} becomes equal to V_{inv}; then it drops linearly until time $t = t_{off}$, when V_{in} becomes equal to the threshold V_{th} of the pull-down. Afterward $I_{pd} = 0$ since $V_{in} \leq V_{th}$. Now we can estimate I_l to be $I_{pu} - I_{pd}$, as shown in Fig. 6.3.

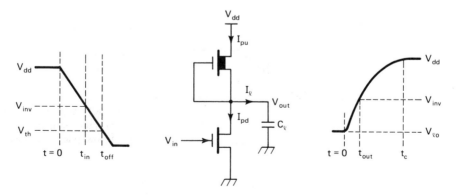

Figure 6.2 Low-to-high transition in an *n*MOS inverter circuit.

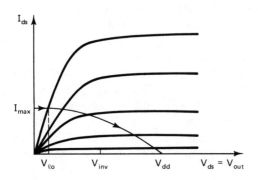

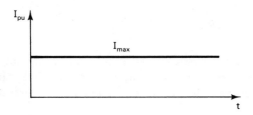

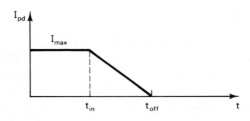

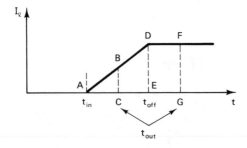

Figure 6.3 Approximation for the pull-up, pull-down, and load currents during low-to-high transition.

Depending on the magnitude of the current, I_{max}, t_{out} could be less than t_{off}, meaning that the output V_{out} could exceed V_{inv} before the pull-down is turned off or $t_{out} \geq t_{off}$, meaning that the charging process continues with pull-up current to raise its value to V_{inv}. If $t_{out} < t_{off}$, we have

$$C_l(V_{inv} - V_{lo}) = \tfrac{1}{2}(t_{out} - t_{in})\left[I_{max}\left(\frac{t_{out} - t_{in}}{t_{off} - t_{in}}\right)\right] \tag{6.8}$$

Note that the right-hand side in the current integral is equal to the area of the triangle ABC in Fig. 6.3. Thus

$$t_p = t_{out} - t_{in} = \left[2C_l\left(\frac{V_{inv} - V_{lo}}{I_{max}}\right)(t_{off} - t_{in})\right]^{1/2} \tag{6.9}$$

If however, $t_{out} \geq t_{off}$, we have

$$C_l(V_{inv} - V_{lo}) = \left[\tfrac{1}{2}(t_{off} - t_{in}) + (t_{out} - t_{off})\right]I_{max} \tag{6.10}$$

The right-hand side now represents the charge equaling the area of the triangle ADE and the rectangle $DEGF$. Simplifying, we get

$$t_p = t_{out} - t_{in} = C_l\left(\frac{V_{inv} - V_{lo}}{I_{max}}\right) + \tfrac{1}{2}(t_{off} - t_{in}) \tag{6.11}$$

The quantity $(V_{inv} - V_{lo})/I_{max}$ has the dimension of a resistance, and setting this equal to R_{pu}, the pull-up resistance, we have

$$t_p = \left[2R_{pu}C_l(t_{off} - t_{in})\right]^{1/2} \qquad t_{out} < t_{off} \tag{6.12}$$

$$t_p = R_{pu}C_l + \tfrac{1}{2}(t_{off} - t_{in}) \qquad t_{out} \geq t_{off} \tag{6.13}$$

When the input voltage is a step voltage, we have $t_{off} = t_{in}$ and obviously $t_{out} \geq t_{off}$, and the propagation delay t_p simply becomes the time constant $R_{pu}C_l$ of the pull-up circuit. In case the voltage is a slowly rising ramp, the term $t_{off} - t_{in}$ becomes dominant in the equation and t_p becomes less dependent on the time constant of the circuit. This means that the output voltage change essentially follows the input voltage change. If the transistor sizes are such that I_{max} is quite large and t_{out} is less than t_{off}, the propagation delay depends in square-root fashion on the time constant of the circuit.

A similar procedure will yield the charging times t_c. In this case it is quite justifiable to assume that $t_c \geq t_{off}$ and we have

$$C_l(V_{dd} - V_{lo}) = \left[\tfrac{1}{2}(t_{off} - t_{in}) + (t_c - t_{off})\right]I_{max}$$

or

$$t_c = C_l\left(\frac{V_{dd} - V_{lo}}{I_{max}}\right) + \tfrac{1}{2}(t_{in} + t_{off}) \tag{6.14}$$

The quality $(V_{dd} - V_{lo})/I_{max}$ again has the dimension of a resistance and we will denote this by R'_{pu}. Furthermore, if the input is a step voltage, we have $t_{in} = t_{off} = 0$ and

$$t_c = R'_{pu}C_l \qquad (6.15)$$

Thus the charging time is again a time constant of the pull-up circuit. Note that $R'_{pu} > R_{pu}$, implying that $t_c > t_p$, which confirms the intuitive notion that the total charging time must be greater than the propagation time. To get a very rough idea of the delay of the inverter, the distinction between the propagation and charging delay can be ignored by simply taking $R'_{pu} = R_{pu}$ and making this resistance be equal to the effective pull-up load resistance, as discussed in Chapter 2.

    Note that in estimating the currents we made the assumption that the pull-down current up to time t_{in} is the same as the pull-up saturation current. This is an overestimation, leading to a possibly lower value of the load current, resulting in an overestimation of the delays. To make a conservative estimate for the delay, the procedure above is quite acceptable. Similarly, we will see next that underestimating the pull-down current will lead to a conservative estimate of the discharge/propagation time.

6.2.2 High-to-Low Output Transition

Let us assume now that the input waveform is a rising ramp, as shown in Fig. 6.4, crossing V_{th} at $t = t_{th}$, V_{inv} at $t = t_{in}$, and becoming equal to V_{dd} at $t = t_{vdd}$. The output voltage starts at V_{dd}, falls to V_{inv} at $t = t_{out}$, and settles to the low value V_{lo} at $t = t_d$, the discharge time. Also, note that $I_l = I_{pd} - I_{pu}$.

    Referring to the $I_{ds}-V_{out}$ characteristic in Fig. 6.3, we note that during the output transition from V_{dd} to V_{inv}, the pull-up is in the linear portion of the characteristic curve. We will, however, approximate the current by its saturation current I_{max1}. Note that the current does not start flowing until time t_{th}. This overestimation will lead to a lower value of I_l resulting in higher estimated values for the delay. As we noted earlier, I_{max1} is given by $(\beta_u/2)(|V_{dep}|)^2$.

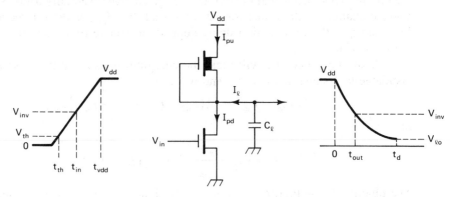

Figure 6.4   High-to-low transition in an nMOS inverter circuit.

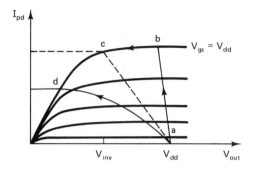

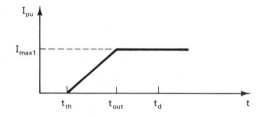

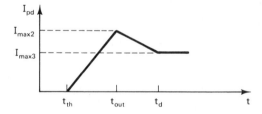

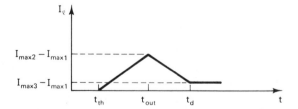

Figure 6.5 Approximation for the pull-up, pull-down, and load currents during high-to-low transition.

The estimation of the pull-down current is not that straightforward. Referring to the characteristic curves in Fig. 6.5, we note that the pull-down starts in the saturation region with V_{out} near V_{dd} at point a on the characteristic curve. If the input transition is very fast, the current rises very rapidly and the operating point moves to b on the characteristic curve for $V_{gs} = V_{dd}$. Then it takes the path bcd. On the other hand, if the input changes slowly, the pull-down current will always be almost equal to the pull-up current and the operating point will follow the path ad. In practice, this can never happen since the pull-down current has to be larger than the pull-up current by at least the amount necessary to sink the charge on the load capacitance. The proposed approximation for I_{pd} is therefore as follows. The current starts flowing at $t = t_{th}$, rises linearly up to time $t = t_{out}$, the time at which V_{out} becomes equal to V_{inv}, following the path ac on the characteristic curve to a value

$$I_{max2} = \beta_d \left(V_{dd} - V_{th} - \frac{V_{inv}}{2} \right) V_{inv} \qquad (6.16)$$

This underestimates the pull-down current for a fast transition, lowering the value of the load current I_l. This will result in an overestimation of delay. The current then drops linearly to the steady-state pull-down current with output voltage equal to V_{lo}. We can approximate this time to be $t = t_d$. The current in the pull-down is now given by

$$I_{max3} = \beta_d \left(V_{dd} - V_{th} - \frac{V_{lo}}{2} \right) V_{lo} \qquad (6.17)$$

The pull-down current will eventually be equal to the pull-up saturation current I_{max1}, but we will assume that the pull-down current will be I_{max3} for all time $t \geq t_d$.

The load current $I_l = I_{pd} - I_{pu}$ can now easily be estimated: It starts at $t = t_{th}$, rises linearly to $I_{max2} - I_{max1}$ at $t = t_{out}$, then drops linearly to $I_{max3} - I_{max1}$ at $t = t_d$, and thereafter remains constant.

We can write the charge conservation equations as

$$C_l(V_{dd} - V_{inv}) = \tfrac{1}{2}(t_{out} - t_{th})(I_{max2} - I_{max1})$$
$$C_l(V_{inv} - V_{lo}) = \tfrac{1}{2}(t_d - t_{out})(I_{max2} - I_{max3}) + (t_d - t_{out})(I_{max3} - I_{max1})$$

which yields

$$t_{out} = 2C_l \left(\frac{V_{dd} - V_{inv}}{I_{max2} - I_{max1}} \right) + t_{vdd} \left(\frac{V_{th}}{V_{dd}} \right) \qquad (6.18)$$

If the input is a step function, $t_{vdd} = 0$ and $t_{in} = 0$. Let us set

$$R_{pd} = \frac{V_{dd} - V_{inv}}{I_{max2} - I_{max1}} \qquad (6.19)$$

We then have $t_{out} = 2C_l R_{pd}$, which is the same as the propagation delay t_p. In deriving an expression for t_d, we will assume that $I_{max3} = I_{max1}$, which is reasonable. Also, as before, assume that the input is a step voltage. Then

$$t_d = t_{out} + (t_d - t_{out}) = 2C_l \left(\frac{V_{dd} - V_{inv}}{I_{max2} - I_{max1}} + \frac{V_{inv} - V_{lo}}{I_{max2} - I_{max3}} \right)$$
$$\approx 2C_l \left(\frac{V_{dd}}{I_{max2} - I_{max1}} \right) \qquad (6.20)$$

ignoring V_{lo} in the presence of V_{dd}. If we let

$$R'_{pd} = \frac{V_{dd}}{I_{max2} - I_{max1}}$$

we get

$$t_d = 2C_l R'_{pd} \tag{6.21}$$

which is again a time constant for the pull-down circuit. Since $R'_{pd} > R_{pd}$, the discharge delay is larger than the propagation delay. Again to get a very rough idea of the delay of the inverter, we would equate R'_{pd} and R_{pd} to the effective pull-down resistance.

If R_{pd} is equal to the channel resistance $R_\square$ of a minimum $2\lambda \times 2\lambda$ size transistor and C_l corresponds to $C_{g\square}$, the gate capacitance of a similar transistor, we will then denote t_d by the symbol δ. Taking $R_\square = 8.3 \times 10^3 \, \Omega$, $C_l = 4.4fF$, we have $\delta = .08$ ns for the minimum-size transistor. All discharge delay computation can now be expressed normalized with respect to δ. Thus if $R_{pd} = nR_\square$ and $C_l = mC_{g\square}$, we have

$$t_d = mn\delta$$

Since the gate capacitance $C_g = \varepsilon WL/T_{ox}$, we can express t_d as

$$t_d = 2C_l \left\{ \frac{V_{dd} - V_{lo}}{\beta_d [(V_{dd} - V_{th} - V_{inv}/2)V_{inv} - (V_{dd} - V_{th} - V_{lo}/2)V_{lo}]} \right\} = \left(\frac{C_l}{C_g} \right) \tau_d \tag{6.22}$$

where

$$\tau_d = \frac{2(V_{dd} - V_{lo})}{(\mu/L^2)(V_{inv} - V_{lo})[V_{dd} - V_{th} - \frac{1}{2}(V_{inv} + V_{lo})]}$$

The quantity τ_d is determined by processing and circuit parameters. Thus Eq. (6.22) emphasizes the fact that the discharge delay is directly proportional to the ratio of the load capacitance to the driving capacitance or that the essential function of the drive transistor is to sink the charge of the load capacitance. The time taken to sink this charge determines the total discharge delay. A large drive transistor, that is, one with small R_{pd}, will reduce t_d. By a similar argument, we can show that the charge time t_c is directly proportional to the size of the pull-up transistor. These ideas are utilized in the design of nMOS "superbuffers," as we will see later.

6.3 Delay in a CMOS Inverter

The analysis technique presented in the preceding section can be directly extended to estimate the delays in a CMOS inverter. The method essentially consists of obtaining a simplified expression for the pull-up and pull-down currents over the relevant time duration and then equating the current integral to the charge lost or gained by the load capacitance. However, a much simplified analysis can be made by recognizing that the current in a CMOS inverter flows only during the transition and that the static component of either the pull-up or pull-down current is practically negligible if the input voltage changes fast, for example, for a step input voltage. We can therefore approximate the current during the discharge phase to be a double-sided ramp of duration t_d of peak magnitude equal to the saturation of the n-channel transistor with $V_{gsn} = V_{dd}$. Similarly, during the charging phase, the current is a ramp of duration t_c of peak magnitude equal to the saturation current of the p-channel transistor with $V_{gsp} = -V_{dd}$. Thus we can write

$$C_l V_{dd} = \tfrac{1}{4}\beta_n (V_{dd} - V_{thn})^2 t_d \tag{6.23}$$

$$C_l V_{dd} = \tfrac{1}{4}\beta_p (-V_{dd} + |V_{thp}|)^2 t_c \tag{6.24}$$

If the output of the inverter is connected to the input of an identical inverter, we can write

$$C_l = C_\Box (W_p L_p + W_n L_n) \tag{6.25}$$

where $C_\Box$ is average gate capacitance per unit area of the channels and is given by ε / T_{ox}, and the expression in parentheses in Eq. (6.25) denotes the total area occupied by the n- and p-transistors. If the p- and n-channels have identical dimensions, we can let $W = W_p = W_n$ and $L = L_p = L_n$. Then we have

$$t_d = \frac{8L^2 V_{dd}}{\mu_n (V_{dd} - V_{th})^2} \tag{6.26}$$

$$t_c = \frac{8L^2 V_{dd}}{\mu_p (-V_{dd} + |V_{thp}|)^2} \tag{6.27}$$

Taking $V_{th} = 0.2V_{dd}$ and $V_{thp} = -0.4V_{dd}$, we have

$$t_d = 12.5 \times \frac{L^2}{\mu_n V_{dd}} = 12.5\tau_e \tag{6.28}$$

$$t_c = 22 \times \frac{L^2}{\mu_p V_{dd}} = 22\tau_h \tag{6.29}$$

where $\tau_e = L^2/\mu_n V_{dd}$ is the transit time of an electron through the n-channel pass transistor and τ_h is the transit time of a hole to pass through the p-channel transistor. Taking $\mu_n = 8 \times 10^{-2}$ m^2/V $\cdot$ s, $L = 4.0$ μm, and $V_{dd} = 5$V, we have $\tau_e = .04$ ns. Taking $\mu_n = 2 \times \mu_p$, we have $\tau_h = .08$ ns. Thus the estimated values of t_d and t_c are 0.5 ns and 0.88 ns, respectively. Thus the transit times give a good measure of the switching speed. Because the electron mobility is about twice the hole mobility, there is some asymmetry in switching speed for the pull-up and pull-down phases. This asymmetry can be removed by roughly doubling the width of the p-channel transistor as is done in many commercial circuits.

Some timing waveforms for inverters obtained by SPICE simulation will now be presented. Figure 6.6(a) shows the input and output waveforms of an nMOS

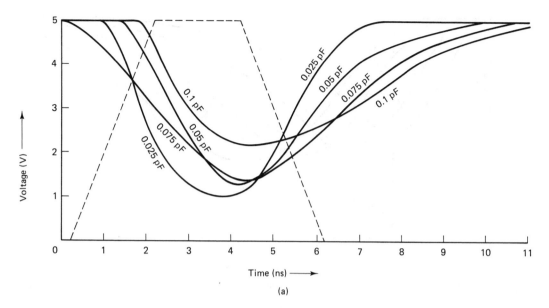

(a)

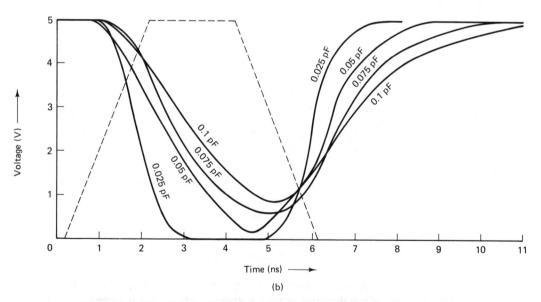

(b)

Figure 6.6 The nMOS and CMOS inverter timing waveforms with varying load and ramp input voltages. (a) nMOS inverter: input versus output voltage for different load capacitances. (b) CMOS inverter: input versus output for different load capacitances. (c) nMOS inverter: output voltage for varying ramp input. (d) CMOS inverter: output voltage for varying ramp input.

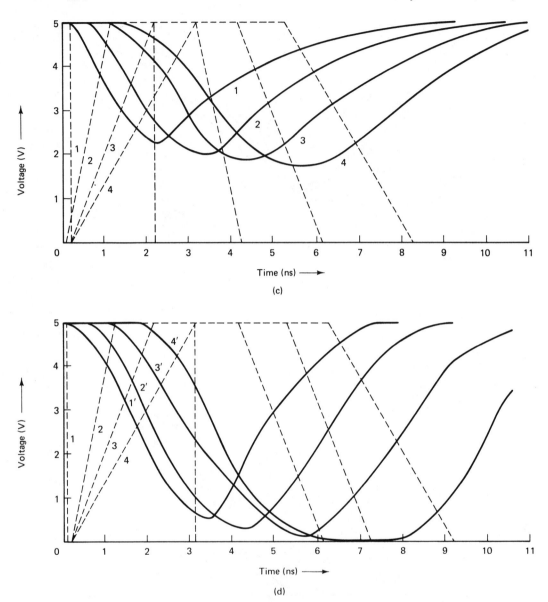

Figure 6.6 (*Cont.*)

inverter with varying load capacitances. Figure 6.6(b) shows similar curves for the CMOS inverter. Note that the output voltage swing is much larger for a CMOS inverter and has more symmetric characteristics than the *n*MOS inverter. Also note that the charging time via pull-up is longer than the discharge time via pull-down. For the CMOS inverter both pull-up and pull-down were chosen to have identical dimensions and therefore the *p*-channel transistor provided a higher resistance and

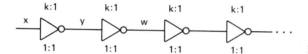

Figure 6.7 Chain of inverters.

hence a longer time constant. Parts (c) and (d) of Fig. 6.6 show the output waveforms with fixed load capacitances (0.025 pF for nMOS and 0.075 pF for CMOS) and a ramp input voltage of varying rise/fall times.

6.4 Delay through an Inverter Chain

Consider a cascade of identical inverters as shown in Fig. 6.7. Assume that the voltages at points x, y, and w are initially at 0 V, V_{dd}, and 0 V, respectively. If x now changes from 0 to V_{dd}, the discharge delay for the first inverter will be t_d and the charging delay for the second inverter will be t_c, so that the total delay between points x and w, called the *pair delay*, is $t_d + t_c$. A similar delay is involved between successive pairs of inverters.

For nMOS inverters, we have seen that the delay t_d can be expressed as $m \times n \times \delta$. If the inverter has a ratio k, we can assume that $R_{pu} = k \times R_{pd}$ and hence $t_c = k \times t_d$. So the pair delay can be expressed as $(k + 1)t_d$, where $t_d = m \times n \times \delta$. If we assume that $n = 1$, that is, R_{pd} corresponds to the smallest size transistor, and $m = 3$, that is, $C_l = 3C_{g\square}$, which includes gate capacitance plus wiring capacitance, we have $t_d = 3 \times \delta$. Furthermore, if $k = 4$, the pair delay is equal to $15 \times \delta$. Taking $\delta = 1.0$ ns the pair delay is about 15 ns, making the maximum operating frequency of the inverter chain to be 66 MHz. If the inverters are of CMOS type, as we have shown in the preceding section, the pair delay is approximately 1.4 ns, yielding a high operating frequency. Including wiring capacitance which might reduce the speed by a factor of 5 – 10, the operating frequency comes in the range of 70 – 140 MHz.

6.5 Delay in Steering Logic

Whenever a signal is steered through a pass transistor or a transmission gate, a delay is encountered. In typical use, the voltage V_{ds} across the transistor is very small and the circuit acts like a resistor R_{pass} in series with load capacitance C_l, as shown in Fig. 6.8. The value of R_{pass} depends on the type of the switch; for example, for the transmission gate it is approximately half of that for a pass transistor since it consists of two pass transistors connected in parallel. The time constant of the circuit $R_{pass} \times C_l$ gives a pretty good estimate of both the high-to-low and low-to-high propagation delays since this, as we noted in Section 6.1, corresponds to the time for C_l to charge to 63% of its final value. Horowitz (1983) shows that better estimates

Figure 6.8 Equivalent circuit of a pass transistor.

for these quantities are $0.63\,R_{pass}C_l$ and $0.79R_{pass}C_l$, respectively. For an nMOS $2\lambda \times 2\lambda$ pass transistor, $R_{pass} = 8 \times 10^3\,\Omega$ and $C_l = 0.01$ pF. Hence the time delay is about 0.08 ns, which is about the transit time of an electron for the n-channel transistor. Similarly, for a p-channel pass transistor the total delay is close to the transit time of the p-channel transistor.

Now, consider the example of a cascade of pass transistors as shown in Fig. 6.9(a). The pass transistor could be n- or p-type or could be replaced by a transmission gate. The equivalent electrical circuit is shown in Fig. 6.9(b), which is a transmission line with distributed parameters. The mathematical analysis of the total delay through such a network is rather complex. We could simplify the circuit by assuming that all the resistances and capacitances are the same and that the

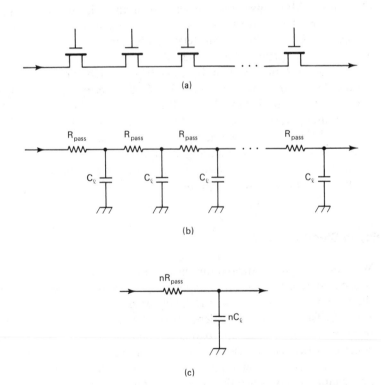

(a)

(b)

(c)

Figure 6.9 Chain of pass transistors and the equivalent circuits.

resistances could be lumped together into one resistor of value $n \times R_{\text{pass}}$, and similarly the capacitances could be lumped together into one capacitance $n \times C_l$. Thus the equivalent time constant of the circuit is $n^2 R_{\text{pass}} C_l$. And the total delay of the cascade increases as the square of the number of pass transistors. Adding one additional pass transistor to a cascade of n switches will increase the delay by $(2n + 1)R_{\text{pass}} C_l$.

The foregoing method considerably overestimates the delay. Horowitz (1983) has shown that a more accurate estimate is obtained if the effective time constant of the circuit is taken to be

$$\sum_k R_{ko} C_k = \tau_{\text{eff}}$$

where o is the output node, k denotes the input nodes of all the switches, and R_{ko} is the resistance of the path common to node k and node o. Thus with $n = 4$, we have

$$\tau_{\text{eff}} = R_1 C_1 + (R_1 + R_1)C_2 + (R_1 + R_2 + R_3)C_3 + (R_1 + R_2 + R_3 + R_4)C_4$$

If all the resistances are equal to R_{pass} and all the capacitance are equal to C_l, the high-to-low propagation delay is $6.3 R_{\text{pass}} C_l$ and the low-to-high propagation delay is $7.9 R_{\text{pass}} C_l$, both of which are considerably lower then $16 R_{\text{pass}} C_l$ obtained from the first method. In general, for arbitrary n, we can write

$$\tau_{\text{eff}} = \frac{n(n + 1)}{2} R_{\text{pass}} C_l$$

This sets the limit to the number of pass transistors that can be used in a cascade, after which the signal needs to be restored by a driver. A practical rule of thumb is that a cascade of not more than three or four switches should be used in any steering logic.

6.6 Superbuffers and Driving Large Capacitance

A "superbuffer" is a specially designed nMOS drive circuit that eliminates the basic asymmetry in ratio-type logic, which is unable to provide equal amounts of currents during the rising and falling transitions. A typical inverting superbuffer is shown in Fig. 6.10. Consider the output stage first: The pull-down or driver transistor is designed with a ratio that provides a low R_{pd}, so that the discharge time is small. To reduce the value of the delay during charging which starts when x becomes 0 V, the gate of the depletion transistor is connected (rather than being tied to its source) to $\bar{x}$, the output of the first inverter. The saturation current for the depletion-mode

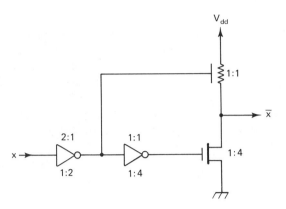

Figure 6.10 Inverting nMOS superbuffer.

output is given by

$$I_{ds} = \frac{\beta}{2}(V_{gs} - V_{dep})^2 \tag{6.30}$$

If the gate and source were tied together, $V_{gs} = 0$. But now V_{gs} is determined by the rising transient at the output of the first inverter. The average value of V_{gs} could be taken to be $0.8V_{dd}$, which means that I_{ds} is approximately four times larger than the current of a standard inverter. Thus the currents for both charge and discharge transients are approximately the same, providing symmetrical voltage waveforms across the load capacitance.

Note that the input inverter has a ratio $k = 4$, but it is designed with a lower pull-up resistance so that it can drive the larger capacitance at the two gates. The inverter in the middle similarly has a lower pull-up resistor to drive a higher gate capacitance of the pull-down transistor of the output stage.

A noninverting superbuffer circuit is shown in Fig. 6.11. Since the output stage of the superbuffer is the same as that of the inverting superbuffer, its driving capabilities are identical to that of the inverting type.

To drive a really large capacitance, say, an off-chip capacitance of 100 pF, a superbuffer is not enough. A similar situation may exist in driving a large number of inputs to many cells, where the input capacitance adds up to form a big load capacitance. If the load capacitance of 100 pF is driven by a driver having $C_g = 0.1$ pF, the total delay will be 1000 τ_d [see Eq. (6.22)]. Instead, if the driving is done in several stages each having a gate capacitance 10 times higher than the previous driver (i.e., $0.1C_g$, $1C_g$, $10C_g$, $100C_g$, and finally 1000 $C_g = 100$ pF), the total delay will only be $40 \times \tau_d$. If the ratio of the gate capacitances of the succeeding stages is s, we need $\log_s(C_l/C_g)$ stages. The delay per stage is $s\tau_d$. Hence

$$\text{total delay} = s\tau_d \log_s\left(\frac{C_l}{C_g}\right) \tag{6.31}$$

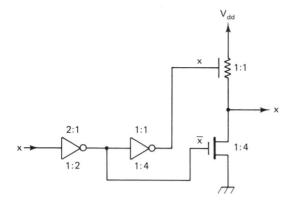

Figure 6.11 Noninverting superbuffer in n MOS.

An interesting question is to determine how many stages will produce optimal delay and how big each stage should be compared to the preceding stage. A large load capacitance C_l is driven by several stages of drivers—the first stage having a pull-up $4:1$, pull-down $1:1$, and gate capacitance C_g. Each succeeding stage is f times larger. For example, the second stage has a $4:f$ pull-up and a $1:f$ pull-down, and so on. What value of f will produce optimum delay, and how many stages will be required to produce the delay? (See the Exercises.) The answer to this question is that s is equal to e, the base of the natural logarithm, whose value is approximately 3. The total minimum delay is about $3\tau_d \ln_3(C_1/C_g)$ (Mead and Conway, 1980).

A superbuffer in CMOS merely consists of an inverter or a pair of inverters with a low L/W ratio for its p- and n-channel transistors so that it can deliver a large current to its load during transition. Ratios as low as $1/120$ have been used in a CMOS input pad to drive signals in the chip.

The quantity f above is sometimes also referred to as *fanout* (in conventional logic design terminology, it means the number of outputs to which the gate is connected). A suitable value of f is usually determined by a lot of experimentation. For a particular process, plots of fanout versus propagation delay or rise/fall time are obtained by simulating an inverter chain of various fanouts over typical and extreme process conditions. For CMOS, f is taken to be the factor by which the width of the n- and p-channels are increased for the succeeding stages of the inverter chain. These plots assume standard channel length, no interconnect load, and a standard p-channel to n-channel width ratio of the inverters, which is taken to be 3 to optimize drive. Typical plots are done at room temperature and at higher temperature. These plots show that the propagation delay or rise/fall time increases linearly with fanout and is higher for higher temperature. The designer selects the fanout that will produce the desired delay or rise/fall time.

To illustrate how the fanout information is used for device sizing (Sutton, 1985), recall from Chapter 5 that first we need to know the equivalent width of the "load inverter" given by

$$W_{eq} = W_p + W_n + 0.8W_{sd} + 0.15L_{interconnect}$$

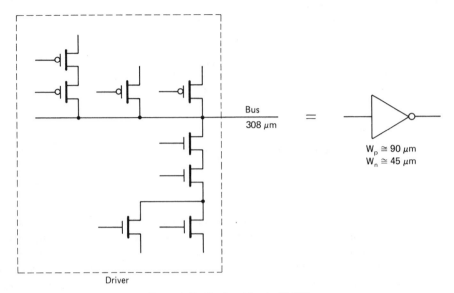

Figure 6.12   Device sizing in CMOS.

Let us assume that $W_{eq} = 308\ \mu$m (see Fig. 6.12) and that the desired fanout is 5 for a propagation delay of the order of 5 ns. This yields $W_{driver} = W_{eq}/5 = 61.6\ \mu$m. Assume that default W_p/W_n is 3. Then we have

$$W_p(\text{effective}) = 0.75W_{driver} = 46.2\ \mu\text{m}$$

$$W_n(\text{effective}) = 0.25W_{driver} = 15.4\ \mu\text{m}$$

Assume now that the driver is a complex gate and has a 2-high p-stack (two p-channel transistors in series) and a 3-high n-stack. This yields

$$W_p = W_p(\text{effective}) \times p\text{-stack} = 92.4\ \mu\text{m}$$

$$W_n = W_n(\text{effective}) \times n\text{-stack} = 46.2\ \mu\text{m}$$

where p-stack and n-stack denote the height of the p- and n-channel transistors in series. Thus, as shown in Fig. 6.12, the driver circuit along with its load of $W_{eq} = 308\ \mu$m can be sized to an inverter load of approximately $W_p = 90\ \mu$m and $W_n = 45\ \mu$m.

Let us now determine the size of a driver that drives a number of gates which have already been sized, as shown in Fig. 6.13, over a bus of known lengths, say 500 μm metal and 300 μm polysilicon. Let us also assume that the design default value

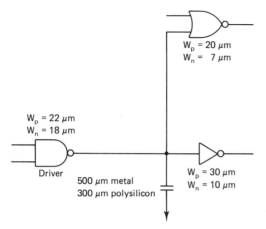

W_p = 20 μm
W_n = 7 μm

W_p = 22 μm
W_n = 18 μm

Driver

500 μm metal
300 μm polysilicon

W_p = 30 μm
W_n = 10 μm

Figure 6.13 Sizing a driver in CMOS.

for W_p/W_n is 2.5/1, and that fanout has been determined to be 6. We have

$$W_{eq} = W_p + W_n + 0.8W_{sd} + 0.15L_{interconnect}$$

$$= (20 + 30) + (7 + 10) + 0.8(0) + 0.15(500 + 300)$$

$$= 183 \; \mu m$$

$$W_{driver} = \frac{W_{eq}}{6} = 31 \; \mu m$$

$$W_p(\text{effective}) = \left(\frac{2.5}{3.5}\right) W_{driver} = 22 \; \mu m$$

$$W_n(\text{effective}) = \left(\frac{1}{3.5}\right) W_{driver} = 9 \; \mu m$$

Assume that the driver has one p-channel transistor and a stack of two n-channel transistors. Then

$$W_p = W_p(\text{effective}) \times p\text{-stack} = 22 \; \mu m$$

$$W_n = W_n(\text{effective}) \times n\text{-stack} = 18 \; \mu m$$

6.7 Delay in Logic Networks

In this section we estimate the delay in several elementary logic circuits. We will make the assumption that if the circuits consist of a number of stages, the total delay is equal to the sum of individual stage delays. This will lead to delay figures more conservative than the actual delay in the circuit. To estimate the delay for a complex logic network, the delay for the path having the largest number of stages,

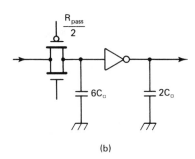

(a)

(b)

Figure 6.14 Delay computation for (a) dynamic nMOS latch, (b) dynamic CMOS latch.

called the *critical path*, will be estimated. Once we have the delays for the critical path, the maximum operating frequency is estimated to be equal to $f = 1/T$, where T is the sum of the rising and falling delays in the critical path. We will look at some examples to illustrate the approach.

Dynamic latch.   Consider the nMOS dynamic latch shown in Fig. 6.14(a). The load capacitances of the input and output of the inverter are assumed to be $3C_{g\square}$ and $C_{g\square}$, respectively. The falling delay t_f at the output equals the rising delay of the pass transistor plus the discharge delay t_d of the inverter. Similarly, we can estimate the rising delay t_r at the output. Thus

$$t_f = 0.79R_{pass} \times 3C_{g\square} + t_d$$

$$t_r = 0.63R_{pass} \times 3C_{g\square} + t_c$$

Taking $t_d = 2\delta = 3.2$ ns, $k = 8$ (the inverter ratio), $t_c = 8\delta = 25.6$ ns, $R_{pass} = 8 \times 10^3\ \Omega$, $C_{g\square} = 0.1$ pF, so we have $t_f = 3.4$ ns and $t_r = 25.8$ ns, resulting in $T = 29.2$ ns and $f = 34$ MHz. These figures have been used only to illustrate calculation procedure.

To estimate the delay in a CMOS dynamic latch, as shown in Fig. 6.14(b), we note that R_{pass} for the transmission gate is half, but the capacitance values are approximately doubled. So we can write

$$t_f = 0.79R_{\text{pass}} \times 3C_{g\square} + 12.5\tau_e \tag{6.32}$$

$$t_r = 0.63R_{\text{pass}} \times 3C_{g\square} + 22\tau_h \tag{6.33}$$

where τ_e and τ_h are the transit time of electrons and holes, respectively, as defined earlier. Taking $\tau_e = 1$ ns and $\tau_h = 2$ ns, we have $t_f = 15$ ns and $t_r = 35$ ns, giving $T = 40$ ns, yielding an operating frequency of $f = 25$ MHz.

Shift register.   If dynamic latch and two-phase nonoverlapping clocking are used to build the shift register, the clock width of ϕ_1 or ϕ_2 cannot be less than T for the dynamic latch and must have an interclock gap of at least T to allow for the clock of opposite phase. Thus the maximum operating frequency of an nMOS shift register is about 17 MHz and that of a CMOS shift register is about 12.5 MHz. These values are, of course, very conservative. Commercial shift registers run much faster.

Complex gate.   To estimate the delay in a complex nMOS logic gate, we need to model the pull-down network as a series–parallel connection of pull-down switches and then replace it by a single pull-down transistor. We also need to make the pull-up of appropriate size to guarantee that the ratio of the gate network is either 4 or 8. Once an "equivalent" inverter circuit has been formulated with effective pull-up resistance R_{pd}, the procedure of Section 6.2 can be applied to estimate the delay. The conclusion is that the time constants $R_{\text{pu}}C_l$ and $R_{\text{pd}}C_l$ will determine the rising and falling delays of the gate, respectively.

In estimating the delay for a complex CMOS gate, the procedure of Section 6.3 needs to be modified in the following way: First, the transient current will rise and fall less sharply due to the increased resistance of the pull-down or pull-up paths and/or increased effective gate capacitance. An approximation will be a ramp duration t_d or t_c. Second, an effective β_n and β_p have to be estimated, which, as a first approximation, could be taken as the β of the series-parallel pull-down and pull-up pass transistor networks, denoted as β'_n and β'_p, respectively. We can then write Eqs. (6.23) and (6.24) as

$$C_l V_{dd} = \tfrac{1}{4}\beta'_n (V_{dd} - V_{thn})^2 t_d \tag{6.34}$$

$$C_l V_{dd} = \tfrac{1}{4}\beta'_p (-V_{dd} + |V_{thp}|)^2 t_c \tag{6.35}$$

A third problem arises is estimating the load capacitance C_l for both nMOS and CMOS gates. In general, the output of a complex gate may be connected to an arbitrary member of gate inputs. Such a load can be modeled more accurately by a linear RC tree network (Horowitz, 1983), whose analysis is beyond the scope of this book. Ignoring the load resistance, the capacitance C_l can be approximated as

$$C_l = \left(\sum_i W_i L_i \right) C_\square$$

where i ranges over all transistors of width W_i and length L_i to which the output and the gate are connected. In typical circuits the wiring between circuits takes up about 15% of the total gate capacitance. An example of delay estimates for an nMOS PLA is given below.

Critical path delay estimate for *nMos* PLA. The problem is to determine the critical path and the worst-case delay for an nMOS PLA represented by the following set of equations:

$$z_1 = ab + \bar{a}bc$$
$$z_2 = ab$$
$$z_3 = a + \bar{b}c$$

The PLA with clocked inputs can be generated using the PLA generator software and the layout is shown in Fig. 6.15. The critical path for the PLA is to be obtained and the worst-case delay for the critical path is determined by three different methods and the results are compared.

The critical path for the PLA is first obtained by using a CRYSTAL timing analyzer. The path connecting the input node a and the output node z_1 is indicated by CRYSTAL as the critical path. CRYSTAL estimated this delay as 11.2 ns. Then a spice deck was created from the PLA's circuit descriptor file and input to the SPICE program. The voltages at the three output nodes and the input clock signal were traced while setting the inputs to different values. SPICE estimated the worst-case delay for the critical path between node a and node z_1 as 7.8 ns. The vast difference in the delay estimation between CRYSTAL and SPICE is due to the fact that CRYSTAL gives an extremely conservative value.

Now we compute the delay for the critical path by hand for comparison with the delays estimated by CRYSTAL and SPICE. The hand computation is done as follows. First for each gate in the critical path, the load capacitance and the resistance values are calculated and then the delay for the gate is estimated. Then the delays for all the gates are summed up to obtain the delay for the critical path. The calculation is given below. The gates in the critical path are shown in Fig. 6.16. The load capacitance due to each layer is calculated from the area of the layer and represented in terms of unit gate capacitance $C_{g\square}$. Then the delay is given as

$$t_c = x \times r \times \tau$$
$$t_d = x \times s \times \tau$$

where τ equals $C_{g\square}R_\square$; r and s denote the length-to-width ratios of the pull-up and pull-down transistors, respectively; x is the total effective load capacitance on the gate in terms of units of $C_{g\square}$; and the value for τ is assumed as 0.3 ns. For gate G_1, the load capacitances due to polysilicon, diffusion, and channel are $2.2C_{g\square}$, $3.25C_{g\square}$, and $8C_{g\square}$, respectively. So the total load capacitance on the gate is $13.45C_{g\square}$. The charging delay t_c for gate G_1 is calculated as 4.035 ns. For gate G_2, the load capacitances due to polysilicon and channel are $2.65C_{g\square}$ and $3C_{g\square}$, respectively. So

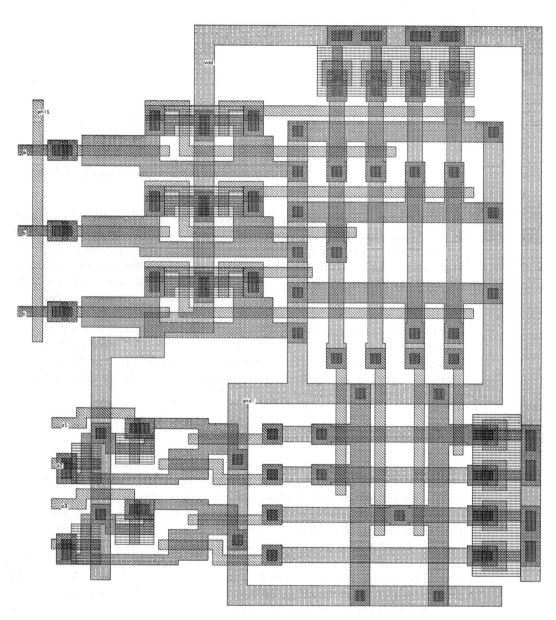

Figure 6.15 PLA layout.

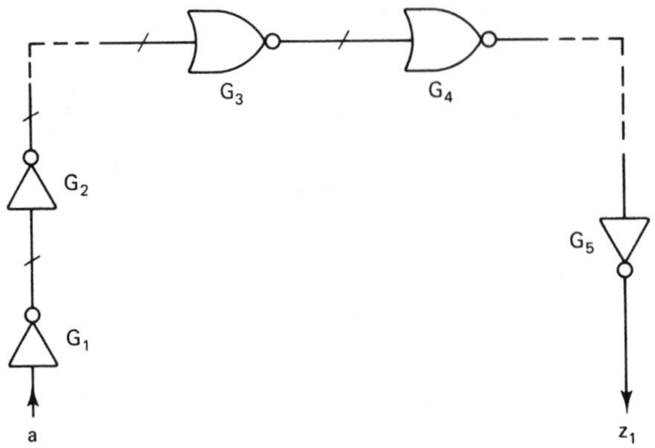

Figure 6.16 Critical path in the PLA.

the total load capacitance on gate G_2 is $5.65C_{g\square}$, and the falling delay t_d is found to be 0.423 ns. The load capacitances for gate G_3 are due to polysilicon, metal, diffusion, and channel and the values are $1.4C_{g\square}$, $2.975C_{g\square}$, $3.75C_{g\square}$, and $2C_{g\square}$, respectively. The charging delay for gate G_3 due to a total load of $10.075C_{g\square}$ is calculated as 3.02 ns. Similarly for gate G_4, the load capacitances due to polysilicon, diffusion, metal, and channel are $0.75C_{g\square}$, $2.25C_{g\square}$, $2.1C_{g\square}$, and $4C_{g\square}$, respectively. The total load is $9.1C_{g\square}$ and the falling delay t_d is 1.365 ns. Finally, for gate G_5, the load capacitances are due to polysilicon and diffusion layers and they are calculated to be $1.15C_{g\square}$ and $3.75C_{g\square}$, respectively. The total load is $4.9C_{g\square}$ and the rising delay is calculated to be 1.47 ns. The total delay for the critical path is the sum of the delays for all the gates and it is 10.313 ns, which is comparable with the estimates given by CRYSTAL.

6.8 Power Estimates

Estimates of power are required for several reasons: avoiding metal migration in power and ground routing, and understanding the design trade-off between power versus area and power versus performance.

The special requirement of power and ground routing is that every active component of the circuit, such as an inverter, must have a direct connection to it. The power and ground lines are never run in polysilicon since it has large resistance, carrying considerable voltage drops. Diffusion is used only as a "jumper" connection to cross the V_{dd} line over the ground and is never used over a long distance. Both the power and ground lines are in metal and the width of the lines must be less than the limit imposed by the metal migration "phenomenon," which will thin out portions of the wires, ultimately becoming an open circuit. A common strategy to route power and ground lines is to use interdigitation as shown in Fig. 6.17, where $D_1, D_2, \ldots, D_n$ are active devices. The interdigitated structures may be looked upon as two planar trees woven together so that their branches near the "leaves" meet at

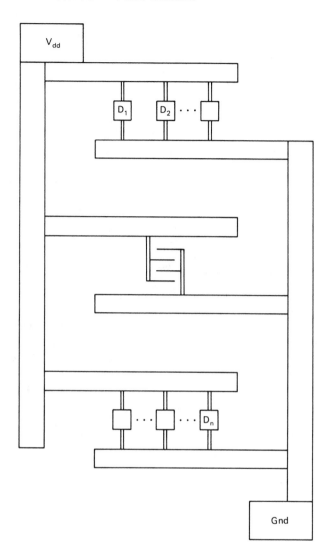

Figure 6.17 Common strategy for power and ground line distribution.

the active devices. The width of the metal lines is calculated as follows: Calculate the maximum average current drawn by all the active devices connected to a supply line at the lowest "level" of the tree. Add these currents and then calculate the required width of the power and ground lines as follows. The maximum current rating for a metal line is about 1 mA/μm. For $\lambda = 2.0$ μm, metal will carry about 2.0 mA/λ. The minimum width of metal is 3λ; it is usually made 4λ wide to fit contact cuts, which gives about 8 mA at the lowest level. At the next highest level of the tree, the width must correspond to the sum of all currents flowing into all its lower level branches, and so on. The width is usually increased in increments of 4λ (8λ, 12λ, 16λ, etc.) to accommodate roughly increments of about 8 mA at each higher level.

The figures above can be used to estimate the number of active devices that can be fed by a power line. For example, if the active devices are nMOS inverters, we can approximate the static current to be the saturation current of the pull-up, which is about $0.24W/L$ mA (see the Appendix A). For a standard inverter with ratio 4, $L/W \approx 4$, the current is 0.06 mA. Thus a maximum of about 100 pull-ups can be supplied by a 3λ-wide metal line. This fact has to be used in the context of a specific circuit. For example, most PLA generators automatically increase the width of the V_{dd} line and the Gnd line between the AND plane and the OR plane, depending on the total current demand. If a shift register circuit is designed using dynamic latches, a maximum of 200 pull-ups can be supplied by a 3λ-wide line since only alternate inverters will have current flowing from V_{dd} to Gnd lines. We are assuming, of course, that $L/W = 4$, although the ratio of the inverter is 8; that is, L/W of the pull-down is taken to be 2. If we take $L/W = 1$ for the pull-down, the total current in the inverter will be reduced approximately by a factor of 2, and a maximum of 400 pull-ups can be supplied by a 3λ-wide power line.

The examples above also illustrate the fact that two completely different layouts of inverters having the same ratio k may have quite different current requirements. Consider two inverter circuits: the first has $k:1$ as the pull-up ratio and $1:1$ as the pull-down ratio. For the second, the ratios of pull-up and pull-down are reversed. The inverters have the same ratio k, but the second circuit will burn about k times more power. But the advantage of the second inverter is that it will handle about a k times larger load capacitance for the same amount of delay. More power means higher speed, but it also means more silicon area.

For a CMOS circuit, there is no static power dissipation; there is only switching power dissipation during transition (Seitz et al., 1985). Consider a CMOS inverter. Its power dissipation can be approximated as $C_l V_{dd}^2 f$, where C_l is the load capacitance and f is the frequency of switching. The average current is $f C_l V_{dd}$ since for each transition an approximately $C_l V_{dd}$ amount of charge is required. Half of the energy dissipated per cycle, $C_l V_{dd}^2$, is spent in the p-transistor and the other half in the n-transistor as a rough estimate. Using the expression of C_l as given in Eq. (6.25), we can write the power dissipation for a CMOS inverter P_s as

$$P_s = V_{dd}^2 f C_{g\square} (W_p L_p + W_n L_n) \qquad (6.36)$$

As a rough figure, if we take $C_l = 0.2$ pF, $V_{dd} = 5$ V, and $f = 20$ MHz, we find an average current of 0.02 mA and an average power consumption of about 0.1 mW. Thus 13λ-wide power and ground lines can carry approximately 300 CMOS inverters at 20 MHz.

The power formula $C_l V_{dd}^2 f$ is applicable to any arbitrary CMOS gate or chip. For CMOS design, the input/output pads are rather large and determine the bulk of the chip capacitance. Thus a good approximation of the power consumption of the chip can be obtained from knowledge of the pad capacitances, the operating frequency, and the supply voltage.

6.9 Summary

In this chapter we have provided a simplified model of the operation of an inverter based on the work of Terman (1983) and have developed an analysis procedure to estimate the delay in the circuit. The delay can be expressed as a simple formula in terms of the RC time constants of the circuit. The analysis procedure is then extended to simple logic circuits, shift registers, and data paths to estimate the performance of the circuit. The analysis provides insight into the nature of delay and forms the basis for simulation programs written for large and complex circuits. The chapter concludes with a discussion of power estimates for both nMOS and CMOS circuits.

REFERENCES

Some of the references cited at the end of Chapters 2 and 5 are also relevant for this chapter. In particular, Terman (1983) has been used heavily. The Mead–Conway text (1980) is also a good source of information on delay. The treatment of device and driver sizing is based on a private communication from James A. Sutton (1985).

Horowitz, M., "Timing Models for MOS Pass Networks," *Proc. IEEE Int. Symp. Circuits and Systems*, pp. 198–201, 1983.

Mayo, R. N., J. K. Ousterhout, and W. S. Scott (eds.), "1983 VLSI Tools: Selected Works by the Original Artists," *Rep. UCB/CSD 83/115*, Computer Science Division, University of California at Berkeley, Berkeley, Calif., March 1983.

Mead, C., and L. Conway (Eds.), *Introduction to VLSI Systems*. Reading, Mass.: Addison-Wesley, 1980.

Seitz, C., A. H. Frey, S. Mattison, S. D. Rabin, D. A. Speck, and J. L. van de Snepscheut, "Hot-Clock nMOS," *Proc. 1985 Chapel Hill Conference on Very Large Scale Integration*, (Ed. H. Fuchs), Chapel Hill, NC.: pp. 1–17. Computer Science Press, 1985.

Sutton, J. A., II, private communication, Harris Corp., Melbourne, Fla., 1985.

Terman, C. J., "Simulation Tools for Digital LSI Design." Ph.D. dissertation, Massachusetts Institute of Technology, Cambridge, Mass. (MIT/LCS/TR-304), Sept. 1983.

EXERCISES

6.1 A control signal originates from a minimum-size inverter (pull-up $4:1$, pull-down $1:1$). A large capacitance load C_l has to be driven by the circuit shown in Fig. 6.7. Each stage is f times larger than the preceding stage but has the same ratio. The gate capacitance of the first driver is C_g. How many stages of inverter would you use? What should be the value of f for minimum delay, and what is the expression for this minimum delay? Compute this delay if the load is (a) a bonding pad of 140 μm; (b) a metal bus 7 mm long and 16 μm wide; and (c) a package pin with 5 pF capacitance.

6.2 First, lay out the standard $4:1$ and $8:1$ ratio inverters that you will use very frequently in your design. Similarly, lay out the superbuffers. Using a timing simulator (such as CRYSTAL; see Mayo et al., 1983), obtain a plot of the total delay versus the length of the polysilicon line ($2 \times \lambda$ wide), diffusion line ($2 \times \lambda$ wide), and metal line (4λ wide) that both the inverter and the inverting superbuffer would drive. Make at least two observations, one at small values of lengths where the superbuffer delay is larger than the inverter delay, and one at relatively large values of lengths where the superbuffer delay is less than the inverter delay. From these curves find the critical lengths for each type of material for which the delays for the inverter and the superbuffer are the same. Repeat the assignment with an inverter pair (first inverter $8:2/2:4$ and second inverter $4:2/2:4$) and compare the results to those of the noninverting superbuffer. Compare your observed results with delays computed by the model discussed in this chapter. What conclusions can you draw from these experiments regarding the situations where a superbuffer should or should not be used?

6.3 Calculate the delays in the circuits described in Exercises 5.1 and 5.2.

6.4 Develop a procedure to compute the peak and average current in a CMOS inverter taking into account the sizes of both the p- and n-channel transistors.

6.5 Obtain the layout of an nMOS and a domino CMOS PLA for the expression of z_1, z_2, and z_3 (see Section 3.24), including the input drivers and the clocks. Determine the critical paths in these PLAs and estimate the delays. Then compare with the observed values using CRYSTAL. From these estimate the maximum frequency of operation of the PLAs.

7

Systems Design

7.1 Introduction

This chapter is concerned with functional, subsystem, and system design. The approach will be to present several interesting circuits and systems as test cases to illustrate the fundamental design issues. An overview of the design process is presented next.

A typical design undergoes roughly the following stages. A design starts as a top-level designer's mental model, presented in the form of a schematic diagram of interconnected subsystems (i.e., memory, ALU, control, I/O, bus, etc.). This system-level specification is then decomposed into functional building blocks which are defined by logic equations, state diagrams, or a precise verbal specification of the input/output behavior. The functional modules are then first transformed into logic diagrams and then, together with timing and sequencing information, into gate- or switch-level structures for the target technology, and finally to a layout representation for the target technology. The design process is thus a multilevel translation scheme as depicted in Fig. 7.1. At each level a set of constraints similar to the syntax and semantic rules of a program compilation process have to be observed. For example, the syntax between the gate level and the layout level is the set of design rules described in Chapter 5. The semantic constraint at this level is simply the equivalence of the behavior of the logic circuit extracted from the layout with electrical and timing characteristics that realize the intended behavior. The behavioral equivalence is necessarily the semantic constraint at all levels, but the language and the primitives by which this equivalence is expressed are different from layer to layer.

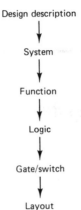

Design description

System

Function

Logic

Gate/switch

Layout   **Figure 7.1** Multilevel design process.

Top-down design is still an idealized approach. The success of complex design will depend on the existence of a set of very powerful automated design aids that will do a job as good as human beings can do. In most practical designs, a combination bottom-up/top-down approach is used. The design starts with some approximate top-level floor plan which implements the system-level architecture. It is then hierarchically decomposed by stepwise refinement to the point at which some preliminary idea is formed about the functional modules or the "leaf cells" of the design, which perform the most primitive functions necessary to design the system. These cells are defined with a view toward using them in an array of one or two dimensions so as to maximize the regularity of interconnection, area utilization, and simplicity of clocking and control structures. At this point the designer starts at the bottom level laying out cells and charting a precise strategy for layer selection and routing a global wiring such as power, ground, synchronization signal, and control lines. An array of cells usually forms a functional module with a well-defined interface of input/output connections, geometrical constraints, and signal types and timing relations associated with its terminal connection. The composition process is carried upward until subsystems are laid out, keeping the overall global strategy of distribution of power, ground, clock, and control intact. The total system is then designed as an interconnection of subsystems, a control subsystem that drives the entire system and input/output pads to communicate with the external world. At every stage of design, verification tools such as design rule checking, electrical rule checking, and dynamic simulation programs are used to validate the integrity of the design.

In this chapter we present several design examples to illustrate the general principles noted above. First, we discuss two typical student design projects completed in a VLSI design course. Then we discuss the design of several nonnumeric processors for pattern matching, lexical comparison, sorting, stacking, and queuing functions. We then have a section on the design of a barrel shifter typically used in an arithmetic data path. The design of an arithmetic processor using precharge logic forms the concluding section of the chapter.

7.2 Variable Register Delay

This section will present the design of a variable register delay (VRD) chip designed by Donovan (1982). The VRD performs the function of delaying a data stream by a selectable number N of clock cycles. This type of function can be used to synchronize one data stream with another data stream in a synchronous pipeline machine. A typical example is when one field of a data word must be processed and the remaining fields of the data word must be delayed until the results of the processing are available.

7.2.1 Functional Specification

The VRD is designed to provide up to 17 ($2 \le N \le 17$) delays for an 8-bit data path at a clock rate up to 10 MHz, which means a new data word is input every 100 ns. The selection of N is done by 4-bit "select" lines and is considered as an initialization process before the chip starts accepting valid input. The function would fit into a 24-pin package with two power supplies (V_{dd} and Gnd), eight data inputs, eight data outputs, four select lines, and two clock inputs. A functional specification of VRD is shown in Fig. 7.2. The size and power consumption of the chip should be within limits of standard student design projects, which are approximately 1000 × 1000λ^2, and less than 1 W, respectively.

7.2.2 Functional Design

The functional block diagram of the VRD is shown in Fig. 7.3. The data (8 bits) flow through a series of registers in order to effect the delay. The selection logic determines where the data enter the register stream. This is done by enabling the 2 : 1 multiplexer for a given register to accept data from the data bus instead of the previous register. In this way, the data effectively bypass the previous registers and are not delayed by them.

For an example where only two register delays are required, the data would experience one delay at register 1 and then the data would be output to the data bus. The select logic would enable register 2, and the data would experience the second delay going through the register 2 output. For three register delays, the data would pass through register 1, be enabled to register 3, pass through register 3, and then

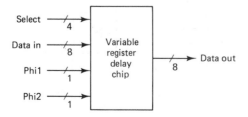

Figure 7.2   External functional interfaces for VRD.

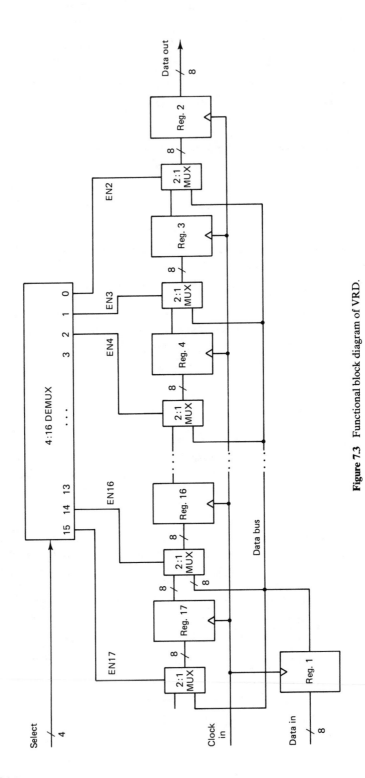

Figure 7.3 Functional block diagram of VRD.

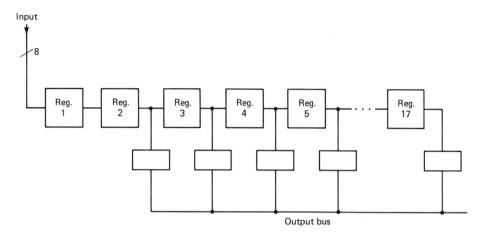

Figure 7.4   Register output select approach.

pass through register 2. The maximum number of delays is 17, when the data are enabled into register 17, and the minimum number of delays is two.

Several alternative design approaches were considered. They are discussed here to illustrate the process of selecting the most suitable design after consideration of the key VLSI design parameters discussed in the preceding section.

A FIFO (first-in-first-out) memory can be used as a VRD. The disadvantage is that a random memory access with complete control logic (read/write, control, wraparound address decoder) is required. The memory cell could be static (large area) or dynamic with refresh capability (see Chapter 8).

The register output select approach is shown in Fig. 7.4. As in Fig. 7.3, the data are shifted from left to right. One of the output registers is enabled onto the output bus. The input select approach is chosen over the output select because of speed/size advantages. The input bus or output bus has to be driven by super-buffers. For the input select approach, the input bus is driven from one 8-bit source (register 1) which uses eight superbuffers. The output select approach would have required sixteen 8-bit sources or 128 superbuffers.

A fourth design alternative is shown in Fig. 7.5. This design contrasts with the register input select approach in separating the storage and multiplexing operations. This causes routing to take a lot of space and will use a large number of superbuffers to drive the signals through the multiplexer circuit.

7.2.3 Circuit / Logic Design

The mixed notation circuit/logic diagram for the VRD is shown in Fig. 7.6. The figure shows the register string for one bit as it flows from the input pad to the output pad. Figure 7.7 shows the NOR-type decoding logic which generates the enable signals EN2, EN3, . . . , EN17. The enable signal enables the data bus into one register in the register string and disables the output of the previous register. The

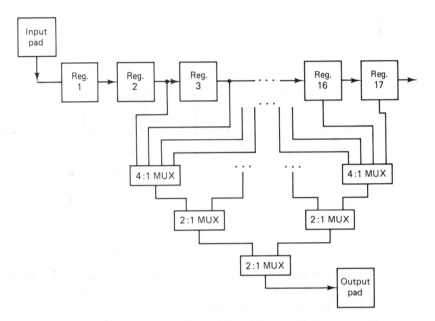

Figure 7.5   Alternative design approach for VRD.

data then shift to the right through the remaining registers until they are sent to the output pad.

The main advantage of this implementation is that only a regular cell structure can be identified. Besides I/0 pads and superbuffers, only two basic cell types are used: a register cell with a 2-to-1 multiplexer duplicated $8 \times 16 = 128$ times, and a control decoder cell duplicated 16 times.

7.2.4 Layout Design

The hierarchical buildup of the VRD chip will start at the lowest level of cell design. The basic storage cell, shown in Fig. 7.8, called VR1BY1, is composed of a register and a $2:1$ multiplexer. (This uses a subcell consisting of an inverter with a clocking transistor.) If the enable register i line is high, the data from the input bus are passed to the first inverter of the register. If this line is low, the data from the previous stage (register $i + 1$) enter from the left and pass through two inverter cells as the clock lines enable the data. When the enable register i is high, the data from the register $i + 1$ are disabled using the phi2 clock line of register $i + 1$. This clock line is forced low when the enable register i is set so that the data from register $i + 1$ do not conflict with the data on the enabled input data bus. The control signal for this is denoted "gated phi2" in Fig. 7.9.

At the next level of hierarchy, a cell called TOPVR4BY2 is built consisting of an array of 4×2 basic storage cells with four clock drivers at the bottom. This cell is shown in Fig. 7.9. The four rows of basic storage cells are aligned so that V_{dd} and

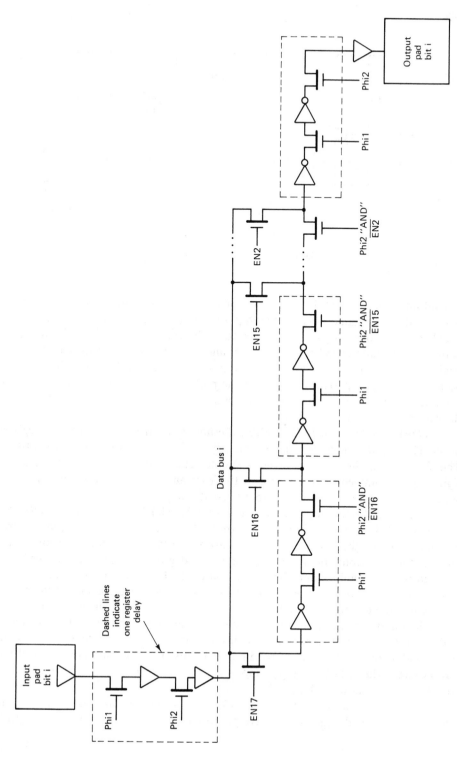

Figure 7.6 Circuit diagram for enabling logic and data path for one bit.

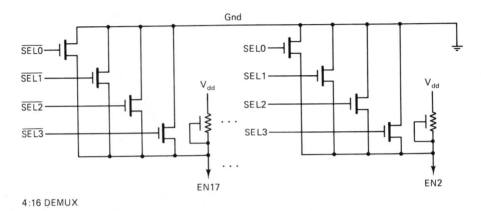

4:16 DEMUX

Figure 7.7 The 4 × 16 demultiplexer.

ground lines can be shared. To do this, the cells in rows 2 and 4 are flipped upside down with the ground line at the top of the cell. A very similar cell, called BOTVR4BY2, consisting of a 4 × 2 register cell array for data bits 5 through 8, is built. The key difference between this cell and the top cell is that its superbuffer is modified to include the gated phi2 signal as sketched in Fig. 7.10, which runs over both bottom and top arrays. Note that the pull-down transistors controlled by the "enable" input effectively transform the output inverter and the second inverter of the superbuffer into two NOR gates.

TOPVAR4BY2 and BOTVR4BY2 are now stacked to form an 8 × 2 register array. The decoding logic for each enable register is placed on top of TOPVR4BY2. The skeleton for the decoding logic is simple. The structure allows either a select signal or its complement (derived directly from an input pad or via an inverting superbuffer) to gate the decoder output "selout" to ground.

The hierarchical design continues to form a VR8BY2 by stacking TOPVR4BY2 and BOTVR4BY2 (four VR8BY2's form a VR8BY8) and finally joining two VR8BY8 cells to give the 8 × 16 register array. To this array, the eight input and eight output superbuffers are added, the V_{dd} and ground buses are run vertically through the superbuffers, and the connection to the horizontal power grid is made. The width of the power lines is determined by calculating the maximum power dissipation at each point. Note that the width increases as the bus nears the power pad. The input data buses also run horizontally across the chip in metal. The clock and control lines run vertically in polysilicon. Note that the clock lines are buffered in the middle of the chip due to speed requirements. This completes the heart of the VRD design. The rest of the chip connects this to the outside world. This final layout with I/0 pads is shown in Fig. 7.11.

It is common in VLSI designs to talk about an initial *chip floor plan*. For the VRD problem, the designer could have proposed a floor plan as shown in Fig. 7.12,

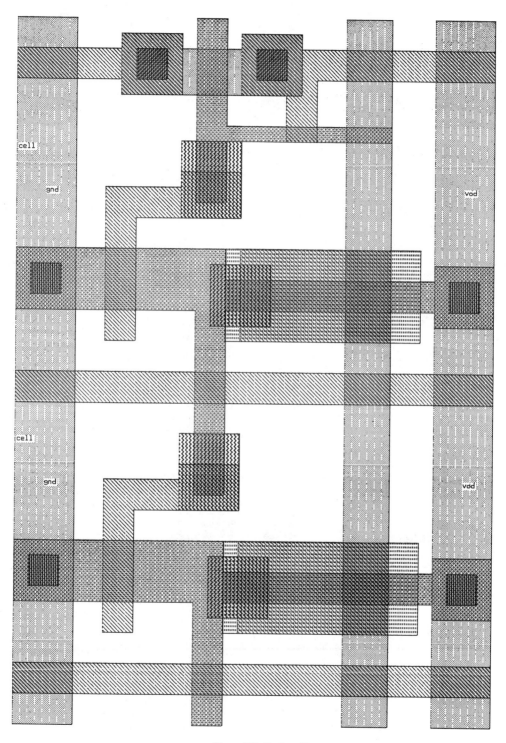

Figure 7.8 Basic cell.

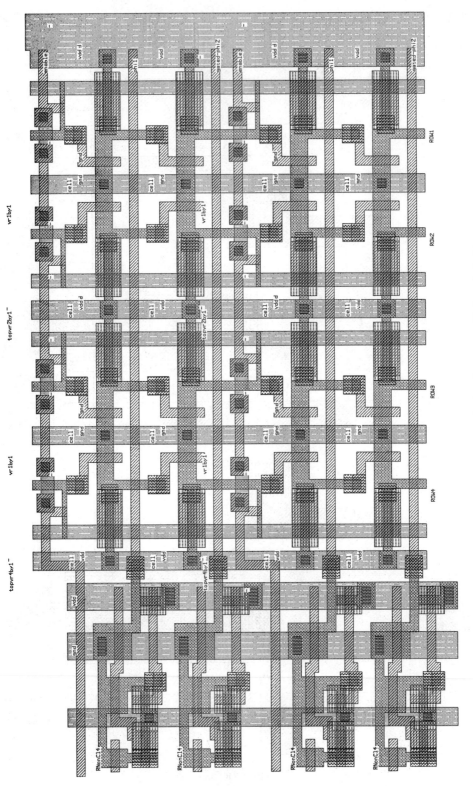

Figure 7.9 TOPVR4BY2: 4-bit by 2 register array with four clock drivers.

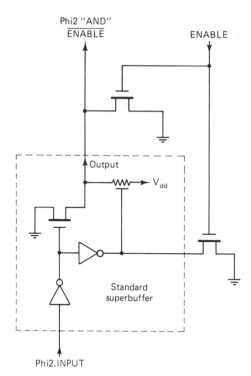

Phi2 "AND"
$\overline{\text{ENABLE}}$

ENABLE

Output

V_{dd}

Standard
superbuffer

Phi2.INPUT

Figure 7.10 Modification to a standard
superbuffer.

which is structurally very close to the final layout of Fig. 7.11. The problem is that a
floor plan cannot be conceived unless the designer has gone through a rough
bottom-up design process, at least mentally, and has had some experience in
designing the basic cells. It is still a good idea to start with a preliminary floor plan
which might be modified as the design evolves. The floor plan defines a good
objective function and serves as a good documentation aid as the design proceeds.

7.2.5 Timing

The select line decoding path is not intended to be a dynamic path and has a time
constant of approximately 100 ns. The select lines must be initialized by several
clocks before data can be pipelined through the chip. The data are latched by the
VRD at the falling edge of ϕ_1 as a stable-ϕ_1 signal. After the selected number of
clock delays, the data are set to output as stable-ϕ_1 signals. All the enable signals are
gated-ϕ_2 signals. The worst-case timing analysis for the case of low-to-high transition
shows that the time constant of the register-to-register delay and the time constant to
charge the data bus are not critical factors. Their values are approximately 2.5 ns

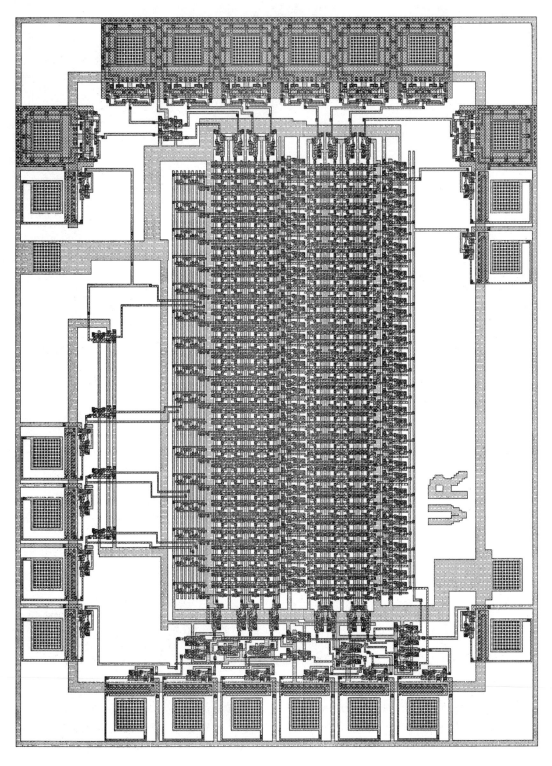

Figure 7.11 The VRD chip.

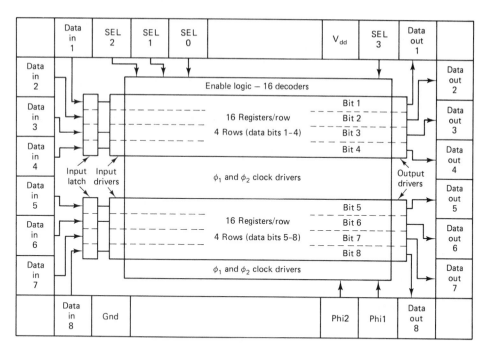

Figure 7.12   Functional floor plan for VRD.

and 2.3 ns, respectively. The critical factor is the clock distribution delay, which is about 20 ns. The distribution delay does not cause a clock skew problem because the clock delay is the same for each register in a given bit path; but it puts a constraint on the hold time of data on the input pad, which must be at least 20 ns. The propagation delay at the output pad is 40 ns. The minimum clock period is therefore 60 ns. The large time constant for the clock distribution is due to the fact that the clock lines run in polysilicon. A different basic cell design would allow the control to be run in metal parallel to the metal power line. This will increase the basic cell size but will decrease the clock period. A similar improvement can be made if the gated clock lines are run in metal.

7.2.6 Power

The V_{dd} and ground layout of VRD is shown in Fig. 7.13, which is an interleaved structure of two planar trees. The chip uses about 75 mA at 5 V, which is 375 mW. The internal main bus is 24λ wide and carries 48 mA, the I/O pads (16λ wide) carry 25 mA, the vertical register bus (16λ wide) carries 31 mA, the horizontal clock bus (6λ wide) carries 10 mA, and the horizontal register bus (4λ wide) carries about 5 mA. All of these currents are below the metal migration limit. The chip size is $900 \times 1200\lambda^2$ with $\lambda = 2.5$ μm. The chip was tested and worked correctly at 10 MHz.

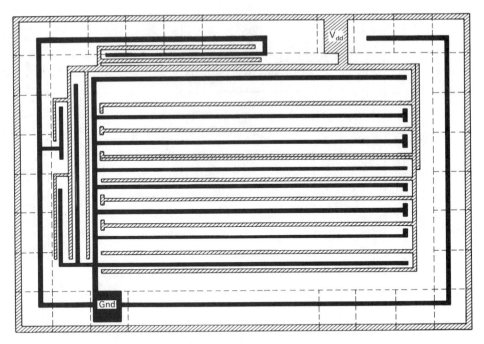

Figure 7.13 Chip floor plan: the V_{dd} and Gnd lines in VRD.

7.3 Interconnection Network

This section presents the design of interconnection networks. We then include a discussion of the design of a barrel shifter, which is a special kind of interconnection network.

    The function of an interconnection network is to provide a set of paths or links from one set of input terminals to another set of output terminals. The interconnection networks are used in several applications, such as parallel machine architecture for communications between processors and memory modules, data manipulation and data alignment networks, data communication networks, and so on. There is a large body of literature on this subject. Some of the well-known network structures discussed are the crossbar switch, the Benes network, the flip network, the perfect shuffle network, the omega network, the barrel shifter, and data manipulation networks. There are different design requirements and modes of operation for interconnection networks: The communication may be synchronous or asynchronous, the control may be centralized or decentralized, the switching method may employ circuit switching or packet switching, and the network topology can be either static (with dedicated links) or dynamic (with programmable interconnection). A dynamic topology could be blocking or rearrangeable. A blocking network is not capable of connecting all the inputs to all the outputs since this may result in a conflict in the use of a switching element. A rearrangeable network is nonblocking.

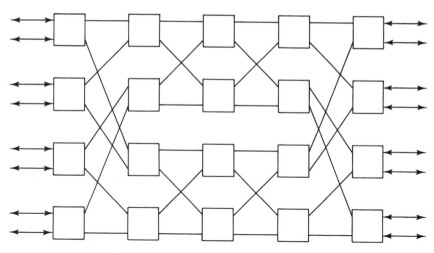

Figure 7.14 An 8 × 8 Benes network.

An obvious solution to implementing an arbitrary permutation of n input lines onto n output lines without blocking is to use a *crossbar switch*. It is usually implemented in the form of a matrix of switches, the switch at the (i, j) position connecting the ith input to the jth output—if a control signal s_{ij} is set on. Bringing out n^2 separate control lines into the array and programming these to realize a particular permutation is the major disadvantage of this design.

The Benes network uses fewer than n^2 number of switches and is rearrangeable and nonblocking. As an example, an 8 × 8 Benes network is shown in Fig. 7.14, where each square box represents a 2 × 2 switch which can pass the inputs straight or permuted onto the output. Thus the network has a possible 2^{20} states, although only 8! permutations are actually realized. In general, it will need $\log_2(n!)$ control signals, which must be brought inside the network and programmed to realize a specific permutation. This again is a major disadvantage of the Benes network. There is a much simpler solution to this problem for at least small values of $n \le 8$. This network, called OPIN for Our Programmable Interconnection Network (Richie and Gatt, 1982), is described next.

7.3.1 Functional Description of OPIN

OPIN is an 8-bit rearrangeable nonblocking interconnection network which has distributed control. OPIN can provide a bidirectional connection link from any of the eight inputs (I_0-I_7) to any of the eight outputs (Z_0-Z_7). The interconnection links are determined by three bytes of control memory. The control memory is loaded into OPIN via a *ld* (load) control input. The direction of data flow is determined by the *dir* (direction) control input.

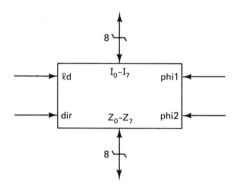

Figure 7.15 OPIN symbol.

OPIN has two control inputs (ld, dir), two nonoverlapping clock inputs (ϕ_1, ϕ_2), and 16 bidirectional data inputs [(I_0-I_7), (Z_0-Z_7)]. The OPIN symbol is depicted in Fig. 7.15.

The significant parts of OPIN consist of an input bus (INBUSS) straddled by eight multiplexer cells. Bidirectional superbuffers (drivers) are incorporated to drive the signals through the chip. Figure 7.16 is a diagram of OPIN, showing the most significant parts.

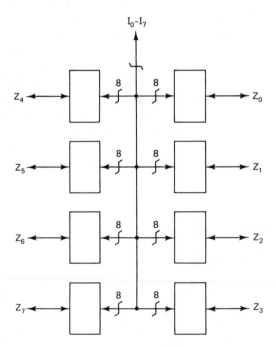

Figure 7.16 Functional diagram for OPIN.

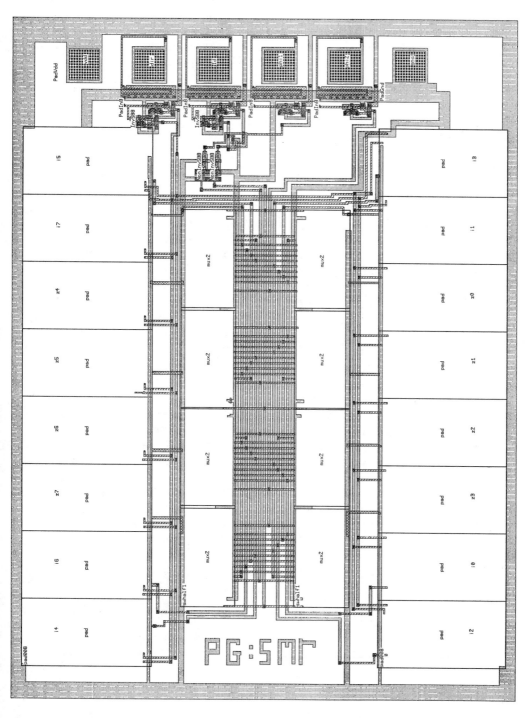

Figure 7.17 OPIN floor plan.

227

The data flow is bidirectional and asynchronous. Figure 7.17 shows a complete OPIN diagram with an expanded INBUSS. The data flow is in either direction through the MUX2 cells. OPIN consists of five major logic cells. These logic cells are MUX2, control word load logic, direction control logic, INBUSS, and pads and superbuffers.

7.3.2 Circuit and Layout Design

In this section we discuss the details of each of the major logic cells of OPIN. A fully expanded Cifplot of OPIN is given in Fig. 7.18.

MUX2 cell.   MUX2 is the fundamental building block of OPIN. MUX2 consists of an 8-to-1 multiplexer and a 3-bit shift register. Figure 7.19 depicts the logic diagram of MUX2. The area of MUX2 was minimized by straddling the 3-bit shift register around the multiplexer as shown in Fig. 7.20.

Control word load logic.   The 3-bit shift register used for the MUX2 control employs the following signals: ϕ_1^*ld, ϕ_1^*ld', and ϕ_2. The functions of these signals are depicted in Fig. 7.21. These signals are generated from the two nonoverlapping clock inputs and the ld input. The ϕ_1^*ld and ϕ_1^*ld' signals are generated by the pass transistor version of an AND gate. Three superbuffers drive these control signals through the circuit. The control logic is depicted in Fig. 7.22.

Direction control logic.   The bidirectional capability of OPIN is controlled by the dir control input. The dir input is not latched, thus retaining flexibility and high-speed operation. The dir signal defines data flow in the $(Z$ to $I)/(I$ to $Z)$ directions. Both dir and dir' are driven to the 16 tristate pads by superbuffers.

Input bus.   The input bus, called INBUSS, consists of 13 vertical metal lines. Eight of these lines are the eight inputs $(I_0–I_7)$, two are Gnd lines, and the other three are the ϕ_1^*ld, ϕ_1^*ld', and ϕ_2 control lines. INBUSS is used to deliver all the needed signals to the eight MUX2 cells, which are straddled around the INBUSS.

Pads and superbuffers.   OPIN utilizes 22 pads. Of these, 16 are bidirectional with superbuffers, to yield driving capabilities in both directions. The dir and dir' control lines determine the direction for each pad. The superbuffers drive all the signals through the OPIN circuit.

Of the remaining six pads, four are input pads. These four input pads are used for the ld, dir, ϕ_1, and ϕ_2 inputs. The last two pads are the V_{dd} and Gnd pads.

V_{dd} and Gnd routing.   A V_{dd} and Gnd map is shown in Fig. 7.23. Both V_{dd} and Gnd are laid around the perimeter of the OPIN circuit, to supply the pads. V_{dd} and Gnd are also interleaved inside the circuit to supply power to the shift register

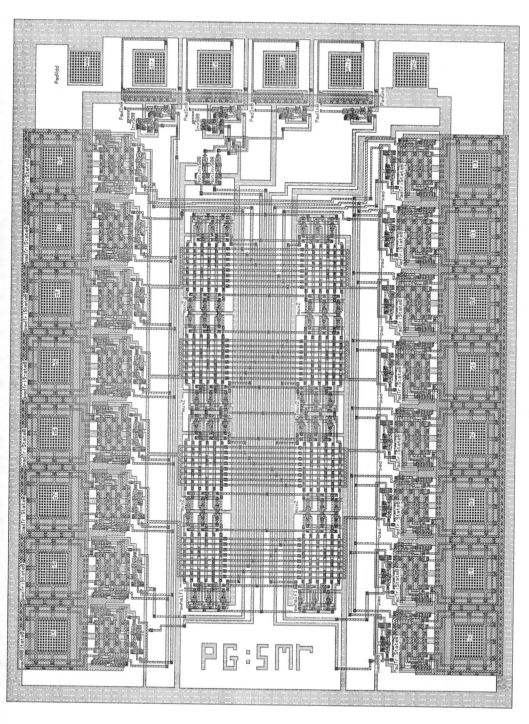

Figure 7.18 OPIN chip.

229

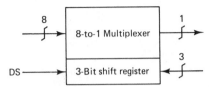

Figure 7.19 MUX2 logic symbol.

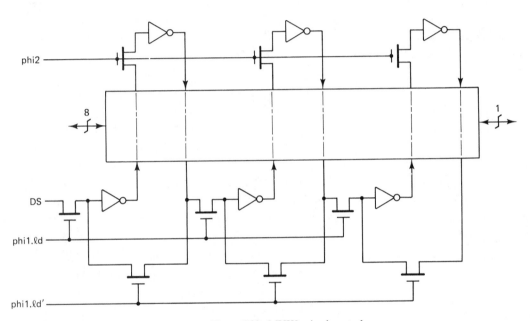

Figure 7.20 MUX2 mixed control.

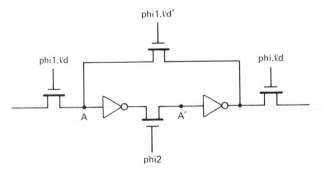

Figure 7.21 Shift register stage.

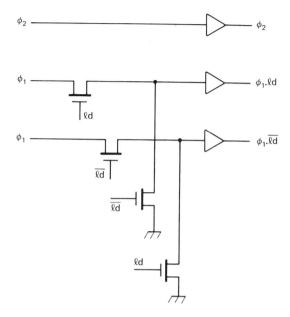

Figure 7.22 Control logic.

cells. Finally, V_{dd} is also required by the superbuffers, which have been added to each NTriState8 pad (Newkirk and Mathews, 1983). The V_{dd} and Gnd line widths can be determined from the maximum dc current needed. The maximum V_{dd} width in OPIN is a conservative 24λ.

7.3.3 OPIN Operation

Our Programmable Interconnection Network is simple to operate. Four states can be identified for OPIN. These states are determined by the logic levels of the *ld* and *dir* control inputs and are identified in Table 7.1. Both S0 and S1 are asynchronous data flow states. S0 defines data flow from *I* to *Z*, while S1 defines data flow from *Z* to *I*. Thus the *dir* control input allows data flow in the $(ZI)/(IZ)$ directions.

S2 is the load control state. When the *ld* input is high in conjunction with ϕ_1, a single control byte is shifted into the eight 3-bit shift registers. The control data are

TABLE 7.1 OPIN OPERATION

State	ld	dir	(I_7-I_0)	(Z_7-Z_0)
S0	0	0	Input	Output
S1	0	1	Output	Input
S2	1	0	MUX2 control	X
S3	1	1	Not allowed	

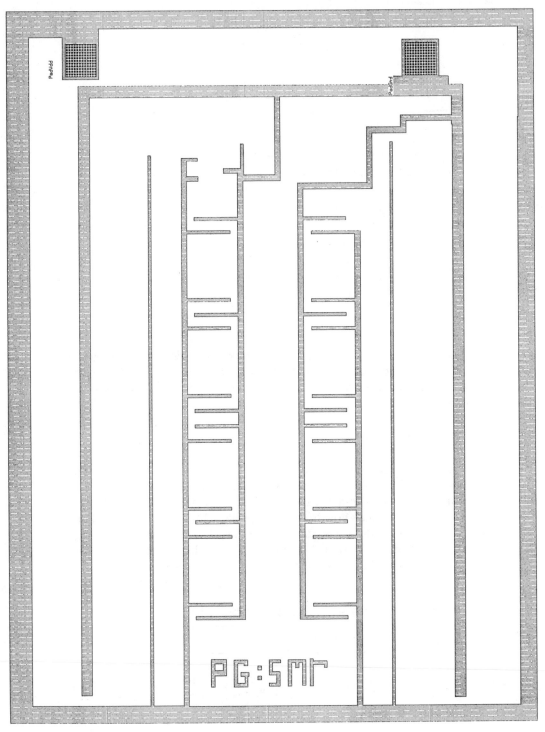

Figure 7.23 V_{dd} and Gnd routing for OPIN.

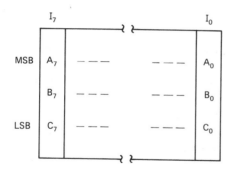

Figure 7.24 Control word format.

loaded from the inputs (I_7-I_0). Each of the eight shift registers receives 1 bit of control from its corresponding input. Thus if the *ld* input is high during three clock cycles, all eight 3-bit shift registers will contain control words which define the current interconnection links. The format of the eight 3-bit control words is depicted in Fig. 7.24.

Each of the eight 3-bit control words (A, B, C) specifies the interconnection link between the ith input and the input specified by the binary equivalent of ABC. For example, if $A_3B_3C_3 = 101$, an interconnection link between Z_3 and I_5 is generated. This is a very simple control scheme!

This control scheme allows for 2^{24} arbitrary sets of interconnection links called *bijections*. The bijections include eight factorial bidirectional single interconnection links, and any combination of single-connection and multiconnection (party) links. Due to the structure of OPIN, party links are available only in the I-to-Z direction.

A party connection is defined as a single-source multiple-destination link. Multiple source-to-single destination links are also possible to specify; however, such links can cause a short circuit. The user is responsible for ensuring that such conflicts do not occur. These conflicts will occur any time that party connections are specified in conjunction with a high signal on the *dir* control line. Thus the possible nonconflicting bijections are 2^{24} or 16,772,216 party and single I-to-Z connections and 8! or 40,320 single Z-to-I connections.

S3 is a not-allowed state. Under this state, undefined values will be loaded into the eight 3-bit shift registers.

When OPIN is first powered up, the shift registers do not necessarily contain valid data. Since the shift register is 3 bits wide, it will require three clock cycles to clear up unknown data with shift registers. This does not pose a significant problem, since it is unlikely that OPIN will be used during the first three clock cycles after system power-up. The chip works at 10 MHz, consuming 174 mW (35 mA at 5 V) with a 50% duty cycle of the clock. The signal propagation delay from I_0 to Z_0 is 140 ns.

7.4 Barrel Shifter

A barrel shifter is typically used for data alignment. In particular, it has been used in mantissa alignment, postnormalization, and rounding operations in floating-point arithmetic. The barrel shifter consists of a $2n$-bit input bus $[A = (a, \ldots, a_n)$, $B = (a_{n+1}, \ldots, A_{2n})]$ and an n-bit output bus $C = (b_1, \ldots, b_n)$. A shift constant s ($0 \le s \le n$) or a control input is specified to the shifter, which gives the number of bits from the B-bus present in the output C-bus. In other words, C gets $(a_{s+1}, a_{s+2}, \ldots, a_{s+n})$. An eight-input four-output barrel shifter is sketched in Fig. 7.25. If $s = 0$, then C equals A; if $s = 4$, then C equals B. If B is absent, the barrel shifter becomes an $n \times n$ shifter and a shift constant s implies that $b_{s+1} = a_1$, $b_{s+2} = a_2, \ldots, b_n = a_{n-s}$, and the first s bits of C (i.e., $b_1 b_2 \cdots b_s$) are filled with all 0's or all 1's. The barrel shifter is said to have "wraparound" features if rather than filling with 0's and 1's, we have $b_1 = a_{n-s+1}$, $b_2 = a_{n-s+2}, \ldots, b_s = a_n$.

In this section we present several designs for a barrel shifter. We will also discuss the design of a precharged bidirectional shifter which allows much higher speed of operation.

An $n \times n$ barrel shifter implements only a subset of permutation operations of a cross-bar switch. An $n \times n$ barrel shifter with wraparound features can be designed with only n control inputs as illustrated in Fig. 7.26. The design can be described as follows. Control lines of transistors providing a path from input a_i to output b_i ($1 \le i \le n$) are connected together to a line marked "shift 0." Similarly, the control lines of transistors connecting a_i to b_{i+1} ($1 \le i \le n - 1$) and a_n to b_1 are tied together to a line marked "shift 1." This construction can be carried through for "shift 2," "shift 3," . . . , "shift n" (not shown in Fig. 7.26) to complete the circuit. Note that only one of the control shift lines is on at any time. If the "shift s" line is on, the barrel shifter realizes a total shift of s bits. Thus these control signals could be generated at the output of a 1-out-of-$(n + 1)$ decoder.

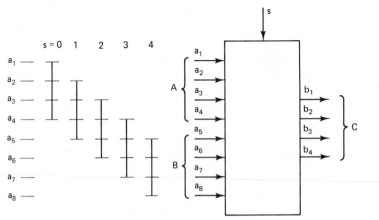

Figure 7.25   Eight-input four-output barrel shifter.

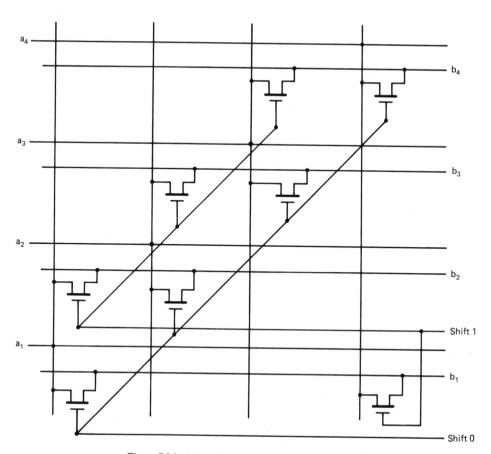

Figure 7.26 A 4 × 4 barrel shifter with wraparound.

Mead and Conway (1980) described an interesting adaptation of the 4 × 4 barrel shifter to design an 8 × 4 barrel shifter. This is illustrated in Fig. 7.27. The input and output lines run horizontally as before but the input lines are shuffled such that the lines (a_i, a_{i+n-1}) and $(b_i, \text{shift } (i-1))$, $(1 \le i \le n)$, form a group of lines running close together. The lines (a_1, b_1) and $(a_{2n}, \text{shift } 0)$ form two special groups on the top and the bottom of the array. An array of switches controlled by shift control lines is used as follows:

Control Line	Connections
Shift 0	$a_1 \to b_1, a_2 \to b_2, a_3 \to b_3, a_4 \to b_4$
Shift 1	$a_2 \to b_1, a_3 \to b_2, a_4 \to b_3, a_5 \to b_4$
Shift 2	$a_3 \to b_1, a_4 \to b_2, a_5 \to b_3, a_6 \to b_4$
Shift 3	$a_4 \to b_1, a_5 \to b_2, a_6 \to b_3, a_7 \to b_4$
Shift 4	$a_5 \to b_1, a_6 \to b_2, a_7 \to b_3, a_8 \to b_4$

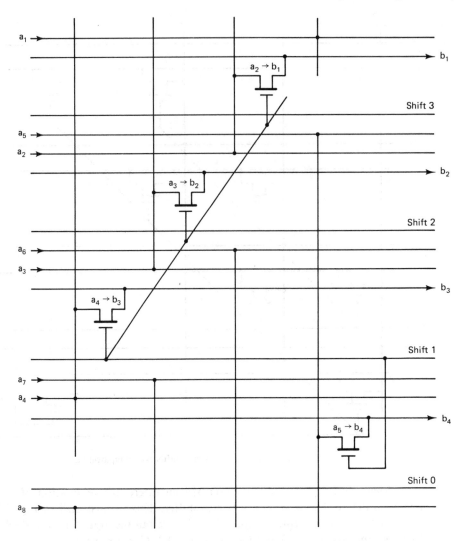

Figure 7.27   Circuit diagram of an 8 × 4 barrel shifter.

Figure 7.27 shows only the transistors and the connections to shift one operation. Note that the vertical lines connecting input and output exhibit an interesting pattern of split.

The major disadvantage of the barrel shifter circuit is that it is basically a signal steering circuit with long data and control lines which have large capacitance associated with them, and is therefore very slow. The barrel shifter circuits described above in nMOS can be directly transformed into CMOS by replacing each pass transistor with a transmission gate whose n-channel path is controlled by some shift i signal and whose p-channel path is controlled by the complement of signal shift i.

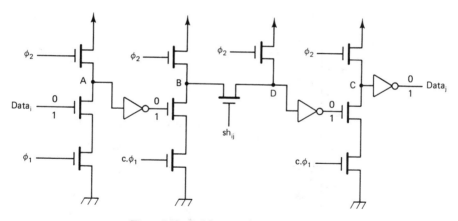

Figure 7.28 Path in a precharge barrel shifter.

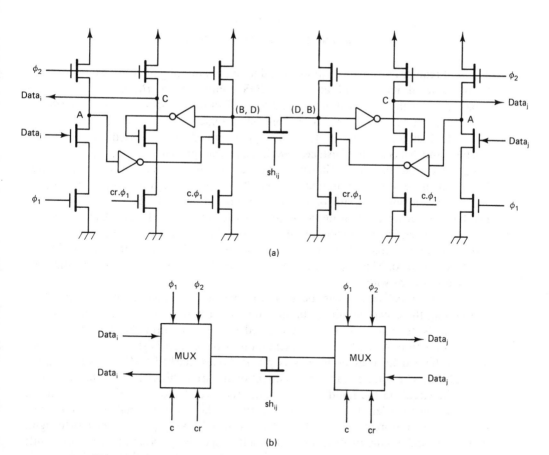

Figure 7.29 (a) Bidirectional path in the precharge barrel shifter; (b) schematic representation.

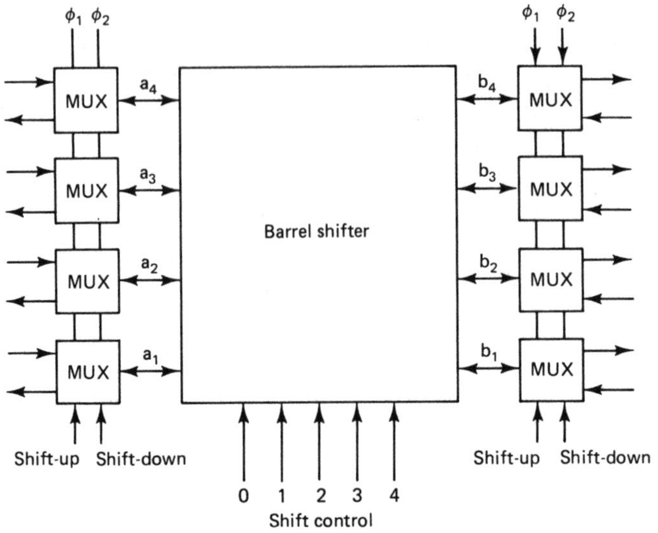

Figure 7.30   Precharge bidirectional barrel shifter.

This will require considerably more area due to the doubling of control lines and due to the general increase of area in CMOS compared to nMOS. The increased capacitance of the lines will have an adverse effect on the speed. A better approach to designing a barrel shifter is to use precharge logic, as described next.

The bidirectional barrel shifter using precharged logic is based on a design by Eustace and Macomber (Eustace, 1984). The basic principle of the design can be understood from the circuit shown in Fig. 7.28. Let us assume that the data in some line $Data_i$ have to be transmitted to some jth output $Data_j$ when the control signal sh_{ij} is on, closing the path of a control switch. Points A, B, C, and D are all precharged during ϕ_2. During ϕ_1, if the gated clock $c.\phi_1 = 1$, then $Data_j = \overline{Data_i}$. This can be verified by tracing the signal values in the path connecting the input to output for all possible values of input. Note that if no data are switched onto $Data_j$, the output is zero. This indicates that the barrel shifter has the feature of filling the emptied positions with 0's.

An identical circuit can be used to reverse the direction of the data flow. Combining these two circuits, a bidirectional data path as shown in Fig. 7.29(a) is obtained. Note that $cr.\phi_1$ is now the gated signal to control the data flow in the reverse direction. A schematic representation of the circuit is shown in Fig. 7.29(b). The precharged logic at both ends is put in boxes marked "MUX," standing for multiplexer. The total circuit for a bidirectional barrel shifter is now shown in Fig. 7.30. The circuit in the middle is essentially the circuit shown in Fig. 7.26 with the modification to delete the wraparound feature. The control signals $c.\phi_1$ and $cr.\phi_1$ can now be interpreted to stand for "shift-down," taking place from left-to-right, and "shift-up," taking place from right-to-left, respectively. Note that the entire shift operation takes place during a single phase of the clock and that the shift signals

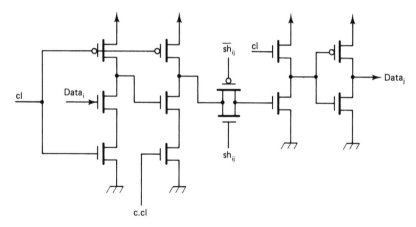

Figure 7.31   CMOS domino circuit for a path in the barrel shifter.

could be either gated-ϕ_1 or stable-ϕ_1 signals. The input and output are precharged valid ϕ_1.

A precharged CMOS circuit using domino logic can also be used to build a barrel shifter. The basic circuit is an adaptation of Fig. 3.88 and is shown in Fig. 7.31. This circuit performs the task of the circuit of Fig. 7.28.

7.5 Nonnumeric Processors

In this section we discuss several examples of hardware for nonnumeric processing. By nonnumeric processing is meant those specialized operations that are not suited for a classical ALU. Some examples of such operations are: string processing, sorting, priority functions, database search and updates, storage management functions, and problems related to graphs and computational geometry. The idea is to use a few simple functions in a regular array of one or two dimensions that reduces communication cost between cells and allows a high degree of parallelism or pipelining on several data streams. Such architectures have been called "cellular" architecture in the literature in the past. More recently, they have been classified as "systolic" architecture by Kung (1979), and a large body of literature has grown on systolic algorithms encompassing a large number of applications.

We present circuits of some of the algorithms, whereas for others only functional descriptions will be provided. These descriptions will provide a basis on which VLSI design projects for students can be formulated.

7.5.1 Pattern Matching

The pattern matching problem consists of searching a large text to find the occurrence of one or more *patterns* starting from a specified position in the text (anchored-mode operation) or from any position in the text (unanchored-mode

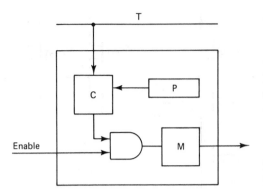

Figure 7.32 Basic pattern matching cell
for parallel operation.

operation). The pattern may have "don't care" or "wild card" characters anywhere
in the pattern or might even have a special character signifying a "variable-length
don't care" character. The well-known linear algorithms of Knuth (1973) and Boyer
and Moore (1977) can only handle patterns having no wild-card character.

The hardware algorithms use a one-dimensional array of simple cells in which
a pattern is stored and the text to be searched is applied to the array either in serial
or in parallel fashion (Mukhopadhyay, 1979). The basic cell for parallel or "broadcast text" operation is shown in Fig. 7.32. The cell consists of a register to hold a
pattern character P, a comparator circuit C which produces a 1 output if P equals
the broadcast character, and an output latch M which is set to 1 if the cell is
"enabled" or "anchored" by the previous cell and the comparator produces a match
bit. A cascade of n such cells, as shown in Fig. 7.33, will form the basic pattern
matching machine. A 1 output from the nth cell will indicate the occurrence of the
pattern during the previous n cycles in the text, provided that the first cell was
enabled during the first cycle. Figure 7.34 shows the nMOS circuit diagram of the
basic cell. The pattern bit p is shifted into the recirculating-type storage cell during
ϕ_1, when the shift signal sh is on. The comparison with a text bit t takes place
during ϕ_2. If this cell is iterated eight times in the vertical direction, as shown in Fig.
7.35, the character comparator is formed. The bottom dashed part of the circuit has
been added to allow wild-card characters. The latches F (the fixed-length don't
care), V (the variable-length don't care), and A (anchor) are loaded along with the
pattern. When both F and V are set to 1, the output of the cell remains 1 for all

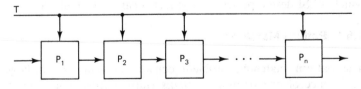

Figure 7.33 Pattern matching machine.

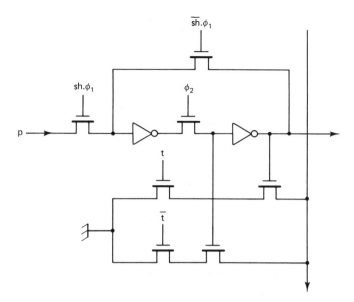

Figure 7.34   The nMOS circuit of the basic cell.

clock times following the first appearance of 1 at the input of the cell, which is equivalent to matching with a variable-length wild character. For fixed-length don't care operation, only the F flip-flop is set. Curry et al. (1983) have introduced a "case" flip-flop and modified the logic to distinguish between upper- and lowercase letters.

The basic cell, described above, can easily be modified to handle serial text input by incorporating a storage register T similar to the recirculating-type storage register P and delaying the output of the stage by an additional clock cycle with a latch. The pattern is loaded into the n cells before the text is applied at the leftmost cell and then shifted right to the last cell of the cascade. The schematic diagram of the serial pattern matching machine is shown in Fig. 7.36. The anchor flip-flop A can be used to determine subpatterns of a given pattern. For example, by setting the ith anchor flip-flop to 1, the subpattern $P_i P_{i+1} \cdots P_n$ can be detected.

Both the serial and parallel versions of the algorithms have been implemented in nMOS (Curry and Mukhopadhyay, 1983; Curry et al., 1983). The serial version was implemented first because broadcasting the text characters to all the cells could increase the power requirement for the chip and hence decrease its speed. However, the text characters only need to be stable during one phase of the clocking and the clock signal has to be broadcast to every cell. As long as the clock skew is within the tolerance limits of the performance of the chip, broadcasting text characters along with the clocks does not involve an additional time penalty. Using a broadcast cell means that the area per match cell is drastically reduced. The implemented chip allows 8 bits per character and up to 64 characters. Any arbitrary number of patterns can be searched for concurrently as long as the total length does not exceed

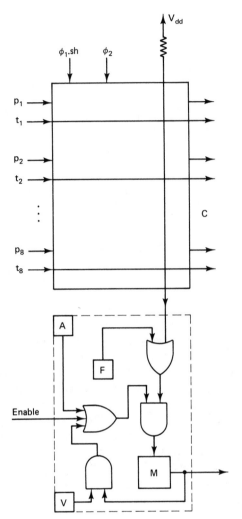

Figure 7.35 Circuit for one character with "don't care" logic.

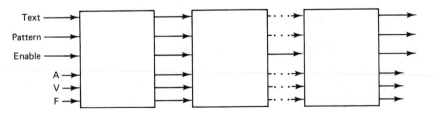

Figure 7.36 Basic pattern matching cell for serial operation.

64 characters. To allow reconfigurability of the cells for a multiple pattern search in parallel, the role of the anchor bit was redefined. The end of the pattern is recognized by checking if the next cell indicates the beginning of the pattern by its anchor flip-flop A being set to 1. If so, and if the current cell has a match, an output decoder circuit will inform the outside world of the exact location where the match occurred. It also informs whether there is a multiple match or a single match. In the case of a multiple match, the location of the rightmost matching cell is reported. The broadcast cell with all the logic for wild-card characters and end-of-pattern recognition has a dimension of $30\lambda \times 1077\lambda$ compared to $100\lambda \times 768\lambda$ for the serial cell. The complete chip, called PAM, contains 64 cells split into two groups of 32 cells with superbuffers placed in the middle to restore the levels of clocks and text information. The maximum operating speed of the chip is 2.5 MHz, drawing a current of 140 mA at 5 V.

7.5.2 Lexical Comparator

A current research project at the University of Central Florida is concerned with the design of a string processor that will support string primitives in a general-purpose computer through incorporating pattern matching and other hardware algorithms (Curry and Mukhopadhyay, 1983). One of the subsystems that will be used in this machine is a lexical comparator. Unlike the simple comparator that formed the basis for the pattern matching algorithms of the preceding section, a lexical comparator determines whether a character C1 is lexically greater than, less than, or equal to a second character C2. Lexical comparison needs two functions: a simple comparison function and a function that determines a leading 1, sometimes also called a *priority function*. We are given C1 $= (a_1 a_2 \cdots a_n)$ and C2 $= (b_1 b_2 \cdots b_n)$. We have to find a basic cell that can be used to build a lexical comparator for an arbitrary value of n. The proposed solution will look as shown in Fig. 7.37. The outputs of the ith cell, z_i and y_i, are defined as

$$z_i = \begin{cases} 1 & \text{if } (a_1 \cdots a_i) \geq (b_1 \cdots b_i), 1 \leq i \leq n \\ 0 & \text{otherwise} \end{cases}$$

$$y_i = \begin{cases} 1 & \text{if } a_j = b_j \text{ for all } j, 1 \leq j \leq i \leq n \\ 0 & \text{if } (y_{i-1} = 1 \text{ and } a_i \neq b_i) \text{ or } (y_{i-1} = 0) \end{cases}$$

with $z_0 = y_0 = 1$. By observing the outputs (z_n, y_n) we can make the following conclusions:

z_n	y_n	Conclusion
1	1	C1 = C2
1	0	C1 > C2
0	0	C1 < C2
0	1	Not possible

The logic diagram of the basic cell is shown in Fig. 7.38. The comparison function is computed at point R by the exclusive-OR gate E. The four pass

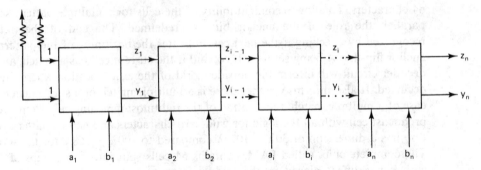

Figure 7.37 Structure of the lexical comparator.

transistors enclosed within the dashed box compute the priority function. If $a_i = b_i$, y_{i-1} propagates to the output y_i, point P is connected to the ground point G, making the output of the NOR gate N equal to 0, and the path through the pass transistor T is open. This enables z_{i-1} to propagate to z_i. If, however, $a_i \neq b_i$, the priority network connects y_{i-1} to P and y_i to ground G. If now $b_i = 1$ and $a_i = 0$, the transistor T will conduct and ground the output z_i. On the other hand, if $b_i = 0$ and $a_i = 1$, the value of z_i will remain unchanged. Note that as soon as the output

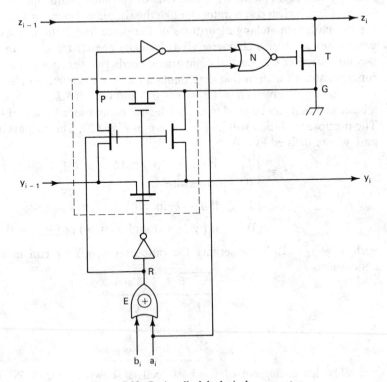

Figure 7.38 Basic cell of the lexical comparator.

of the ith cell becomes $(z_i = 1, \ y_i = 0)$ or $(z_i = 0, \ y_i = 0)$, these values are maintained for all cells $i + 1$ through n.

A nonnumeric logic unit (NLU) has been designed (Crystal and Hendry, 1983) incorporating the foregoing circuit. The unit has as input the characters C1 and C2 from two character strings. A special terminating character t is preloaded into a register and the circuit makes the following checks: C1 = t, C2 = t, C1 = C2, or C1 > C2. We will not discuss the details here except to comment that the critical delay appears in the carry propagation path of the priority network when the output of the exclusive-OR gate is low. The signal passes through one pass transistor in each stage, and after each four stages, a superbuffer has to be used to boost the signal level. The large capacitance of the z output lines contributes to the delay.

To improve the speed of the lexical comparator, precharge techniques can be used (Eustace, 1984). A "negative" logic is more suitable in which rather than propagating a 1, a 0 signal is propagated. The priority circuit is shown in Fig. 7.39 and only the y outputs are shown. The outputs of the circuit $(\hat{x}_1, \hat{x}_2 \cdots \hat{x}_n)$ indicate the position of the leading 1 in $(x_1 x_2 \cdots x_n)$ and are used to control the propagation of the z signal, as discussed earlier.

The operation of the circuit can be explained as follows. The outputs y_i are precharged to 1 during ϕ_2. During ϕ_1 if the signal $x_i = 0$ and $y_{i-1} = 0$, a zero will propagate to the output y_i. The propagation of a 0 signal means discharging the chain of the precharge pass transistor during ϕ_1. If 0 propagation takes place over a number of stages, the time constant of the pull-down circuit will be large. The part of the circuit in the dashed box helps reduce this discharge time by a feedback process which grounds the y output solidly if y is already low. In other words, it removes the charge in a time determined by the time constant of the circuit in the dashed box rather than by the time constant of the entire propagation chain. If $x_i = 1$, the 0 signal propagation stops and x_i becomes 1, indicating that a leading 1 has occurred. Once y_i output is 1, the outputs $y_{i+1}, \ldots, y_n$ will always precharge during ϕ_2 or by propagation during ϕ_1. Note that the signals $(x_1, \ldots, x_n)$ have to be stable during ϕ_1.

7.5.3 Stack

The stack, also known as a last-in-first-out stack, consists of an array of shift registers, $S_1, S_2, \ldots, S_n$, with three basic operations: PUSH, the register array is shifted right one place $(S_{i+1} \leftarrow S_i, 1 < i \leq n - 1)$, the input data get into S_1, and S_n is lost; POP, the register array is shifted left one place $(S_i \leftarrow S_{i+1}, 1 < i \leq n - 1)$, S_1 is read as data out, and S_n becomes undefined; and NOP, or no operation, when the registers refresh their values $(S_i \leftarrow S_i, 1 \leq i \leq n)$. A proper stack must be able to generate error or warning signals if a full stack is pushed or an empty stack is popped. The design of an nMOS stack subsystem has been fully described in Mead and Conway (1980). We will discuss briefly a similar CMOS design.

It is possible to design a stack using two separate left and right shift registers each using recirculating-type cells. These two registers can be combined with a

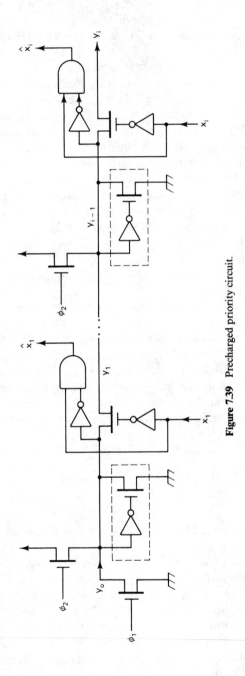

Figure 7.39 Precharged priority circuit.

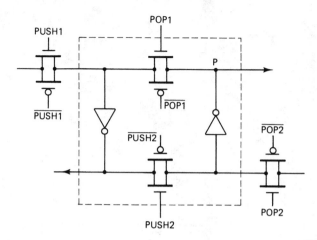

Figure 7.40 Basic stack cell.

considerably reduced number of transistors. The basic cell is shown in Fig. 7.40. Note that the recirculating memory cell is enclosed in the dashed box and the value stored in this cell is represented by the value at point P. An array of n such cells will form an n-bit stack. New information is brought into the stack during ϕ_1 and the popped output data are obtained in complemented form during ϕ_2; an additional ϕ_1-clocked memory element is needed to convert the output data to be of the same type as that of the input. The pairs of signals (PUSH1 and PUSH2) and (POP1, POP2) can never by high at the same time.

The signals POP1 and PUSH2 are on during NOP operation in synchronism with ϕ_1 and ϕ_2, respectively, while both the shift signals are off. To start a PUSH operation, PUSH1 should be brought high at ϕ_1 and the POP1 signal must be suppressed, that is, brought to zero level. This should be followed by PUSH2 becoming high during ϕ_2. Similarly, to implement a POP operation, POP2 should be raised high at ϕ_2, with simultaneous suppression of PUSH2, followed by POP1 becoming high during ϕ_1.

A PUSH followed by a POP operation in two consecutive cycles can do nothing much except perhaps lose the last bit of information. In any practical application, there has to be at least one NOP operation between two such operations.

The basic stack cell in CMOS is shown in Fig. 7.41. An n-bit w-word deep stack can now be built by iterating this cell n times in the horizontal direction and then repeating the entire row in the vertical direction w times. One of the rows can be dedicated for overflow/underflow detection by shifting a 1-bit from the left side at the beginning of the stack operation. If this 1 emerges from the right side of the stack, overflow must have taken place. Similarly, if this 1 emerges from the left-hand side of the stack during a POP operation, an underflow has taken place.

Because of the regular structure of the array, the layout strategy of the global control, power, and ground lines is well defined for the stack system. The floor plan will have the array of stack cells in the middle and control drivers on the top and

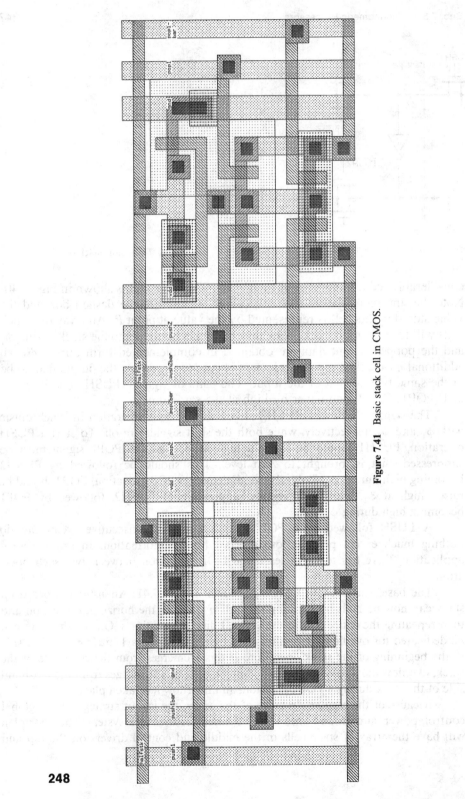

Figure 7.41 Basic stack cell in CMOS.

bottom of the array. The data in/out could be on either the left or right side of the array. Note that if the input and output are taken at opposite sides, the system can be used as a FIFO (first-in-first-out) queue. The design of a complete CMOS stack system will be an interesting student project.

The design of the stack described above uses global signals such as PUSH and POP. Guibas and Liang (1982) describe a hardware stack that uses no global signals but needs more circuits and time.

7.5.4 Sorting

Sorting has been recognized as a fundamental operation since the early days of computers. There is a vast literature on this topic and a large number of algorithms for sorting have been developed. In this section we describe a VLSI design for a hardware sorting device based on an algorithm called *weavesort* developed by the author (1981). The VLSI design is based on the work by Carey, Hansen, and Thompson (1982), who also developed this algorithm independently. Armstrong and Rem (1982), Chung et al. (1980), Leiserson (1979), and Miranker et al. (1982) have developed similar hardware sorting algorithms.

Weavesort.   The basis for the weavesort algorithm is the alternating sorting algorithm (Mukhopadhyay and Ichikawa, 1972), also called an odd–even transposition sort (Knuth, 1973). The algorithm can be described with the help of Fig. 7.42, which shows a linear array of n cells (for the moment assume that n is even) numbered $1, 2, \ldots, n$. Each cell contains a register that can hold a key. Initially, each of the cells contains a special value "*" (the maximum or the minimum value of the key). The registers are connected in shift register form so that a right shift or push operation means that the content of cell i moves to cell $i + 1$ ($i = 1, 2, \ldots, n - 1$), the content of cell n is lost, and if necessary a new element can be inserted at cell 1 from the left. Similarly, a left shift or pop operation means that the content of cell $i + 1$ moves to cell i. The content of cell 1 is recognized as the output from the array and the special value "*" enters cell n. Pairs of adjacent cells i and $i + 1$ ($i = 1, 3, 5, \ldots, n - 1$) share logic which can compare the contents of cells i and $i + 1$ and swap, if necessary, according to the particular sorting criterion employed. One input (output) step of the machine consists of a right shift (or a left shift) operation followed by a comparison-swap operation. The operation of the sorting machine can be described as follows. The data or the sequence of n keys is inserted into the array by n input step operations. The sorted sequence is extracted from cell 1 by n output step operations immediately following the input operations. An example is shown in Table 7.2. The unsorted key sequence are the numbers 63, 5, 17,

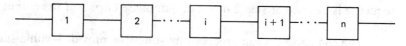

Figure 7.42 Sorting machine.

TABLE 7.2 EXAMPLE OF WEAVESORT OPERATION

Initial configuration:			*	*	*	*	*	*	
Input:	63	Step 1:	63	*	*	*	*	*	Shift
			63	*	*	*	*	*	compare-swap
	5	Step 2:	5	63	*	*	*	*	
			5	63	*	*	*	*	
	17	Step 3	17	5	63	*	*	*	
			5	17	63	*	*	*	
	11	Step 4:	11	5	17	63	*	*	
			5	11	17	63	*	*	
	2	Step 5:	2	5	11	17	63	*	
			2	5	11	17	63		
	32	Step 6:	32	2	5	11	17	63	
			2	32	5	11	17	63	
Output:	2	Step 1:	32	5	11	17	63	*	
			5	32	11	17	63	*	
	5	Step 2:	32	11	17	63	*	*	
			11	32	17	63	*	*	
	11	Step 3:	32	17	63	*	*	*	
			17	32	63	*	*	*	
	17	Step 4:	32	63	*	*	*	*	
			32	63	*	*	*	*	
	32	Step 5:	63	*	*	*	*	*	
			63	*	*	*	*	*	
	63	Step 6:	*	*	*	*	*	*	

11, 2, and 32. The keys are sorted in ascending order. Notice that the last two complete output steps are unnecessary. At the end of the fourth step, two left shift operations would yield the correct result. Similarly, the first input step can be replaced by a right shift only.

If n is odd, it is possible to add a fictitious $(n + 1)$th cell, which will always contain the element "*", and a nonexistent comparison-swap operation between the nth and $(n + 1)$th cell can be defined. In practice, this nth cell need only have shifting capability.

There are a number of important observations that can be made from the example above. First, unlike the alternating sorting procedure, all the processors are busy at all steps and we need only $(n/2)$ processors. The odd–evenness is simulated by the temporal movement of the keys to the left or right. Second, at the conclusion of an input or an output step, the odd-numbered cells (i.e., 1, 3, 5) always contain the minimum of the sequence of the keys to the right. Finally, the number of cells containing "*" shrinks and grows linearly with a period of n steps. As far as sorting is concerned, any comparison involving "*" as an argument is a useless operation. It is possible to initiate a new sorting operation from the right side of the array if the sequence is inserted at step 2 of the output phase (step 3 of the output phase if n is odd).

The comparison-swap processor must now be provided with a state flip-flop F which will specify whether or not the keys being compared come from the left or the right side of the array. If $F = 1$ and if the partial ordering condition of the sorting process is not satisfied, a swap takes place; if $F = 0$, a swap takes place under the reverse condition, that is, when the condition is satisfied. Thus, for our ascending-order sorting problem, if the $\le$ relation between cells i and $i + 1$ is satisfied, a swap does not take place if $F = 1$. If $F = 0$, a swap will take place under this condition. The F flip-flops of the $(n/2)$ processors could be connected in the form of a shift register. The value $F = 1$ can be propagated right along with the key during the input phase from the left at half the speed of the key propagation. The value of F is complemented at step 2 of the output phase and propagated to the left

TABLE 7.3 TWO-WAY WEAVESORT OPERATION

	Input/Output	Step	Contents of cells 1, 2, 3, 4				Step	Input/Output
S_i	75	1	75	*	*	*		
			75	*	*	*		
	9	2	9	75	*	*		
			9	75	*	*		
	22	3	22	9	75	*		
			9	22	75	*		
	37	4	37	9	22	75		
			9	37	22	75		
Sorted S_1:	9	1	37	22	75	*		
			22	37	75	*		
	22	2	37	75	*	46	1	46 : S_2
			37	75	*	46		
	37	3	75	*	46	33	2	33
			75	*	46	33		
	75	4	*	46	33	19	3	19
			*	46	33	19		
			46	33	19	23	4	23
			46	33	23	19		
			*	46	33	23	1	19 : sorted S_2
			*	46	33	23		
S_3:	55	1	55	*	46	33	2	23
			55	*	46	33		
	7	2	7	55	*	46	3	33
			7	55	*	46		
	15	3	15	7	55	*	4	46
			7	15	55	*		
	29	4	29	7	15	55		
			7	29	15	55		
Sorted S_3:	7	1	29	15	55	*		
			15	29	55	*		
	15	2	29	55	*	*		← At this time a
			29	55	*	*		fourth sequence S_4
	29	3	55	*	*	*		can be inserted.
			55	*	*	*		
	55	4	*	*	*	*		

along with the key sequence coming from the right, again at half the speed of the key propagation. The example shown in Table 7.3 illustrates the idea. Let the sequence to be sorted be $S_1 = (75, 9, 22, 37)$, $S_2 = (46, 33, 19, 23)$, and $S_3 = (55, 7, 15, 29)$. S_1 and S_3 are applied from the left and S_2 is applied from the right.

The sets of unsorted key streams are like the threads in a loom. The machine manipulates them in alternating periods to *weave* out a sorted sequence—hence the name *weavesort*. For a proof of correctness of the algorithm, the reader is referred to the original paper (Mukhopadhyay, 1981).

The fact that cell 1 always contains the minimum key means that the sorter can also be used as a *priority* or *sorted stack*. Inserting a key in the sorter corresponds to a *push* operation, and the *pop* operation reads the minimum element in the list. This fact is exploited in the design of the record-sorting stack (RESST) described below.

RESST chip (Carey, Hansen, and Thompson, 1982).   The chip allows the (key, record-pointer) pair to be pushed and popped. Each key is an 8-bit pointer value. The algorithm works in two phases. During ϕ_1, data are pushed in or popped out, and the storage cells are either cleared or refreshed. Associated with the key, a ninth bit indicates whether the cell contains data or is empty. The empty cells are cleared and the nonempty cells are refreshed. The ninth bit is set to 1 for empty cells. The empty cells are always kept to the left side of the array. During ϕ_2, the actual compare and exchange operations take place. The storage circuits for the key and pointer information for two consecutive cells and their associated compare/exchange logic are built into one single cell called a COL cell. The RESST chip uses 16 such cells and the overall structure is shown in Fig. 7.43. The COL cell contains five cell types: CE, TOPCE, RP, CBUF, and PH12SIG, and has the structure depicted in Fig. 7.44. The symbols key $[i]\langle j \rangle$ and recPtr$[i]\langle j \rangle$, where $i = 0, 1, \ldots, 31$ and $j = 0, 1, \ldots, 7$, denote the jth bit of the ith word of the key and record storage, respectively.

The most important cell is CE, which contains the storage and compare/exchange logic for the pair of key bits and is shown in Fig. 7.45. The control signals POP1, PUSH1, and HOLD1 are gated ϕ_1, and EXCH2L and EXCH2 are gated ϕ_2. The top part of the circuit shows the register cells: PUSH1 reads in data BDIN to the first cell and delivers the content of the first to the second cell, at storage nodes p and q, respectively. POP1 delivers the information at node r to BDOUT and s to point p of the first cell. If no exchange is necessary, the EXCH2L signal comes on

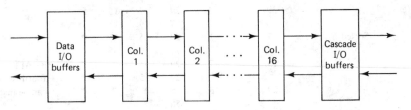

Figure 7.43   RESST chip.

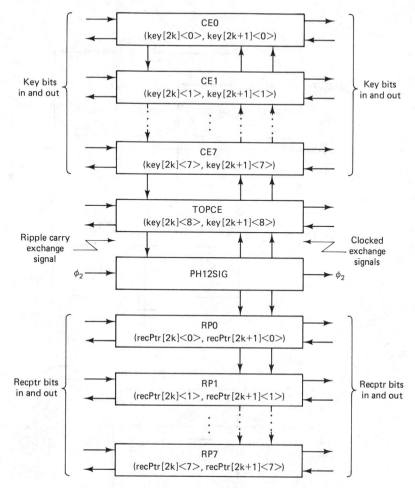

Figure 7.44 COL cell structure.

during ϕ_2 and establishes the new value read into the cells at points r and s, respectively. If exchange is necessary, EXCH2 signal comes on during ϕ_2, transferring p to s and q to r. If no data are read during ϕ_1, HOLD1 and EXCH2L signals refresh the information followed in the storage cells. Note that POP and PUSH are mutually exclusive, and so are EXCH2 and EXCH2L. The bottom part of the circuit generates the global signal EXCHIN/EXCHOUT. It consists of a comparator circuit (equivalence gate) with output M and logic to implement:

$$\text{EXCHOUT} = \text{EXCHIN} + \overline{M} \cdot \text{rightL}$$

The EXCHIN signal is initialized to 0 at the topmost cell. If the match signal $M = 1$, it propagates EXCHIN to EXCHOUT. If there is a mismatch, exchange is necessary if the right bit of the storage cell is 0, that is, rightL is 1. The signal

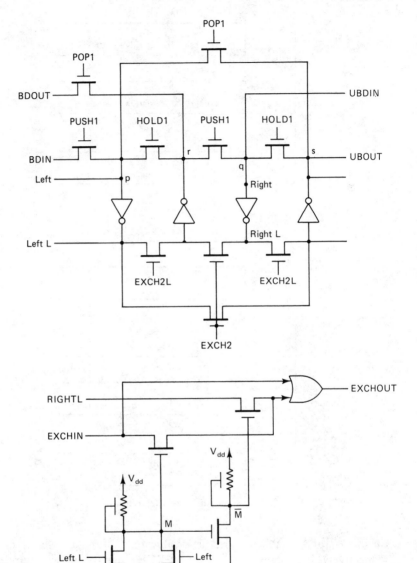

Figure 7.45 Compare/exchange cell structure.

EXCHOUT propagates from least to most significant bit position (top to bottom). If the EXCHOUT signal has value 1 from the most significant digit, the cell PH12SIG will then generate the global EXCH2 and EXCH2L signals, which are distributed over to both CE and RP cells via superbuffers.

The communication between adjacent cells takes place connecting UBDIN of the ith cell to BDOUT of the $(i + 1)$th cell and connecting UBDOUT of the ith cell to BDIN of the $(i + 1)$th cell.

We will describe very briefly the functions of other cells. The RP cell is identical to the CE cell except that it does not need the comparator and exchange signal logic. The TOPCE cell is identical to the CE cell with a clear signal which causes the most significant bit of all keys to be set to 1. The CBUF is simply a superbuffer cell to boost the EXCHOUT signal after every three or four stages. The PH12SIG is a cell to generate superbuffered gated-ϕ_2 signals EXCH2 and EXCH2L based on the value of the EXCHOUT signal coming from the TOPCE cell. There is a similar circuit used to generate the gated-ϕ_1 POP, PUSH, and HOLD signals. A few other control signals and a set of I/0 pads from the standard cell library complete the list of circuits used in the RESST chip. The total size of the chip is $3358 \times 4468 \ \mu m^2$, it has a worst-case power dissipation of about 0.2 W, and it can operate at about 1.2 μs of cycle time.

7.6 Arithmetic Processors

We conclude this chapter with a discussion of the design of an arithmetic-logic unit (ALU). There are varieties of ALU design with different capabilities, speed, and architecture. Our discussion will be concerned with a specific bus architecture to illustrate the use of precharge circuits in the design of carry chain and bus transfer logic.

The general structure of ALU can be depicted as shown in Fig. 7.46. The ALU performs arithmetic operations such as add or subtract; logical operations such as AND, OR, or EXOR; and bit manipulation operations such as shift or rotate on two operands coming from two input buses called the ABUS and BBUS, and delivers the output to a result bus CBUS. The operands could be derived from a variety of sources, such as the memory, the register bank of the processor, or the data path, via a set of multiplexers set by the selector control input signals. Several multiplexer circuits in both nMOS and CMOS have been presented in Chapter 3. By putting some additional logic at the output of the multiplexer, both the ABUS or BBUS could be made to assume a constant 1 or 0 value, or the complement of the selected bus could actually be applied to ALU. The operation of the ALU proceeds in two phases. First, the operands are latched into the ALU register or input latches. Then the actual operation and carry propagation take place. If we use two-phase clocking, these two phases can be done during ϕ_1 and ϕ_2. As we will see soon, sometimes more than one cycle may be necessary.

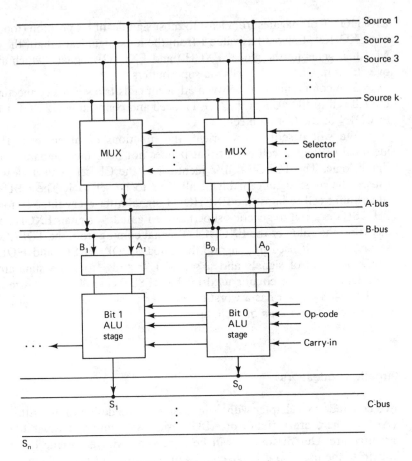

Figure 7.46 ALU schematic.

Since the carry propagation chain forms the critical delay path in the ALU, a precharge technique is used to design the carry logic, such as the circuit used in the machine OM2 (Mead and Conway, 1980). A similar circuit is shown in Fig 7.47 for the ith-stage adder. The input data are latched during ϕ_1, while the output C_{i+1} is precharged to 1. A negative logic is used to represent the existence of a carry signal by 0. During ϕ_2, the two functions $P = A_i \oplus B_i$ and $G = AB$, called the *propagate* and the *generate* functions, control the carry propagation. (The circuits to qualify these signals with ϕ_2 are not shown in Fig. 7.47.) If $G = 1$ and $P = 0$ (meaning that $A = B = 1$), C_{i+1} must be a 0, which is produced by pulling down the precharge line. If $G = 0$ and $P = 1$ (meaning that $A \neq B$), C_{i+1} equals C_i, and if both G and P are 0 (meaning that $A = B = 0$), the precharge holds C_{i+1} to a 1 value. The $G = P = 1$ condition never occurs.

The circuit above only shows the carry bit logic. The sum bit S_i can easily be generated by using an additional EXOR gate and an inverter since $S_i = P \oplus C_i$. The Stanford nMOS cell library (Newkirk and Mathews, 1983) contains a number of

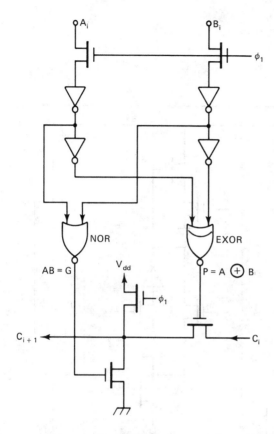

Figure 7.47 Precharged carry chain.

cells based on the original design by Johannsen of Caltech, which was used to build a bit-slice arithmetic logic unit.

The second major source of delay in the ALU comes during the transfer of information from the data bus to ALU and back. Eustace (1984) proposed a three-bus precharge architecture that attempts to alleviate this problem. The scheme is shown in Fig. 7.48. All three buses are precharged high during ϕ_2. Negative logic is used, which means that ABUS, BBUS, and CBUS represent $\bar{A}$, $\bar{B}$, and $\bar{C}$, respectively. Information from the recirculating storage cells R_A and R_B can be read into either ABUS or BBUS during ϕ_1 and applied to ALU at the same time. In a similar fashion, information from any external source selected by the multiplexer can be read into the bus during ϕ_1. At any time, only one such source can sink the precharged line. The ALU operation takes place during ϕ_2 and the result appears in the CBUS during ϕ_1 and is passed on to R_A and R_B during ϕ_1. Nothing happens during the next ϕ_2; that is, the ALU performs no operation. The addition operation thus takes two complete cycles.

The circuit diagram of a generic cell that can be used as either an R_A or an R_B cell is shown in Fig. 7.49. Note now that the control signals Write A and Write B have to be stable during ϕ_1, to avoid any possible charge-sharing problem.

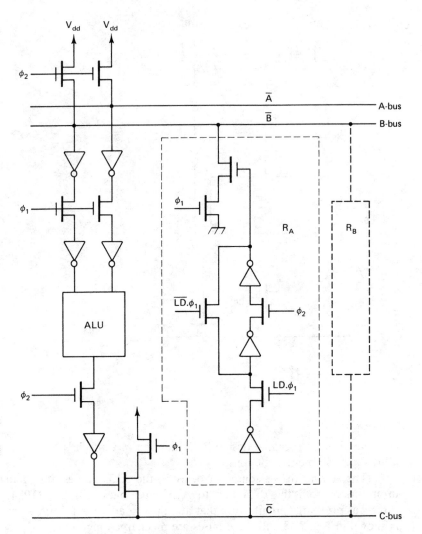

Figure 7.48 Three-bus precharge architecture.

Note that the circuits described above do not allow CBUS to be directly transferred to bus A or B. Such a facility is sometimes useful to speed up arithmetic operations in which intermediate results are used as operands. For example, if the operation is to compute $3 \times R_B$ and put the result in R_A, this can be done by the sequence: $R_A \leftarrow R_B + R_B$; $R_A \leftarrow R_A + R_B$. For this, we need the result of the first addition to be available in ABUS directly. The circuit to do this is shown in Fig. 7.50, which is self-explanatory. Note that the idle ϕ_2 phase of the addition cycle is now effectively utilized and that $3 \times R_B$ takes only two additional phases or one clock cycle to compute.

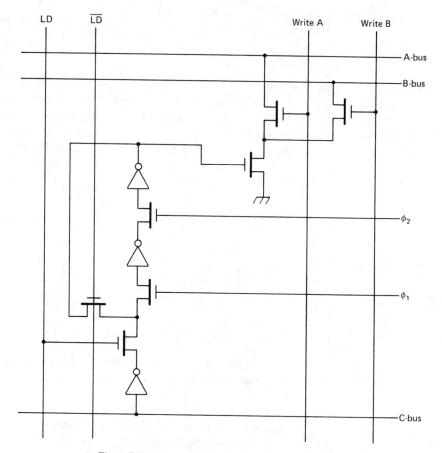

Figure 7.49 Generic cell for the three-bus architecture.

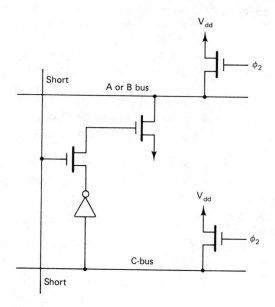

Figure 7.50 Logic for direct transfer of C-bus to A- or B-bus.

259

7.7 Summary

This chapter has presented several examples of designs using a top-down hierarchical approach. Several applications in both nonnumeric and numeric processing have been included. The reader should supplement his or her knowledge of the material in this chapter by additional reading from the literature describing innovative processor architectures such as RISC (Patterson and Sequin, 1981) and MIPS (Hennessy et al., 1981). A good source of information is the paper by Hennessy (1984). The paper by Seitz (1984) is also highly recommended for study.

REFERENCES

Armstrong, P., and M. Rem, "A Serial Sorting Machine," *Comput. Electr. Eng.*, Vol. 9, No. 1, 1982.

Boyer, R. S., and J. S. Moore, "A Fast String Searching Algorithm," *Comm. ACM*, Vol. 20, No. 10, 1977.

Carey, M. J., P. M. Hansen, and C. D. Thompson "RESST: A VLSI Implementation of a Record-Sorting Stack," *Final Project Report CS-248 (MOS-LSI Design)*, University of California at Berkeley, Berkeley, Calif., 1982.

Chung, K., F. Luccio, and C. Wong, "On the Complexity of Sorting in Magnetic Bubble Memory Systems," *IEEE Trans. Comput.*, Vol. C-29, No. 7, 1980.

Crystal, J., and R. Hendry, "Non-numeric Logic Unit," *Design Report CDA 5182*, Department of Computer Science, University of Central Florida, Orlando, Fla., Fall 1983.

Curry, T. W., and A. Mukhopadhyay, "Realisation of Efficient Non-numeric Operations Through VLSI," *Proc. IFIP TC 10/WG 10.5 International Conference on Very Large Scale Integration*, Trondheim, Norway, Aug. 16–19, 1983.

Curry, T. W., D. Kotick, E. Diaz, J. Klages, and A. Mukherjee, "PAM—A Pattern Matching Chip," a Design Report submitted to MOSIS, University of Central Florida, Orlando, Fla., 1983.

Donovan, K. B., "A Variable Register Delay Chip," *Design Report CDA 5182*, Department of Computer Science, University of Central Florida, Orlando, Fla., Fall 1982.

Eustace, R. A., private communication ("Precharged Bidirectional Barrel Shifter and Three Bus ALU Architecture"), 1984.

Guibas, L. J., and F. M. Liang, "Systolic Stacks, Queues and Counters," *Proc. Conference on Advanced Research in VLSI*, Massachusetts Institute of Technology, Cambridge, Mass., Jan. 25–27, 1982, pp. 155–164.

Hennessy, John L., "VLSI Processor Architecture" *IEEE Transactions on Computer*, Vol. C-33, No. 12, Dec. 1984, pp. 1221–1246.

Hennessy, J., N. Jouppi, F. Baskett, and J. Gill, "MIPS: A VLSI Processor Architecture," *Proc. CMU Conf. VLSI Systems and Computations*, pp. 337–346. Rockville, Md.: Computer Science Press, Oct. 1981.

Knuth, D. E., *The Art of Computer Programming*, Vol. 3: *Searching and Sorting*. Reading, Mass.: Addison-Wesley, 1973.

Kung, H. T., "Let's Design Algorithms for VLSI Systems," *Proc. Conference on Very Large Scale Integration: Architecture, Design, Fabrication*, California Institute of Technology, Pasadena, Calif., Jan. 1979.

Kung, H. T., *Advances in Computers*, Vol. 19. New York: Academic Press, 1980. (Also, "The Structure of Parallel Algorithms," *CMU Tech. Rept. CMU-CS-79-143*, Department of Computer Science, Carnegie-Mellon University, Pittsburgh, Pa., Aug. 1979.)

Leiserson, C., "Systolic Priority Queues," *CMU Tech. Rept. CMU-CS-79-115*, Department of Computer Science, Carnegie-Mellon University, Pittsburgh, Pa., 1979.

Mead, C., and L. Conway (Eds.), *Introduction to VLSI Systems*. Reading, Mass.: Addison-Wesley, 1980.

Miranker, G., L. Tang, and C. Wong, "A 'Zero-Time' VLSI Sorter," *IBM Research Rept.*, IBM Thomas J. Watson Research Center, Yorktown Heights, N.Y., 1982.

Mukhopadhyay, A. (Ed.), *Recent Developments in Switching Theory*. New York: Academic Press, 1971.

Mukhopadhyay, A., "Hardware Algorithms for Nonnumeric Computation," *IEEE Transactions on Computers*, Vol. C-28, No. 6, June 1979, pp. 384–394.

Mukhopadhyay, A., "Hardware Algorithms for String Processing," *Proc. IEEE International Conference on Circuits and Computers*, New York, Oct. 1–3, 1980, p. 508.

Mukhopadhyay, A., "Weavesort—A New Sorting Algorithm for VLSI," *Tech. Rept. TR-53-81*, University of Central Florida, Orlando, Fla., 1981.

Mukhopadhyay, A., and T. Ichikawa, "An *n*-Step Parallel Sorting Machine," *Tech. Rept. 72-03*, Department of Computer Science, University of Iowa, Iowa City, Iowa, 1972. (See also "A Survey on Macrocellular Research" by A. Mukhopadhyay, *Rev. R.A.I.R.O.*, Dec. 1972, J-4, pp. 3–42.)

Newkirk, J., and R. Mathews, *The VLSI Designers Library*. Reading Mass.: Addison-Wesley, 1983.

Patterson, D. A., and C. H. Sequin, "RISC-I: A Reduced Instruction Set VLSI Computer," *Proc. 8th Ann. Symp. Computer Architecture*, Minneapolis, May 1981, pp. 443–457.

Richie, S., and P. Gatt, "OPIN: A VLSI Implementation of a Programmable Interconnection Network," *Design Report CDA 5182*, Department of Computer Science, University of Central Florida, Orlando, Fla., Fall 1982.

Seitz, Charles L., "Concurrent VLSI Architectures," *IEEE Transactions on Computers*, Vol. C-33, No. 12, Dec. 1984, pp. 1247–1265.

8

Memory Systems

8.1 Introduction

In this chapter we are concerned with memory systems. We discuss typical nMOS and CMOS memory circuits and discuss general principles to organize these circuits to build systems. We discuss trade-offs with respect to power, speed, and silicon areas. The aim of this chapter is to provide a broad overview of the principles and technology of memory systems. We will have occasion here to refer to some of the circuits presented in earlier chapters. Our discussion will be brief and many of the details will be omitted for the zealous reader to discover in the huge body of literature on semiconductor memory devices.

Basically, all semiconductor storage devices are either dynamic or static. In dynamic device the storage is due to small capacitance, which holds information in the form of the presence or absence of electric charge. In a static device, stored information is due to the state of conduction or no conduction of a transistor. Static devices need more power, but can hold the information indefinitely. The dynamic devices use low power, but need periodic refreshment of the stored charge, which may be lost due to leakage.

8.2 Basic Memory Cell

8.2.1 Four-Transistor Dynamic Memory Cell

We first describe the operation of a basic dynamic storage cell for random access memories (RAMs) consisting of four transistors, as shown in Fig. 8.1(a). Many of

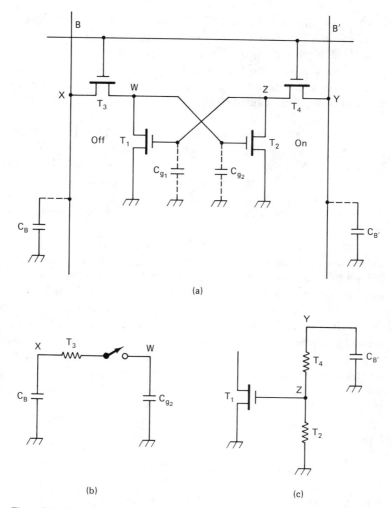

(a)

(b)   (c)

Figure 8.1 (a) Four-transistor dynamic memory cell; (b) equivalent electrical circuit on the bit line B; (c) equivalent electrical circuit on the bit line B'.

the concepts underlying the operation of static storage cells using four or six transistors are based on the operation of this basic cell.

The two "bit" lines B and B' are precharged high before any operation (read, write, or refresh) can take place. The "row select" transistors T_3 and T_4 are turned off during precharge. The bit lines are connected to a "sense" amplifier (described later), which is also turned off during the precharge phase.

Let us assume that T_1 is off and T_2 is on when the row select line is turned on. The charge on capacitor C_{g2} has been holding T_2 on and its value must be high enough to hold the charge between refresh cycles. This charge is restored during the read or refresh operation. But C_{g2} cannot be high compared to the precharged line capacitance C_B because it will create a charge-sharing problem, since C_{g2} and C_B are connected in parallel via T_3. See the equivalent circuit in Fig. 8.1(b). The voltage at node W may drop below threshold to turn T_2 off, destroying the information of the cell, and/or the voltage at node X may drop too low to activate the sense amplifier. Also, to ensure that enough charge is transferred to C_{g2} before the end of the cycle, we need a large W/L ratio for T_3.

Now, consider the B' side of the cell. The charge in this line is discharged via the transistors T_4 and T_2 [see the equivalent circuit shown in Fig. 8.1(c)], driving the bit line B' to low, which is sensed by the sense amplifier. The sum of resistances of T_4 and T_2 must be low enough to ensure that this discharge takes place quickly, but the resistance of T_4 must be large compared to that of T_2 to make sure that the point Z is at a low voltage; otherwise, transistor T_1 will turn on, shorting the stored charge in C_{g2}.

By symmetry, if the cell is storing a complementary bit (T_1 on, T_2 off), the constraints on the sizes of the capacitance C_{g1} and C_B' are similar to the one between C_{g2} and C_B. Similarly, the absolute and relative sizes of transistors T_1, T_2, and T_4 must satisfy criteria similar to those applicable to T_2, T_1, and T_3.

To perform a write operation, let us assume that we wish to write a 0. The I/O circuit driving the bit lines must discharge the B line to low and supply a current needed to sustain a high on bit line B'. Transistors T_3 and T_4 are enabled by the row select line. Transistors T_4 and T_2 will be on, sinking current from B' and holding T_1 off (the ratio of sizes of T_2 and T_4 guarantees that T_1 will remain off). On the other side, the capacitance C_{g2} will discharge through T_3 to the bit line B, which is pulled low. When the voltage on C_{g2} drops below T_2's threshold voltage, T_2 will shut off, allowing C_{g1} to charge up and turn on T_1. At this time, a 0 has been written into the flip-flop.

8.2.2 Six-Transistor Static Memory Cell

The basic six-transistor static CMOS memory cell is shown in Fig. 8.2. The operation of the cell is essentially the same as that of a four-transistor cell except that the cell does not need a refresh operation. The loss of charge at the gate is replenished by a current flowing through the p-transistors. For example, if the flip-flop state is such that the gate voltage at T_5 must be 0, this requires that a current through T_5 maintain the charge at C_{g2}. A current through T_2 at the same time will maintain the gate of T_1 at ground potential, discharging C_{g1}. A similar but complementary situation will exist if the flip-flop state is complemented. The p-transistor sizes could be very small since their only function will be to latch the flip-flop and not to drive the bit lines.

The p-transistors could be replaced by depletion-mode n-channel load transistors yielding a six-transistor nMOS static cell or simply by two resistors, which need only supply enough current to the gate capacitors to restore the leaking charge (McKenny, 1977).

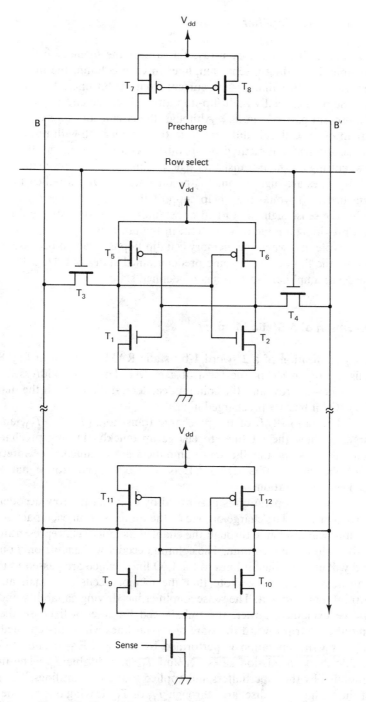

Figure 8.2 Six-transistor RAM memory cell and a sense amplifier.

8.2.3 Sense Amplifier

To build a large memory array, the size of the memory cell must be as small as possible. By adding a sense amplifier in each column, the memory cell size can be reduced since it will only sink current from the bit lines. The driving of the bit lines is done by a specially built flip-flop attached at the end of the bit line, as shown at the bottom part of Fig. 8.2. Although the circuit looks very similar to the flip-flop circuit of the cell, it is different from the memory cell with respect to the sizes of the transistors and the method of operation. During precharge, the sense line is set to low, cutting off the n-transistors. The p-transistors are automatically turned off since the bit lines are high. If one of the bit lines, say B', is pulled low, the p-transistor attached to it would then help support the charge on the high bit line B. At this point the sense signal is turned on, stabilizing the state of the flip-flop even further and providing a large drive current in the bit lines.

Note that when the memory cell flip-flop is disabled because the row select line is low, the flip-flop is in some predetermined state. But when the sense line is low, the sense amplifier flip-flop has no defined state.

8.3 Organization of a Static Memory

The organization of a 2^k-word 1-bit static RAM is shown in Fig. 8.3. The memory cells are organized in the form of a matrix. The k-bit address is decoded by two decoders—the row and the column decoders. Initially, both the matrix bit lines and the I/O bit bus are precharged high.

The ratio W/L of the precharge transistors (T_7, T_8, T_{13}, and T_{14}) are made large to allow the bit lines to charge up quickly. During precharge, the row and column decoders and the sense amplifiers are disabled. The internal timing starts after the chip enable signal CE is active. The control signal $R/\overline{W}$ initiates a read/write operation.

The read operation is performed by enabling the row decoder, which selects a particular row. The charge on one of the bit lines from each pair of bit lines on each column will discharge through the enabled memory cell, representing the state of the active cells on that column. The column decoder will enable only one of the columns and will connect the bit lines to the I/O lines, which are passed to the output via the transmission gate T_{17}. Note that the memory cells are small and can only sink current, not source it. The sense amplifier has driving capability and is now enabled. As we explained earlier, the unbalanced bit lines will cause the balanced sense amplifier to trip toward the state of the bit lines when it is enabled.

A write operation is performed by setting CE = 0 and $R/\overline{W} = 0$. This will enable the transmission gates T_{15} and T_{16} and disable T_{17}. The input data will be amplified by the superbuffers and applied to the I/O bit lines. The precharge on one of the bit lines will discharge through T_{15} or T_{16} leaving one bit line high and one bit line low. The column decoder will select a particular column connecting the bit lines

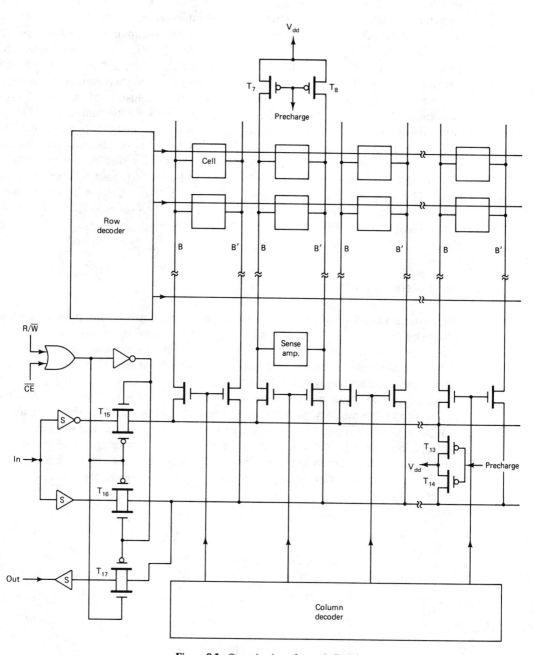

Figure 8.3 Organization of a static RAM.

to I/O lines, thereby discharging one of the bit lines through T_{15} or T_{16}. The row decoder will select a particular row and the information on the bit lines will be written on the cell at the intersection of the row and column, as explained earlier.

The internal timing starts with the chip enable (CE) signal going low. The precharge is turned off immediately after this and is not turned on again until the entire operation is done. The column and row decoders are then activated, followed by the activation of the sense amplifier. At the conclusion of the read/write operation, the sense amplifier is deactivated along with CE becoming low. This is followed by disabling the decoder and then the precharge line comes on again.

The main advantages of the static RAM are that it is nondestructive, does not need refreshing, and is simple to operate. But it requires a rather large area, large dc power, and large support areas for decoding and control (about 50% of the total area). The design of a successful commercial static RAM involves not only memory cells but also innovative circuits for decoding and I/O (Kramer, 1983; Ochii et al., 1982; Meusburger, 1983).

8.4 Dynamic Memory Cells

We have presented several dynamic memory cells in Chapter 3 based on a six-transistor shift register cell. The main advantages of this kind of memory are that it is very easy to design and needs very little support circuitry, resulting in very high utilization of memory area. There are several disadvantages: The access is word sequential; the area/bit is rather large: the basic cell takes up about 5000 μm^2 with $\lambda = 2.5$ μm; and the dc power consumption per cell is about 150 μW. Thus the limit of the size of the memory on a 8 mm × 8 mm chip is about 6K, consuming about 1 W in a 40-pin package. The power can be reduced by increasing the size of the pull-ups, but this will take more area and reduce the maximum operating frequency. An alternative approach is to use two-phase ratioless logic, as shown in Fig. 8.4. The operation of the cell can be explained as follows. During ϕ_1 the data at A are

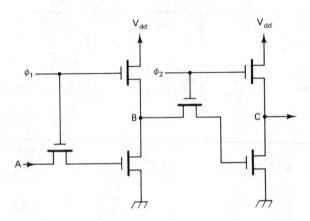

Figure 8.4 Two-phase ratioless register cell.

latched into the gate of the inverter, which sets $B = \overline{A}$. At the termination of ϕ_1, B retains its value in the form of a charge at the output capacitance of the inverter. During ϕ_2 the process is repeated with respect to the second inverter and we get $C = \overline{B} = A$. Note during ϕ_2, there is charge sharing between the node B and the gate input capacitance of the second inverter, and similar charge sharing occurs during ϕ_1 between node A and the input gate capacitance of the first inverter. This makes the circuit a little unreliable, and special care is needed to adjust the capacitance values to ensure correct operation.

In Chapter 3 we discussed a basic dynamic CMOS memory cell (Fig. 3.87) using eight transistors. The circuit leads to a very compact layout (see Chapter 5) because of its symmetry, and it consumes no static power. The register arrays discussed in Section 3.8 also provide examples of dynamic and semistatic memory structures.

To increase the layout densities of dynamic memories, cells that require only a few transistors have been designed. We will describe a few important types. The four-transistor dynamic cell has already been presented in Section 8.2.

To increase the speed of dynamic CMOS RAM, a mixed-mode memory architecture has recently been developed (Altnether, 1983). The decoding circuits are static CMOS circuits with low power consumption, but the memory array consists of dynamic cells. This type of architecture would not be possible with nMOS because the power consumption of the decoding circuit would be too great. Furthermore, to reduce package size and pin count, the row address and the column address are multiplexed, leading to what is called a *page-mode* operation. The row address is fetched first, causing a "page"—one row of memory bits—to be latched into the sense amplifiers. The column address can then be latched to access the desired bit. Since the entire page is residing in the sense amplifiers, data within the page can be accessed simply by changing the column address. This leads to faster access and a smaller effective cycle time. The column address must be stable throughout the cycle time. In another variation of the operation of this type of memory, called *ripple-mode* operation, it is possible to set up the column address for the next cycle in a pipelined fashion during the current cycle, leading to even faster operation.

8.4.1 Three-Transistor Dynamic Memory Cell

The basic principle of operation of the cell can be explained with reference to Fig. 8.5, which shows one bit slice of memory with associated control circuits. The basic bit cell is enclosed in the dashed box. The bit line is a precharged bus which is charged high during ϕ_1. Reading and writing are done during ϕ_2. If Data = 1 and write enable (WE) and the write line (W) are high during ϕ_2, the bus will be pulled low, the transistor T_1 will be on, and T_3 will be off; for Data = 0, the precharged bus will not be pulled to zero and a 1 will be written at the gate capacitance of T_3. The transistor T_3 will be on, but since the read signal $R.\phi_2$ is zero, T_2 will be off, keeping the precharged state undisturbed. To read the stored information, the read signal R is set to 1 (with $W = 0$). If the stored bit is 1, the bus will be pulled low,

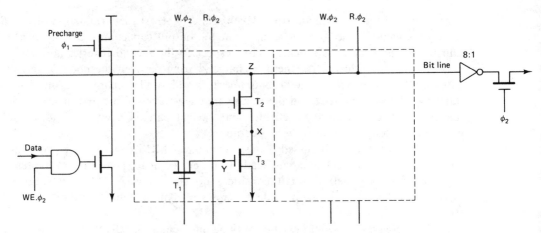

Figure 8.5   Three-transistor dynamic memory cell.

but the output inverter provides a 1 output. If the stored bit is 0, the bus remains charged and a 0 output is obtained.

The pull-down circuit for the bit line is not one of the safe pull-down circuits for precharged logic that we discussed earlier. There is a charge-sharing situation as follows. Assume that the point X is set to zero during the previous read operation; it is then followed by a 0 write operation. If a read operation is now performed, the points X and Z will share charge and will produce an imperfect 1 at the input of the inverter. This problem is not that critical and the circuit works since the capacitance of the bit line is much larger than the capacitance of node X, and therefore the value of node Z dominates.

Since the read/write line is a common bus, the cell refreshes itself in a loop in a complementary fashion (recall that the output of the cell is the complement of the stored information). This cell is therefore sometimes called an *inverting cell*. The refresh period is typically 2 ms at room temperature. The refresh operation consists of reading the whole memory, one word at a time, and writing it back into the memory.

If the memory has to have ω words of n bits each, each bit slice must have ω basic cells, and the bit slices have to be organized as sketched in Fig. 8.6. A good strategy to lay out this cell will be to use polysilicon lines for control and metal lines for the bit and ground lines in the orthogonal directions (see Newkirk and Mathews, 1983, where a number of other cells useful for a three-transistor memory system are described).

An alternative design will be to use ratioed logic rather than precharged logic. If we use only minimum-size transistors in the memory cells, the pull-up transistor must have a ratio of 16 : 1. This design will consume much more power and will be slower than that of the precharged design. In the precharged design, current flows through only one of the pull-down paths during ϕ_2. The estimated speed comparisons are as follows. Let the total capacitive loading of the bit line for the slice with n

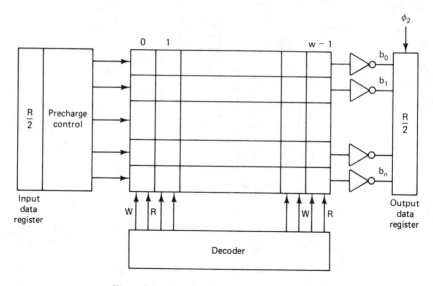

Figure 8.6 Organization of a dynamic memory.

cells be nC, where C is the capacitative load per cell. Let R and r denote the pull-up and pull-down resistances for ratioed design, where $r = R/4$. Then the charging delay is $RnC\delta$, where δ is defined as in Chapter 6. The clock period is twice the charging delay and hence the maximum operating frequency is $f = 1/2RnC\delta$ cycles per second. For precharged logic, the maximum frequency is determined by the pull-down delay, which is $4f$ cycles per second. This, of course, assumes that the precharge time is not larger than the pull-down time. Sometimes more than one precharge transistor can be attached to the bus line in parallel to make sure that the precharge time is smaller than the discharge time. The maximum frequency may even be slightly higher, since during pull-down the transistors sink only the charge stored in nC, but for ratioed design they have to sink this charge as well as the charge flowing through the pull-up resistor. The output inverter has to have a ratio of $8:1$ since the "pull-up" is an enhancement-mode transistor which produces a weak 1 signal ($V_{dd} - V_{th}$). The net effect of this might be to reduce the effective speed of the entire system by half. A suggested cure for this problem is to use both enhancement-mode precharge transistors and a depletion-mode pull-up, which will boost the output to V_{dd}.

One interesting application of a three-transistor memory is in signal processing, where it can be used as a serial–parallel–serial memory to match the high-signal-bandwidth requirement. This principle is illustrated in Fig. 8.7. The memory is configured like an n-word shift register with a simple addressing scheme which drives the read/write lines in the sequence ($R_1, W_1, R_2, W_2, \ldots, R_n, W_n$) cyclically. The input register D_1 converts the serial data to parallel form for memory to read and the parallel data from memory are delivered to output register D_2 for serial output. The shift register action is obvious from the addressing scheme. This addressing scheme was proposed by R. F. Lyon (see Newkirk and Mathews, 1983).

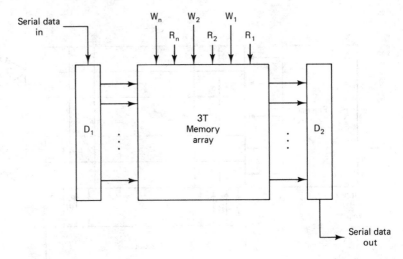

Figure 8.7 Application of three-transistor memory as a serial–parallel–serial memory.

The memory cell discussed above is sometimes classified as a 3-2-1 configuration (three transistors, two addresses, and one bit line). A 3-2-2 configuration is also possible, as shown in Fig. 8.8. To perform a write operation, the data to be written are put in the WData line and the WSelect line is driven high, which charges the capacitance C via transistor T_1 to the appropriate data value. The reading operation is performed in a complementary fashion: the $\overline{\text{Rdata}}$ line is precharged high; if the stored information in C is 1, and the RSelect line is driven high, both transistors T_2 and T_3 will conduct, bringing the precharged line to 0. If the stored information is 0, the precharged line will remain high. Note that the reading is nondestructive, but the

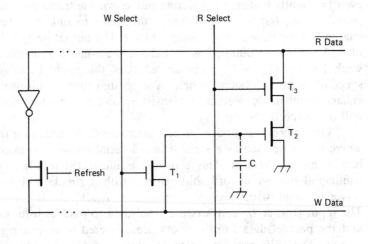

Figure 8.8 A 3-2-2 dynamic memory cell.

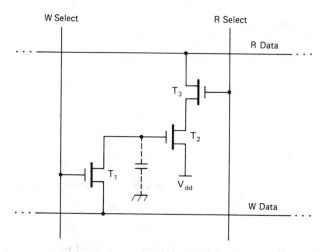

Figure 8.9 A 3-2-2 dynamic cell with true output.

cell needs to be refreshed to restore the loss of charge due to leakage. This can be done by driving both RSelect and WSelect lines high simultaneously on a given column and turning on all the refresh amplifiers for all the rows at the same time. Thus the overhead support circuitry for this type of memory is moderate to high, but the memory array is about four times as dense as static RAM.

A 3-2-2 configuration of the cell which provides true output, rather than complemented output, is shown in Fig. 8.9. The principle of operation is very similar to that of Fig. 8.8 except that one side of transistor T_2 is connected to V_{dd} rather than Gnd, to provide a true output.

Notice that when data are read, selection of the WSelect line turns on all the T_1 transistors in one column at the same time. Therefore, all the WData lines must have correct data for input at the same time. Similarly, during the refresh operation, the entire column is refreshed at the same time.

It is appropriate here to comment on the basic organization of a large memory using dynamic memory cells. Usually, one bit of information is derived from one memory chip. Thus a 4K memory chip has $2^{12} = 4096$ words each of one bit. If the data are to be handled in, say, 16-bit words, 16 such chips are employed. Each chip comes with in-chip decoding circuitry, which is usually word-organized. Thus the 12-bit decoding is done by two separate decoders each handling 6 bits, the memory being organized in a matrix of 64 words of 64 bits each. The periodic refresh operation in Figs. 8.8 and 8.9 thus refreshes the cells of a selected word in one column of the matrix, which has to be repeated for each column at intervals of a few milliseconds.

8.4.2 One-Transistor Dynamic Memory Cell

The memory cell uses only one transistor and as a result provides quite high density. This cell is shown in Fig. 8.10. The information is stored in capacitance C_s, which is built using polysilicon. The "transmission" gate T acts to insulate the stored charge

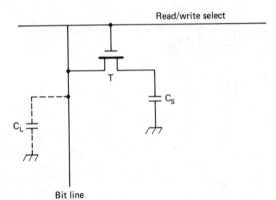

Figure 8.10 A one-transistor memory cell.

Bit line

except during read/write operations. To write, the bit line is charged to the desired level (1 or 0), and the read/write select line is driven high, which charges up C_s via transistor T. To read, the read/write select line is again selected, the bit line is tristated, and the charge in C_s is put in the bit line and is detected by a specially designed sense amplifier. The reading is destructive; after reading, the information must be immediately restored. Note that the same bit line is used for both read and write operations, but the direction of currents is reversed.

Although the cell provides high density and high speed, it has poor noise immunity. The capacitance C_s is usually small compared to the bit-line capacitance C_4, which gives rise to a bad charge-sharing problem. As a result, the output voltage does not have a large enough voltage swing. A special kind of balanced sense amplifier is necessary to read the information out (see Muroga, 1983).

8.5 Read-Only Memory

A read-only memory (ROM) is essentially an AND plane of a PLA. It produces as output n "words" or product terms each of m "variables" by an array of $n \times m$ nMOS switch matrices, as shown in Fig. 8.11. The m bit lines are precharged high during ϕ_1, and a word line is selected. During ϕ_2, the bit lines that have a ground path via a transistor controlled by the selected word line will be discharged to 0 V, signifying a 0 output. If a transistor is not present, a 1 output is produced. Read-only memories have been widely used to store microprograms, code conversion tables, tables of mathematical functions, and general combinational functions (Muroga, 1983).

The structure of a CMOS ROM is shown in Fig. 8.12. It uses a precharge principle as in nMOS, but a single-phase clock is sufficient (if $\bar{\phi}_1$, is replaced by ϕ_1, and ϕ_1 by ϕ_2, the circuit will still work). The output bit lines are precharged low or to 0 during $\bar{\phi}_1$. During ϕ_1, the selected word line goes low (all others remain 1, implying that a precharge decoder to select word lines can be used). If a p-channel

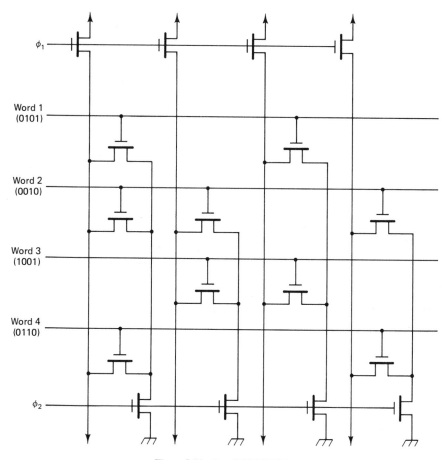

Figure 8.11 An nMOS ROM.

transistor is placed at the intersection of the bit line and the selected word line, a 1 will be transmitted to the bit line; otherwise, a 0 will be transmitted.

The use of precharge techniques in both nMOS and CMOS ROM enhances the speed of operation, but practical limitation will come from the word size and the number of 0's (for nMOS) and 1's (for CMOS) in the word since this determines the maximum resistive as well as capacitive loading of the bit lines. Typically, sense amplifiers are used to boost the signal levels at the bit lines.

Since the pattern of transistors in the switch matrix determines the stored information of the ROM, this pattern can be set during manufacture. Such a ROM is called a mask-programmable ROM. There are also field-programmable ROMs (PROMs), whose switch matrix can be set by the user. There are also erasable and reprogrammable ROMs (EPROMs), whose information can be erased by ultraviolet rays and reprogrammed with new information. These ROMs use a special dual-dielectric gate insulator or FAMOS (Glaser and Subak-Sharpe, 1979).

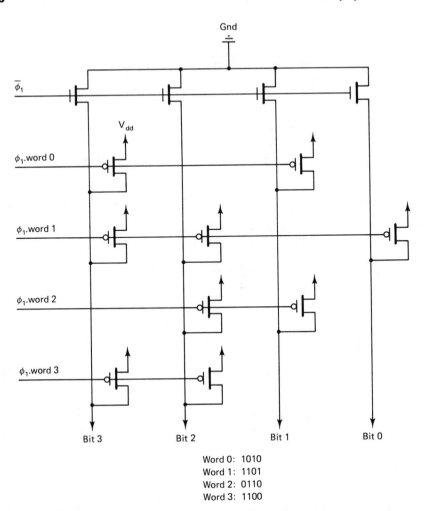

Figure 8.12 A CMOS ROM.

8.6 Summary

This chapter has presented the basic principles of operation of static and dynamic memories in both *n*MOS and CMOS. In general, the CMOS memory has a lower power consumption but takes a larger area and has lower speed than the *n*MOS memory. But with new and innovative circuit design and using mixed-mode memory architecture, the CMOS memories are becoming comparable to *n*MOS in speed. The CMOS memory also has the advantage that it is less immune to "soft error"—a phenomenon caused by bombardment of alpha particles, which destroy the stored information.

The goal of this chapter was to present only the basic classes of memory devices. The memory technology is now highly developed, and the reader is referred to the vast literature on this subject.

REFERENCES

Altnether, J., "A Look at CMOS Dynamic Memory," *BYTE*, Sept. 1983.

Glaser, A. B., and G. E. Subak-Sharpe, *Integrated Circuit Engineering*. Reading, Mass.: Addison-Wesley, 1979.

Kramer, G. A., "A 4K × 1 Static Random Access Memory," *VLSI Memo No. 83-1-38*, Massachusetts Institute of Technology, Cambridge, Mass., Apr. 1983.

McKenny, V. G., "A 5 V-Only 4-K Static RAM," *IEEE International Solid State Circuits Conference*, Session 1: MOS Memories, Feb. 1977.

Meusburger, G., "1.5 V 1 K-CMOS-RAM with Only 8 Pins," *IEEE J. Solid-State Circuits*, Vol. SC-16, No. 3, 1983.

Muroga, S., *VLSI Systems Design*. New York: Wiley, 1983.

Newkirk, J., and R. Mathews, *The VLSI Designer's Library*. Reading, Mass.: Addison-Wesley, 1983.

Ochii, K., K. Hashimoto, H. Yasuda, M. Masuda, T. Kondo, H. Nozawa, and S. Kohyama, "An Ultralow Power 8K × 8-Bit Full CMOS RAM with a Six-Transistor Cell," *IEEE J. Solid-State Circuits*, Vol. SC-17, No. 5, 1982.

9

VLSI Design Tools

9.1 Introduction

The specification and the design of an integrated-circuit chip are complex processes. Verification and validation of a design are an even more difficult challenge. Moore's basic question of what can be done with VLSI is related to a fundamental design issue: how to manage the complexity of VLSI chips. The key to the success of VLSI technology lies in the development of powerful design tools and software systems that help the designer produce an integrated-circuit chip. The area of VLSI design tools is changing rapidly. Most of the design tools that are currently used may become obsolete within a short period. The emphasis in this chapter is therefore not on the description of the tools but on the descriptions of algorithms underlying the tools.

A large number and variety of design tools are presently being used and developed for integrated-circuit chips. In this chapter we discuss some of the basic tools and algorithms used for the development of an integrated design automation system. Excluding the tools dealing with systems or functional specifications, these tools can be broadly classified as design entry or logic capture systems, circuit layout tools (language-based or graphics-based), layout verification tools such as the circuit extractor and static electrical parameter analyzer, simulation tools for functional and timing verification, and routing tools for interconnection and placement. A schematic diagram of the organization and interrelationships of these tools is shown in Fig. 9.1. This diagram broadly describes the major components of a design automation system in a typical design house. The components are: system and functional specification and simulation, the logic capture, the layout, generation of an inter-

278

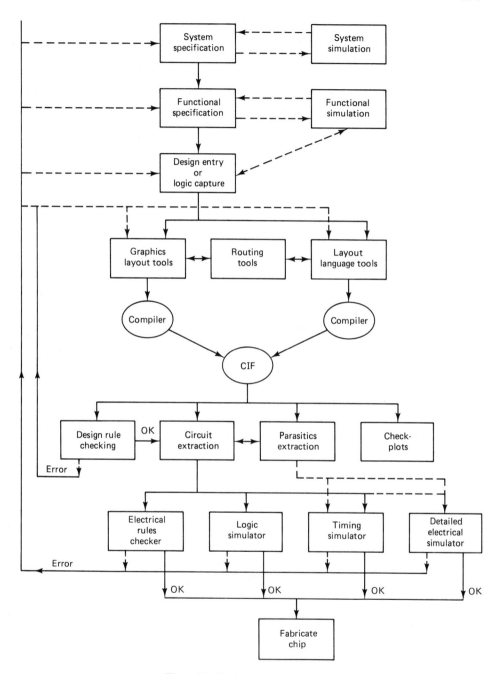

Figure 9.1 Design automation system.

change format, and several simulation tools at low levels of physical design. The design undergoes several levels of verification and change before being sent for fabrication (indicated by the dashed lines). One important component of this system —the test generation system—has been left off this chart and will not be discussed here. Most design automation systems currently in use in universities do not have logic capture or schematic logic synthesis tools. Simple routers are beginning to be available. Tools to extract parasitics and detailed electrical simulation belong to industrial design environments striving for competitive chip performance, although the SPICE simulator has been used in a number of university designs.

Most of the design tools developed at universities use a geometry-based interchange format called *Caltech Intermediate Form* (CIF) as the common denominator for representing circuits. In the next section we present a brief description of CIF.

9.2 Caltech Intermediate Form

The ultimate objective of the VLSI design tools is to produce a description of every geometric object corresponding to the underlying circuits that can be transferred on a masking plate for the processing step. The Caltech Intermediate Form (CIF version 2.0), originally conceived by Sutherland and Ayers at Caltech, is an interchange format to describe integrated-circuit features. This form of description of the circuit is now widely accepted as the standard interchange form between higher-level symbolic or graphic description of a chip and the lower-level descriptions needed for generating the PG files for photolithographic machines as well as for driving output devices such as plotter or video displays. CIF is not intended to be used by the designer for actual layout, but all higher-level representations of the layout are usually translated by computer programs to CIF. CIF is machine-independent in the sense that it is not tied to a specific PG equipment or fabrication technology and strives to provide a standard for communication between the design environments and the processing environments. It can easily be generated, transported, and processed and is now a de facto standard interchange format used by a number of universities and industrial laboratories. A detailed description of this language is presented in Mead and Conway (1980). We provide only a brief synopsis here.

The language uses an abstraction of the circuit as a collection of rectangular geometrical objects in different mask layers. The syntax for the layer specification is

$$LN*$$

where L declares layer, N denotes nMOS process, and * could be a single letter from the set (M, D, P, I, C, B, G), each denoting a layer, as explained in Chapter 4.

For the CMOS process, the layer specification is

$$LC*$$

where * denotes the mask layers of a typical CMOS process such as the one described in Chapter 4 for the bulk p-well process with layers (P, C, M, D, W, G). The language uses a right-handed coordinate system; distance is expressed in units of hundredths of a micron. The most common geometric primitive is a *box*, specified as

$$B\ l\ w\ x\ y\ a\ b$$

when l and w denote the length and width of the box, respectively, (x, y) is the coordinate of the center of the box, and (a, b) specifies that the X-axis of the box is rotated so that it has a slope a/b. In most implementations, this direction parameter is not used, so that the length and width are along the x- and y-axes, respectively. The box is in a layer specified by a layer statement preceding it. Other primitives are polygon (P) specified as a series of points (x, y values), wires (W) with specified width, and round flashes.

The language allows definition of a "cell" or "symbol" which can later be called for repeated use. The syntax for cell definition is

$$(\text{Definition start}) - - - DS \# A/B;$$

$$\langle \text{CIF Statements} \rangle$$

$$(\text{Definition finish}) - - - DF;$$

where # is an identifying number for the cell and A/B is a scaling factor to avoid the use of large numbers. The actual distance equals the distance specified in the CIF statements multiplied by (A × 100)/B microns. The cell can be called by the statement

$$C \# \{\text{transformation}\};$$

where the optional "transformations" could specify linear translation of the origin of the symbol, mirroring in X (multiply X coordinate by -1), mirroring in Y, and rotation.

Another useful feature is the "user extension" command, which starts with "94" in a line. This allows a symbolic name to be assigned to a node or wire used in the circuit extraction phase. The standard symbolic names V_{dd} and Gnd are automatically assigned to wires carrying supply and ground connections. A simple example of a CIF program is shown in Fig. 9.2. This CIF file corresponds to the CMOS inverter described in Chapter 5. This is a compiler-generated CIF which uses mostly the box construct.

In the following sections, we present some of the basic algorithms used in the development of VLSI design tools. We will start with verification tools, of which the design rule checker (DRC) is most heavily used at the initial stages of the design of simple subsystems.

```
DS 1 200 4;
9 pcell;
L CP;   B 24 8 52 68;
L CD;   B 16 16 52 84;
        B 16 16 52 52;
        B 8 16 52 68;
L CM;   B 16 16 52 52;
        B 16 16 52 84;
L CS;   B 32 64 52 68;
L CC;   B 8 8 52 84;
        B 8 8 52 52;
94 vdd 48 84;
DF;
DS 2 200 4;
9 ncell;
L CP;   B 24 8 −12 24;
L CD;   B 16 16 −12 40;
        B 16 32 −12 0;
        B 8 16 −12 24;
L CM;   B 16 16 −12 40;
        B 16 32 −12 0;
L CW;   B 40 88 −12 16;
L CS;   B 32 24 −12 −12;
L CC;   B 8 8 −12 40;
        B 8 24 −12 0;
DF;
DS 3 200 4;
9 cmosinv;
L CP;   B 16 8 48 76;
        B 8 40 52 100;
        B 8 40 52 52;
        B 24 8 76 76;
        B 16 4 72 82;
        B 16 4 72 70;
L CM;   B 40 16 68 20;
        B 40 16 68 132;
        B 16 16 72 76;
        B 12 8 70 88;
        B 12 8 70 64;
L CC;   B 8 8 72 76;
94 in 44 76;
94 gnd 64 16;
94 out 84 76;
C 2 R 1 0 T 80 12;
C 1 R 1 0 T 16 48;
DF;
C 3;
End
```

Figure 9.2 A CIF file.

9.3 Design Rule Checking

A design layout must conform to the design rules laid down by the particular technology. In Chapter 5 we presented the design rules in terms of the photolithographic resolution parameter λ. A *design rule checker* (DRC) is a piece of software or hardware that is used to verify that all the design rules of the technology have been met. The DRC takes as input the CIF description of the design and produces

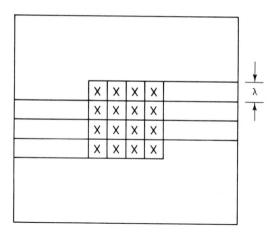

Figure 9.3 The four scan lines in Baker's algorithm.

as output an indication of the geometrical design errors that occur with regard to the given set of design rules.

In this section we describe several methods for design rule checking and compare their relative advantages and disadvantages. We also discuss several approaches to circuit extraction which must be used to analyze and consolidate error reporting by DRC.

9.3.1 Raster-Scan Methods

Baker – Terman algorithm.   Baker at MIT developed the raster-scan method (Baker, 1980; Baker and Terman, 1980). In this method the design layout is represented on a lambda grid and each pixel in this grid contains a bit for each layer, indicating whether or not the layer is present at that point. Baker's algorithm stems from the observation that Mead–Conway design rules are local checks in that they specify only minimum widths and spacings. In the set of nMOS design rules, the largest minimum width or spacing is 3λ. Therefore, there should be sufficient information in a 4×4 "window" to determine whether any design rules are violated in the window. This window scans the bit map such that each pixel appears in every position of the window. For this a buffer storage area equal to three scan lines and four pixels is sufficient, as shown in Fig. 9.3. To discover width and spacing violations of maximum λ, a 3×3 window is adequate, of which there are $2^9 = 512$ possible bit configurations. For the purpose of brevity, rather than analyzing each pattern separately for violations, the patterns are classified using a parameter A, called the *alternation* number, as shown in Fig. 9.4, which equals the number of transitions from 0 to 1 and 1 to 0 in the bit sequence in the perimeter. Thus if the perimeter contains 01011011, then $A = 6$. If the window contains a 0 in the center, it does not lead to an error since no wire passes through the region, although wires of width 1 may "poke" into the area, which is not a violation. Similar arguments lead to the result that patterns with a 1 in the center and having $A = 0$ or 2 do not lead to any width 2 violations. Errors will occur, however, if $A = 4$ or 6, since the pattern now definitely indicates a wire of insufficient width passing through the region.

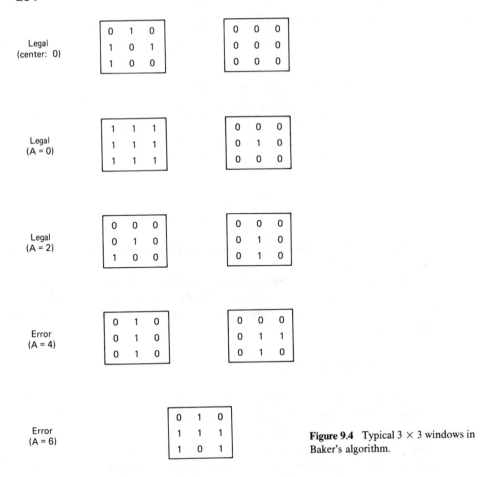

Figure 9.4 Typical 3 × 3 windows in Baker's algorithm.

The main part of the algorithm is table-driven. Two tables are used: one table indicates violations of 3λ rules on a 4×4 window and the other indicates 2λ rules violations on a 3×3 window on the same scan as the 4×4 window. Since the spacing checks are width check on white space, the first table can be used to detect metal space/width or diffusion spacing and the second table is used for polysilicon–diffusion width or polysilicon–polysilicon spacing. A single bit indicating legal or not legal is stored for each possible window. The bit configuration of the window specifies either a 9-bit or a 16-bit address for these tables. This means that a total of $2^9 + 2^{16}$ bits or about 8 kilobytes of storage per layer is needed to store the tables.

Design rules that involve more than one layer, such as contact cut, butting or buried contact, transistors, and ion implantation rules that involve conditional analysis of extension and overlap, are handled by special-purpose codes because the number of possible windows in these cases prohibits the use of tables.

There are several limitations to the raster-scan approach. First, the algorithm reports some false errors since it does not have any knowledge of electrical connectivity of the component parts, as illustrated in Fig. 5.4. It is necessary, therefore, to combine circuit extraction and analysis together with design rule checking. A simpler approach which eliminated most of the false reporting was adopted which examined a 10×10 window for connectivity in the neighborhood of any 4×4 windows which checked an error in the first phase. The additional overhead is not excessive since the number of such windows examined is quite small. Second, the width and spacing violations usually run along the length of a wire which may be long, resulting in a voluminous error report output file. This could be avoided if the algorithm incorporated any kind of global connectivity check. Third, the algorithm scales up poorly when the technology requires a window larger than 4×4. This is due to the fact that all knowledge of the design rules is embedded in the actual code of the DRC in pregenerated tables. Thus when the design rules change, the DRC will probably have to be totally rewritten. Finally, the storage requirement for the algorithm is considerable and is agonizingly slow for designs of even humble size.

Eustace – Mukhopadhyay algorithm.

A second raster-scan DRC algorithm has been developed by Eustace (1981) and Eustace and Mukhopadhyay (1982). This algorithm is based on a simple model of the design rules in terms of a set of deterministic finite-state automata (DFA) which accept the rasterized inputs and produce the error outputs. The algorithm is flexible and technology-independent since the design rules are embedded in the form of transition tables of the DFAs. Changes in the design rules or even the technology can be handled by simply redefining the transition tables without changing the algorithm. The algorithm is easy to implement in special-purpose hardware.

The basic idea of the algorithm can be explained with respect to Fig. 9.5. For each design rule, a deterministic finite automaton (DFA) is defined which takes as its input the raster image of one column at a time starting from the lower left corner. The DFA is a 4-tuple (S, p, t, S_0), where S is a set of n finite states $(0, 1, \ldots, n-1)$ and p is the input symbol at the (i, j) location of the raster imager. The value of p could be any one of the layer names (i.e., p for polysilicon, d for diffusion, etc.) or the layer name of a derived layer as well as symbols such as $\bar{p}, \bar{d}, \ldots$, denoting "no polysilicon" "no diffusion," $\ldots$. S_0 denotes the initial state at the very beginning of the scan process, and t denotes the transition function given below while taking the pixel point β as the input:

$$S_{i,j} = t\left(S_{i-1,j}, S_{i,j-1}, p\right)$$

The states $S_{i-1,j}$ and $S_{i,j-1}$ will also be called the *horizontal state* and the *vertical state*, respectively, and will be denoted as S_h and S_v, respectively. The output is produced with each transition and is either "E," denoting a design rule violation, or null, signifying no violation. When reading a symbol in column 1, $S_{i-1,j}$ will be assumed to be 0 for all j; similarly, when reading a symbol in row 1, $S_{i,j-1}$ will be

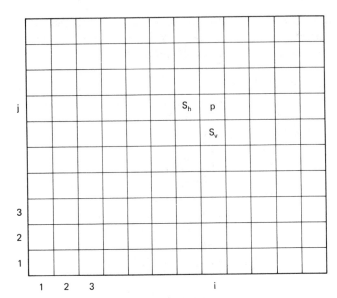

Figure 9.5 Two-dimensional DFA model.

assumed to be 0 for all i, that is, $S_0 = 0$. The raster image will be padded by a column at the right and a row on top with "white" space (dashed squares), that is, pixel points with no layer. This will simplify error reporting.

Let us take an example to illustrate the idea of the algorithm. Figure 9.6 gives the DFA for minimum width check 2λ of a wire in layer ℓ. The first two entries in each row denote the values of the pair (S_v, S_h). The entries in the table denote the next state and possibly an error output, denoted as "/E," corresponding to input $\bar{\ell}$ and ℓ as the pixel being scanned. Certain combinations of (S_h, S_v) do not occur in this table because they correspond to unrealizable situations.

Figure 9.7 shows the states the machine assumes while processing a design. Note that each state has some physical significance. State 0 means that the machine is scanning a region outside the layer, state 1 means that a lower left corner of the layout has just been entered, state 2 means that width 2λ has been satisfied in the vertical direction, state 3 means that width 2λ has been satisfied in the horizontal direction, state 4 means that width 2λ has been satisfied in both horizontal and vertical directions, state 5 indicates white space immediately following a diffused region, state 6 indicates that a corner has been encountered and checks to ensure this does not cause a reverse diagonal width violation, states 7 through 10 indicate a corner and check to ensure that this does not cause a diagonal violation, and states 11 and 12 are special error states that flag errors in the vertical and horizontal directions, respectively.

To appreciate some of the subtleties of the construction of the transition tables, consider the patterns shown in Fig. 9.8. Here the errors correspond to both width violations and diagonal errors, but are recognized only as width violations. Furthermore, the transition table consolidates error reporting so that the width violation report is made only at the initial location and not along the length of the

S_v	S_h	$\bar{\ell}$	ℓ	S_v	S_h	ℓ	$\bar{\ell}$
0	0	0	1	6	0	5	2
0	1	12/E	3	6	1	11/E	7
0	2	12/E	3	6	3	11/E	7
0	3	0	3	6	12	5	2
0	4	0	3	7	2	11/E	8
0	5	0	1/E	7	7	11/E	8
0	6	12/E	3/E	7	9	11/E	8
0	10	0	3	7	11	11	2
0	11	0	1	8	2	11/E	4
0	12	0	1	8	4	5	4
1	0	11/E	2	8	5	5	6
1	1	11/E	7	8	6	11/E	4
1	3	11/E	7	8	7	11/E	9
1	12	11/E	2	8	8	11/E	10
2	0	5	2	8	9	5	4
2	1	11/E	7	8	10	5	4
2	3	11/E	7	8	11	11	6
2	12	5	2	9	2	11/E	4
3	2	11/E	4	9	8	11/E	10
3	4	11/E	4	9	9	11/E	10
3	4	11/E	4	9	11	11	2
3	5	11/E	2/E	10	2	11/E	4
3	6	12/E	4/E	10	4	5	4
3	7	11/E	9	10	5	5	6
3	8	11/E	4	10	6	11/E	4
3	9	11/E	4	10	9	5	4
3	10	11/E	4	10	10	5	4
3	11	11	2	10	12	5	6
4	2	11/E	4	11	0	0	1
4	4	5	4	11	1	12/E	3
4	5	5	6	11	2	12	3
4	6	11/E	4	11	3	0	3
4	7	11/E	9	11	4	0	3
4	8	11/E	4	11	5	0	1/E
4	9	5	4	11	6	12/E	3/E
4	10	5	4	11	7	11	3
4	11	5	6	11	8	11	3
4	12	5	6	11	9	0	3
5	0	0	1	11	10	0	3
5	1	12/E	3	11	11	0	1
5	3	0	3	11	12	0	1
5	4	0	3	12	0	0	1
5	5	0	1/E	12	2	12	3
5	6	12/E	3/E	12	4	0	3
5	9	0	3	12	5	0	1/E
5	10	0	3	12	7	12	3
5	11	0	1	12	8	12	3
5	12	0	1	12	11	0	1/E

Figure 9.6 The DFA for the width 2 check.

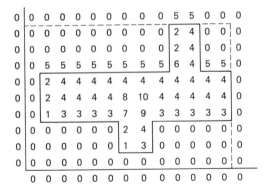

Figure 9.7   Example of the width rule.

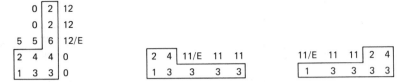

Figure 9.8   Error states.

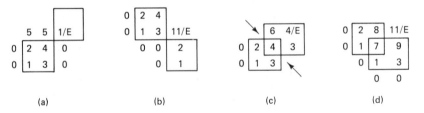

Figure 9.9   Diagonal errors.

wire. Next, consider the patterns shown in Fig. 9.9. In parts (a) and (b), the error combines a diagonal error with a width error to mark the beginning of the lower left corner as well. In part (c), the state 4 recognizes that the 2 width on both horizontal and vertical directions has been satisfied, but $S_h = 6$, $S_v = 3$ imply a diagonal in the direction shown by arrows. Similarly, in part (d) the error message is really due to diagonal violation ($S_h = 8$, $S_v = 9$), but state 11 recognizes this square to be the beginning of the corner of white space. Figure 9.10 shows an example of a design with several design rule violations. Note how the use of "error" states (11 and 12) serves to consolidate redundant error messages.

Figure 9.6 can also be used on any spacing violation of width 2, since a spacing check between two wires is the same as a width 2 check on white space.

In general, the DFA approach for specifying design rules is flexible. Since each state represents some physical characteristics of the design layout, we can, given enough states, describe almost any topological constraint. Unfortunately, the size of the transition tables as well as their complexity grows with the square of the number

0	0	0	0	0	0	0	0	0	0	0	5	5	0	0
0	0	0	0	0	0	0	0	0	0	0	2	4	0	0
0	0	0	0	0	0	0	0	0	0	5	1/E	3	0	0
0	0	5	0	0	0	0	0	0	5	6	12/E	0	5	0
0	0	2	12	0	0	0	0	0	2	4	0	0	2	12
0	0	2	11/E	11	11	11	11	11	2	8	11/E	5	6	12/E
0	0	2	4	3	3	3	3	3	3	7	9	4	4	0
0	0	1	3	0	0	0	0	0	0	1	3	3	3	0
0	0	0	0	0	0	0	0	0	0	0	0	0	0	0
0	0	0	0	0	0	0	0	0	0	0	0	0	0	0
0	0	0	0	0	0	0	0	0	0	0	0	0	0	0

Figure 9.10 Example with several design rule errors.

of states. This constraint may require some type of systematic approach to transition table generation rather than the ad hoc methods used to create the current tables. The simplicity and flexibility of the DFA design-rule-checking algorithm, together with its inherent redundant error consolidation, are its major advantages, but the software implementation is slower than in the Baker–Terman implementation.

9.3.2 The Polygon Method

The polygon or the rectangle method (Yamin, 1972) for design rule checking can be described as a geometry engine. The input is a list of polygons each associated with a specific mask layer and a sequence of commands describing operations on all the objects in one or more layers. Typical commands include MERGE, which combines the polygons in one layer that intersect or touch; and EXPAND, which enlarges or shrinks each polygon of some layer by a specified amount. OR performs the same operation as MERGE, but uses polygons of two different layers; AND produces a third derived layer which is the overlap of two different layers; and EXOR performs the same function as OR with intersecting areas removed. The design rules are expressed in terms of these commands. For example, the spacing check can be performed by doing the EXPAND operation with half the minimum spacing and then doing a MERGE operation to see if any of the polygons are enlarged to touch or overlap with any other polygon, indicating a violation. The transistor areas can be derived as the intersection of polysilicon and diffusion masks. Then by using a sequence of EXOR, EXPAND, and MERGE, it is easy to check the 2λ extension rule as well as the minimum-channel-length rule for nMOS. To check for minimum width, an EXPAND operation with a negative half-width (actually a shrink operation) is performed. A violation is detected if any of the new polygons does not maintain its original "form" with reduced size. To capture the notion of "form" a data structure that represents the polygon as a sequence of line segments of its boundary in a clockwise direction is more convenient. The width check can be stated by saying that the new boundary maintains the original orientation of its line segments and the integrity of its inside area.

The major advantage of the polygon method is that the algebra of polygons can easily be adopted to new design rules or processes. The major disadvantage is that the method is global and involves a time-consuming operation of polygon intersections which is proportional to the square of the number of polygons. Recent DRC programs reduce the number of comparisons by using algorithms that incorporate sorting and windowing schemes (Baird, 1976), which ensure that comparisons between two pieces are within a specified distance of each other. McCreight (1980) has developed an almost $O(n \log n)$ algorithm to solve this intersection reporting problem; see also Bentley and Ottmann (1979). Another disadvantage of the polygon method is that the design rules that are not symmetric in all directions have to be handled by an elaborate sequence of operations.

9.3.3 Corner-Based DRC

The corner-based DRC "Lyra" reported by Arnold and Ousterhout (1982) takes a middle ground between the raster approach, which is based on local checks on bits, and the polygon method, which is global and manipulates larger entities such as polygons. Most design-rule violations of Manhattan geometry layouts can be detected by inspecting small regions around the corners on individual or derived mask layers. Figure 9.11 illustrates the corner points of a transistor. The corners are classified as convex (symbol "a") or concave (symbol "o"). Each rule is specified in two parts. The *context pattern* detects the existence of a corner and the *constraint pattern* will specify what mask features are prohibited in the neighborhood of the corner. For example, Fig. 9.12 specifies the polysilicon spacing rule, which actually represents eight rules corresponding to all orientations of the structures at convex and concave corners.

The design rules in Lyra are specified in a Lisp-like text language at a high level. To give some idea of this specification language, consider the single-layer spacing rule for polysilicon illustrated in Fig. 9.12. It is specified in the form of a

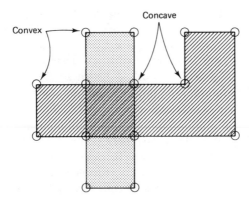

Figure 9.11 Definition of convex and concave corners in Lyra.

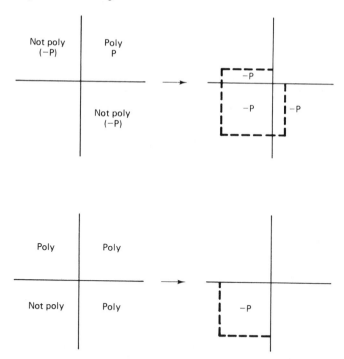

Figure 9.12 Polysilicon–polysilicon spacing rule.

macro "*ss*" having the form

$$(ss\langle\text{layer}\rangle\langle\text{dimension}\rangle\langle\text{text}\rangle)$$

In particular, for polysilicon spacing it has the form

$$(ssP4\text{“}P\_s\text{”})$$

which expands to two rules:

$$\big(\text{rule} \quad (\text{corner: (aP)}) \quad \big(\text{constraints: }\big(\text{e-outside 4 (notP) P\_w}\big)\big)\big)$$

$$\big(\text{rule} \quad (\text{corner: (oP)}) \quad \big(\text{constraints: }\big(\text{outside 4 (notP) P\_w}\big)\big)\big)$$

The first rule applies to a convex corner of a primary layer P (polysilicon), and the constraint part says that the layer specified by the predicate [Fig. 9.13(a)] is in the vicinity of the convex corner. The dimension of this configuration is 4 μm (λ = 2 μm). If the rule is violated, an error message "P_w" will be issued to the user. The second part of the rule applies to a concave corner with configuration "outside" [see Fig. 9.13(b)]. Similar rules can be derived for other spacing and width violations. For a certain error a context pattern is sufficient to detect the error: for example, a missing polysilicon extension in a transistor and the context pattern that will detect the error are shown in Fig. 9.14. The rules for buried and butting contact present some special problems since simple context patterns are not possible to construct

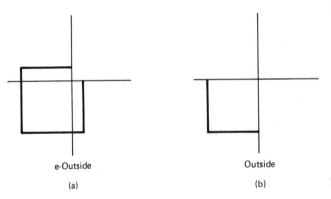

e-Outside

(a)

Outside

(b)

Figure 9.13 Two corner configurations.

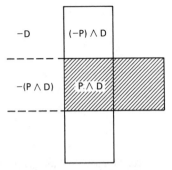

Figure 9.14 Transistor error.

and some amount of global analysis regarding connectivity is necessary. Some rules require reference to corners defined by a composite layer; some rules apply to corners only in the presence of an additional layer at the corner; and so on.

Lyra is easily adaptable to new rule sets and incremental changes in rules. A rules compiler then generates the actual code for checking the mask layers. Lyra works in conjunction with CAESAR—a layout editing system developed by Ousterhout at Berkeley. The major problem seems to be its speed—it is agonizingly slow. An improved implementation is being developed at this time. Another problem seems to be the fact that no precise algorithm is specified to obtain the rule set and to verify its correctness. The rule set has been obtained by enumeration.

9.3.4 Magic's Incremental DRC

The latest Berkeley VLSI design tools distribution (circa spring 1985) incorporates a design rule checker which runs in the background continuously while the designer is editing a design on a graphic workstation. The DRC operates *incrementally*, meaning that the DRC acts on the most recently modified area in the layout and flags layout errors if there are any. Thus when the layout is finished without errors, the DRC is automatically done.

In Magic, mask layers are represented by *tiles*, which are layout segments of maximum width and length in a contiguous region and are made of material of the

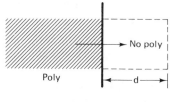

Figure 9.15 Edge-based rule for Magic.

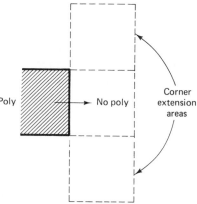

Figure 9.16 Edge rule with corner extension.

same *abstract* layers. Besides the primary layers, such as polysilicon, diffusion, metal, and "white space," the abstract layers allow identification of composite layers such as "transistors" or "contacts." The layout is represented by a small number of *planes*, each plane containing layers that have the most design-rule interactions. The tiles cover an entire plane with a data structure called *corner stitching* (each tile is linked to its four neighboring tiles with four "pointers").

The design rules are applied at the edges between the tiles in the same plane. For example, Fig. 9.15 specifies a rule that requires polysilicon to the left of the edge and no polysilicon to the right of the edge over a distance d. To account for diagonal error, if the edge belongs to a corner of a tile, an additional check is made in the corner extension area, as illustrated in Fig. 9.16. For further information, the reader should consult the original papers (Ousterhout et al., 1983). The Magic DRC also incorporates a "hierarchical checking" principle, which is explained in the next section.

9.3.5 Hierarchical DRC

It is a common experience of university designers using Baker's DRC or Lyra that the first time a layout is DRCed, a voluminous error output may be generated. Much of this volume is attributed to two factors: first, redundant reporting of the same error, such as a width violation across the length of the wire, and second, repetitious reporting of errors. The second type of error is due to the fact that a cell with an error has been used in an array and the DRC has found the same error repeated over the array. Whitney (1981) proposed a hierarchic approach to DRC by incorporating a filter so that each repetitive structure need be checked only once. Each time the structure is utilized, only the interaction between the structure and its surroundings needs to be checked. The algorithm analyzes the hierarchic nature of the design and checks the DRC violations by examining each cell and the interaction of the cells the first time they are encountered.

In order that the hierarchic DRC be efficient, certain conservative design strategies must be followed (Rowson, 1980). First, a high degree of regularity of the design is encouraged. Second, the cells should not be allowed to overlap; to ensure connectivity, abutment must be enough. This may sometimes lead to inefficient layout in the form of redundant extensions of layers at the periphery of the cell. Third, the entire design must be expressible as a composition of basic cells; that is, higher-order cells cannot contain instances of cells and custom wiring. The "custom" part must be encapsulated in the form of cells. Quite often, wires are run over several cells, which is considered bad strategy under hierarchic discipline.

9.4 Circuit Extraction

The *circuit* or *node extractor* is a program that takes a layout description, say a CIF file or its rasterized image, and produces as output the underlying electrical network consisting of the list of transistors and node connectivity information. Usually, other information, such as the length/width ratio of each transistor and the resistance and capacitance of connecting wires, is also derived during the circuit extraction step. Programs that attempt to derive precise resistance and capacitance values are called *parasitics extractors*. The output of the circuit extractor is used by switch-level simulators such as ESIM (Bryant, 1981) to verify functional correctness, whereas parasitics extractors provide data for timing simulators. Parasitics extractors are usually available in industrial design environments, where performance is an important goal. Circuit extraction combined with some parameter extraction such as that used in RSIM (Terman, 1983) or CRYSTAL (Ousterhout et al., 1983) is usually the universities' norm. We present next several approaches to circuit extraction.

9.4.1 Raster-Scan Method

Baker (1980) proposed a simple circuit extraction algorithm based on the raster-scan approach as used in his DRC algorithm. The bit map is examined left to right, top to bottom, with an L-shaped window for each of the following layers: P, polysilicon; M, metal; D, diffusion without polysilicon overlap; and T, diffusion and polysilicon overlapped. The template, shown in Fig. 9.17, has three cells: the current cell, the top cell, and the left cell. Only one scan line of storage per layer is needed. The connectivity of the circuit is derived by following the four simple rules given in Fig. 9.18. If the current cell is 1 and the top and left cells contain 0's, the upper left corner of a new electrical node is located and a new node number is assigned. In other cases, the left or top node number is continued or two different node numbers are discovered, which should belong to the same connected node. In this case they

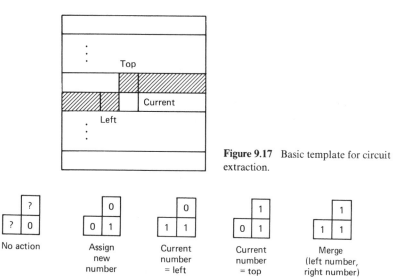

Figure 9.17 Basic template for circuit extraction.

Figure 9.18 Possible template configurations.

are merged into one node, discarding one of the node numbers and replacing all references to this node by the chosen node number. The algorithm above determines the connectivity of all areas except those made by a contact cut. A separate check is made to determine whether the current cell contains a contact cut, metal, and then either polysilicon or diffusion. If this is true, the nodes for the M and P or D are merged into one. A similar approach is used to find connectivity due to buried contact.

The identification of transistors is complicated by two factors: First a T layer does not necessarily imply a transistor, as in a butting or buried contact. This is easily resolved in this algorithm by examining the contact cut or buried layer as explained above. Second, the source and drain could be arbitrarily apart and the transistor does not fit one window. A two-phase algorithm has been proposed. In the first phase, the transistor pieces are located by finding one of the four configurations as shown in Fig. 9.19. An additional bit indicating whether the piece was implanted is also utilized to detect depletion-mode transistors. In the second phase, the pieces are combined to form entire transistors by utilizing the merged node information.

A number of errors can be detected at the end of an extraction process. These errors can be detected by checks to ensure that all node numbers were merged, and checks to ensure that nodes with different names are not connected together and that nodes with the same name are not left unconnected. A number of basic "static" checks are also made at this point: The length/width ratio and threshold drops are used to calculate the effective ratios, which are then compared against a set of

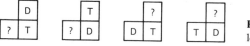

Figure 9.19 Basic windows for transistor location.

acceptable values. More than two threshold drops are flagged for potential errors, transistors which short V_{dd} and Gnd are flagged, transistors whose gate is connected to a constant value such as V_{dd} or Gnd are flagged, depletion-mode transistors are checked so that they are used properly (as pull-up loads, in superbuffers, and as yellow transistors), and a list of "propagate" errors is provided indicating which points cannot be pulled up or down. These checks provide valuable debugging information at the initial stages of design before the design grows to the point that errors become very difficult to track down.

9.4.2 Other Methods

The "polygon method" of circuit extraction is based on the MERGE operation on polygons that we explained in Section 9.3.2. The input is the set of rectangles in primary mask layers. The connectivity relation on polygons is a transitive relation; that is, if polygons Q_1 and Q_2 are connected and Q_2 and Q_3 are connected, then Q_1 and Q_3 must also be connected. The algorithm essentially consists of merging polygons using this transitive relation and assigning unique identifying numbers to each set of connected regions. Special care must be taken to handle channel, contact, or buried regions. The 1λ overlap of polysilicon and diffusion in a buried contact presents a special problem. In the presence of a contact layer overlapping with polysilicon and diffusion, and noting the fact that it abuts with diffusion on only one side, the structure can be ruled out as a transistor region. Similar considerations must apply to polysilicon–diffusion overlap regions in the presence of a buried layer. The depletion-mode transistors must be delineated using implant layer information. Finally, for each maximally connected channel region, the gate, source, and drain regions have to be identified.

Baird (1976) proposed an implementation of a circuit extraction program based on an "edge-based" data structure. Each edge of a polygon on a mask is given a *role*, which identifies the *frontier* number corresponding to the groups of edges connected via common vertices; a *direction* bit, which identifies the edge to be *forward* (inside area is to the left of the edge) or *backward* (inside area to the right of the edge); and a *color*, to indicate in what mask layer it belongs. The edges are kept in a data base in lexicographical order (i.e., sorted by *x*-coordinate and then by *y*-coordinate). The *local environment* of a vertex is defined to be all incident vertices together with their roles. A vertex scan algorithm is then developed to make all connectivity analysis based on their local information. For details, the reader is referred to Baird (1976).

Efficient data structure can simplify the circuit extraction process considerably. The Magic circuit extractor is a good example to illustrate this point. Because of the use of "abstract" layers and "corner stitching" of the tiles, the circuit is already extracted. A tile represents a node corresponding to a connected region in a given layer. Interconnection of nodes can be discovered by examining the corner stitches at the four corners of the tile. For hierarchical design, a problem arises when cells are allowed to overlap because each cell is a collection of separate corner-stitched

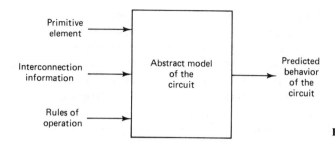

Figure 9.20 Simulator block diagram.

layers. This might lead to the creation of a new transistor (when polysilicon of one cell overlaps with the diffusion of another cell) or a split transistor. These problems have been solved in Magic by restricting overlap and abutment such that the type and number of transistors do not change due to overlap. This is checked by a special design rule incorporated in the incremental DRC.

9.5 Simulation Models and Tools

Simulation tools are software systems that are used to predict the performance and functionality of large integrated circuits. The simulation program sketched in Fig. 9.20 is based on an abstract model of the circuit to be simulated and takes as input the primitive elements, the interconnection rules, and a set of rules of operation and produces as output the predicted behavior of the circuit. The level of abstraction used determines the level of the design process at which the simulator is applicable. The *functional* simulators deal with systems specified in terms of major building blocks and their interconnection. The *register-transfer-level* simulators deal with registers, ALU, buses, I/O, control, and the interfaces that transfer information among these modules. A *gate-level* simulator has to deal with the logic gates that implement the different functional modules, abstracting the circuit behavior from its electrical behavior. The *circuit* simulator produces detailed information regarding the signal levels and delays at every node of the circuit based on electrical models of the circuit. In general, as the level of abstraction goes down, the amount of computation goes up. The circuit simulator takes a huge amount of computation time. The gate-level simulator takes manageable resources and handles reasonable-size circuits, but for several reasons fails to incorporate realistic models for MOS circuits. First, the MOS networks resemble a relay-contact network; they are bidirectional, although the designer intends a direction of data flow. This might lead to "sneak" paths and charge-sharing phenomena that need to be modeled. Second, gate-level simulators do not model "dynamic storage"—the memory due to the storage of charge at a node over several cycles. Third, the relative values of the capacitances and resistances of the circuit nodes of a MOS network play an important role in determining the state of the network which the gate-level simulators cannot take into account. Finally, the gate-level model cannot capture an important MOS structure—precharged logic. In the recent past, several switch-level models have been proposed which describe the logical behavior of MOS circuits (Bryant, 1981; Hayes, 1981) more accurately, and several *switch-level simulators* or *logic simulators* have been developed based on these models (Bryant, 1980; Baker and Terman, 1980). We discuss some of these in this section. Logic simulators, however, do not predict timing or speed characteristics of the circuit. A set of circuit models has recently been explored for MOS circuits which are refinements of the switch-level models incorporating simplified calculations of resistance, capacitance, and time delays (Terman, 1983; Ousterhout et al., 1983). *Timing simulators* based on these models have been developed (RSIM, CRYSTAL) which are used primarily by university designers. We discuss the basic models used in these simulators in this section.

9.5.1 The Switch Model and the Logic Simulators

One of the first switch-level simulators (MOSSIM) was developed by Bryant (1980). The basic model of the network consists of a set of nodes $(n_1, n_2, \ldots, n_n)$ connected by a set of transistors $(t_1, \ldots, t_m)$. Each node could be in one of the states $(0, 1, X)$, where 0 and 1 correspond to low and high voltage levels, respectively, and X denotes an undefined state corresponding to an uninitialized node or a node with a signal value between the two logic values 0 and 1. The nodes are classified as: *input* nodes which provide a strong signal supplied from outside the chip, such as V_{dd} or Gnd; input nodes that are connected to V_{dd} via a pull-up resistor; and a *normal* node, a typical storage node in the circuit which does not generate a signal. A transistor is modeled as a three-terminal (gate, source, drain) bilateral device which could be in any of the three states: 0 (open or nonconducting), 1 (closed or conducting), and X (undefined, intermediate conductance between its conductance when open and when closed). The three types of transistors considered are: *p*-type, *n*-type, and the *d*- or depletion type. The state of the transistor as a function of the gate node state is given in the following table.

Gate State	*n*-Type	*p*-Type	*d*-Type
0	0	1	1
1	1	0	1
X	X	X	1

It is further assumed that the system is two-phase clocked with an interclock gap that allows all the transients to settle and that the circuit does not have any critical races. The *network state* is defined by the values of its node state and the transistor states. The simulation algorithm divides the network into a set of equivalence classes connected by closed transistors. Two nodes are considered to belong to the same equivalence class if they are connected by a path of closed transistors. The equivalence classes are determined by a depth-first search on an equivalent graph representing the network. Nodes in the same equivalence class as an input node are assigned to either a high or a low state depending on whether the input node is V_{dd} or Gnd. If the equivalence class contains no input node but contains a pull-up node, it is assigned a high state. If the equivalence class does not contain an input or a pull-up node, it would not change its state and would remain in a normal state. The simulator is event-driven; it keeps a list of events, each event being either a node-state change or a transistor-state change. For each change of network state resulting from an event, the simulator will iteratively compute the equivalence classes of all nodes in the network until two successive computations yield the same result. This is called an *atomic step*. A complete clock cycle is simulated by breaking the simulation into four *phase* simulations, each phase corresponding to the four epochs of the two-phase clock, and each phase taking several hundred atomic steps. One of the major difficulties with this algorithm is the computation of equivalence classes which contain transistors in the X state. This is done by a combinational exhaustion of all possible 1 or 0 values assigned to all gate nodes in the X state, checking whether the resulting states are the same or different. Furthermore, all nodes affected by this transistor are also assigned to the X state and the effect has to be propagated in the circuit. Since the complete circuit has to be simulated at once (i.e., parts of the circuit cannot be isolated for separate simulation), the algorithm is inefficient. The algorithm has been modified and improved by Bryant (1981). It must be pointed out, however, that no known algorithm can deal with the X states efficiently. When a circuit is powered up, some of the nodes must go to a "void but unknown" state, which is not the same as saying that they are in X states. For example, we know that as the power is turned on, the two states Q and $\overline{Q}$ of a flip-flop must satisfy $Q + \overline{Q} = 1$. But $X + \overline{X} = X$ for whatever definition of $\overline{X}$ we accept.

Baker and Terman (1980) wrote a switch-level simulator using a model very similar to the above. The model used a larger number of node states. One is *input* (V_{dd} or Gnd); this input node has enough "strength" to be unaffected by connections to other nodes. The *driven* nodes are those that are connected to an input node or a driven node by a chain of closed transistors. The *weak* nodes are those that are connected to a pull-up by a chain of closed transistors. A *charged* node is one that maintains its value stored in its associated capacitance. A set of "pseudo" states are also introduced, corresponding to the driven and weak states of the transistor paths containing at least one gate node in state X. During simulation, a state of the node is calculated by interacting its state with the states of its neighboring nodes by iterating the computation with respect to pairs of nodes. For example, a driven high node interacting through a closed transistor with another node will always be driven high except when the second node is driven low or driven X, in which case the resulting node is driven X, and so on. A separate table is prepared to compute the final nodes states connected by an X-transistor. The transistor in the X-state partitions the network and a computation proceeds between the pairs of groups of nodes on both sides of the X-transistor using this table. The simulation algorithm starts with an "event list" of nodes whose values have been changed since the last simulation step. The nodes are chosen from this list in sequence, and for each node its new value is calculated using the interaction tables, and nodes affected by this node are then added to the event list. This calculation is repeated until the event list is empty. Thus the computation is restricted to the neighborhood of the circuit that is active, and is more efficient than MOSSIM in this respect.

The two simulators discussed above have simulated a large variety of nMOS designs, including some having 10,000 transistors. The major criticism of the methods is that the models used are rather ad hoc, not based on a rigorous mathematical structure, and do not account for a large number of practical situations where transistor sizes, their conductances, and the values of possible capacitances vary over a range.

Later works by Bryant and Hayes and Terman (Bryant 1981; Hayes, 1982; Terman, 1983) describe lattice-theoretic approaches to handle a hierarchy of size of transistors and capacitors. These models, however, cannot adequately incorporate the effects of transistors in X states. More recently, Bryant (1984) proposes a more general graph-theoretic approach to the problem. Here we will discuss one of Terman's algorithms to give the reader an idea of this class of algorithm.

Let us first understand the difficulty with the X states. As an example, consider a nine-state simulator which might use the following table to characterize the "strength" and the "logic state" of a node:

		Logic State	
Strength	0	1	X
Driven	DL	DH	DX
Weak	WL	WH	WX
Charged	CL	CH	CX

The letters H, L, and X stand for high, low, and unknown, respectively, and D, W, and C stand for driven, weak, and charged, respectively. Thus a node is in state 1 if its signal strength in one of DH, WH, or CH. A signal A has higher strength than the signal B, written $A \geq B$, if the final value of the signal is A when both signals are connected together. This notion can be formalized in the form of a lattice, as shown in Fig. 9.21. Here λ denotes the null or the "weakest" signal and DX is the strongest signal. Note that certain pairs of signals could be "incomparable" (i.e., CH and CL). In this case, the final signal assumes the value CX, the "lowest upper bound" (LUB) of CH and CL. More formally, the LUB of two signals A and B, written $A \cup B$, is defined to be signal V such that $V \geq A$ and $V \geq B$, and for every signal Z, if $Z \geq A$ and $Z \geq B$, then $Z \geq V$. In Fig. 9.22, point a goes to logic state X since DX is the LUB of DH and DL. But there is ambiguity about the logic state of point b in Fig. 9.23—Is it 1 corresponding to WH, or X corresponding to DX?

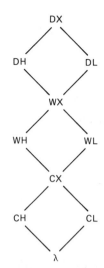

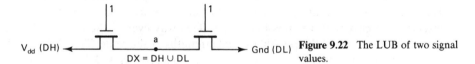

Figure 9.21   Lattice of signal strengths.

Figure 9.22   The LUB of two signal values.

This ambiguity results since the signal at point *c* is undefined; it could be a 0 or a 1, and once *DX* is assigned to point *b*, the *DX* value sticks there since it is the strongest signal in the lattice and overrides *WH*.

To remove this difficulty, Terman (1983) developed an "interval value set" approach, which allows a possible range of signal values to a node. He proposed drawing a line separating the "high" and the "low" signals as shown in Fig. 9.24 and suggested that if the possible signal range completely falls within the high region (i.e., the range [*DH*, *CH*]), its logic value should be 1; if it falls in the low region, its logic value should be 0 (i.e., the range [*CL*, *DL*]); but if the range crosses the line,

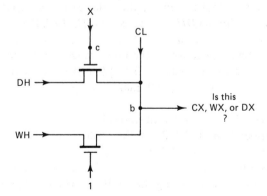

Figure 9.23   Ambiguous signal value.

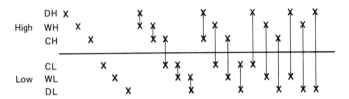

Figure 9.24 Interval value set diagram.

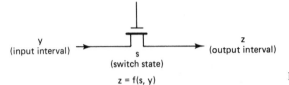

Figure 9.25 Switch function.

the logic value is X (e.g., the range $[DH, WL]$). Note that there are only 21 possible ranges because a signal must assume all possible signal values within the range. When two nodes are merged, a range "algebra" is used to compute the smallest interval that covers all the possible states. This way even if the node goes to an X state, it keeps track of the possible signal strengths at the node. The algebra is based on the following state of a MOS switch: "open," "closed," "unknown," and "weak." A depletion-mode transistor is always in a weak state. For both p- and n-channel transistors, if the gate is at value X, the switch is in an unknown state. The "open" and "closed" states of the switch depend on the 1 or 0 value at the gate and on the type (p or n) of transistor. The output range z of the switch (see Fig. 9.25) as a function of switch state s and input range y is given by the *switch function* f as

$$z = f(s, y)$$

The function f is defined as follows. If s = open, $z = \lambda$ (the null signal strength is considered weaker than CH or CL and corresponds to a logic state X). If s = closed, then $z = y$; that is, input is transmitted unaffected. If s = weak, all the "driven" signals become "weak," that is, replace the letter D in the input range by W in the output range. Thus, the input $[DH, DL]$ goes to $[WH, WL]$. If s = unknown, then if the input range already corresponds to X states (i.e., $[DH, WL]$, etc.), the output range is the same as the input range; if the input range corresponds to logic 1, make the lower value of the range to be λ (i.e., $[DH, WH]$ goes to $[DH, \lambda]$); finally, if the input range corresponds to logic 0, make the upper value of the range to be λ (i.e., $[CL, WL]$ becomes $[\lambda, WL]$).

When a node is connected to two different nodes by switches, its final interval z is obtained as a least upper bound (LUB) of the input intervals z_1 and z_2 as $z = z_1 \cup z_2$, as shown in Fig. 9.26. The LUB computation is done with the lattice shown in Fig. 9.27. If a node is connected to several nodes, a signal interval can be computed for the node by repeatedly applying the LUB on the constituent intervals.

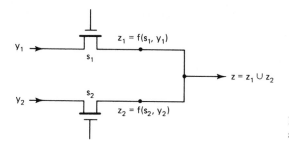

Figure 9.26 Computation of a composite switch function.

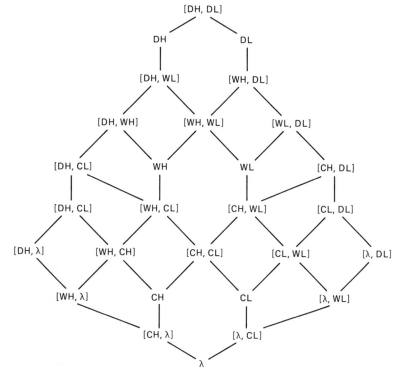

Figure 9.27 Interval value set lattice.

This is finally merged with the contribution of the node, which is taken to be CL, CH, or $[CH, CL]$ if its logic state is 0, 1, or X, respectively.

The calculation of the node's state proceeds as follows. All the input nodes that can be reached from the node via paths of open, closed, or weak transitions are found and their contribution to the node's value is calculated by the procedure discussed above, starting at the inputs and working back toward the node. If no inputs are connected to the node, a group of nodes are identified which are isolated from the inputs. A charge-sharing computation is performed to determine the common voltage of the shorted nodes as the ratio of all capacitors at nodes at logic high, denoted C_{high}, to all the capacitors independent of logic states in this group, denoted C_{total}. Let C_x denote the capacitance of nodes in logic state X. Then the state of the node is given as

$$
0 \quad \text{if} \quad \frac{C_{high} + C_x}{C_{total}} < 0.2
$$

$$
1 \quad \text{if} \quad \frac{C_{high}}{C_{total}} > 0.8
$$

$$
X \quad \text{if} \quad \text{otherwise}
$$

The main simulation algorithm, called the *global simulation algorithm* by Terman, performs these computations in two ways: Nodes that are affected directly by the input change via source/drain connections are evaluated; then nodes that are affected indirectly because of the change of the transistor states induced by the direct changes are evaluated. The second part of the computation is performed by keeping an *event list* that maintains all transistor switches that have a changed state. The algorithm terminates when the event list is empty, at which point the network has settled to a stable state. For further details, the reader is referred to the original reference (Terman, 1983), where a second algorithm, called a *local algorithm*, is also presented, which is more precise but uses a lot more internal states.

9.5.2 Logic-Level Timing Simulators

In this section we briefly discuss the underlying concepts in the logic-level timing simulator, the RSIM (Terman, 1983), which is a commonly used timing simulator for university projects. Another commonly used timing simulator is CRYSTAL (Ousterhout et al., 1983), which will not be discussed (the documentation for CRYSTAL is available with Berkeley VLSI software distribution). The more accurate timing simulator is SPICE, which is also frequently used to determine the performance of small circuits. The main difference between SPICE and these simulators lies in the level of details in modeling transistor operation. SPICE allows a detailed model of a transistor, where parameters such as threshold voltage, supply voltage, temperature, mobilities, and rise/fall times of the waveforms can be varied. RSIM adopts a linear model of the transistor in terms of an effective resistance R_{eff} of the transistor, which is taken to be the average channel resistances of the transistor over its range of operating terminal voltages. As we have seen in Chapter 6, R_{eff} depends on varieties of parameters, including its dimension and the context of its use. Actually, RSIM uses three effective resistances: R_{static} at steady state, R_{dylow} during high-to-low transition, and R_{dyhigh} for low-to-high transition. The transistor is also looked upon as an on–off switch and its effective source-to-drain resistance is described as

$$
R_{ds} = \begin{cases} R_{eff} & \text{(switch closed)} \\ \infty & \text{(switch open)} \\ [R_{eff}, \text{inf}] & \text{(switch in unknown state)} \end{cases}
$$

Note "∞" stands for very high resistance; when the switch is in an unknown state, its resistance is given a range of possible resistance values. A depletion-mode transistor is modeled as a resistance R_{eff} since the switch is always closed. In CRYSTAL, the model is simpler: the transistor is viewed as an on–off switch with a fixed resistance when it conducts.

The second major difference between SPICE and the logic-level timing simulators is the fact that voltages are quantized into one of three values: 0, logic low, corresponding to voltages in the range $[0, V_{low}]$; 1, logic high, corresponding to the range $[V_{high}, V_{dd}]$; and X, corresponding to intermediate or unknown values. In SPICE, a continuum of voltage values corresponding to each operating point defined by the terminal voltages is calculated.

Finally, in logic-level timing simulators, there is much more emphasis in computing the "worst-case" delay or power. This means that the simulators will have to consider all possible input values in the analysis of a circuit configuration. Because of the simplicity of models, such calculations can be done for circuits having a modest number of transistors (about 50,000 transistors) in a reasonable amount of time. SPICE, on the other hand, examines each input condition as a separate analysis situation and the worst-case computation takes an exorbitant amount of time and resources for reasonably large circuits.

Thus the logic-level timing simulators sacrifice accuracy to get some approximate figures quickly and with a reasonable expenditure of time and storage. It is also possible to identify potential problem areas in speed, which can then be analyzed by SPICE for more accurate results. In the following paragraphs, we present a few more details regarding RSIM.

For RSIM, the basic problem of determining the value of a node is handled by deriving an equivalent electrical circuit, called the *Thévenin* equivalent circuit, as shown in Fig. 9.28. This is characterized by a voltage source V_{thev} in the range $[V_-, V_+)$ specifying the possible range voltages the output node may have, and a resistance R_{drive} in the range $[0, \infty]$ in series with the voltage source. Depending on the value of V_{thev} in relation to the quantum levels denoting $0, 1, X$, a logic value is assigned to the node. If this logic value differs from its previous value, a *transition* has taken place, and then the simulator computes the transition time to be $R_{drive}C_{load}$. In Chapters 5 and 6 we have given methods to compute C_{load}. The computation of R_{drive} is based on the equivalent series–parallel resistance network of the drive circuits and depends on the type of transistor (low-to-high or high-to-low).

The main simulation algorithm is "event" based. An event specifies a node in the network, a new logic state, and a time at which the node's value is changed. The list of events, sorted by time, is maintained as an event list. If an input is changed by

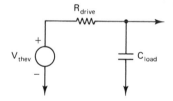

Figure 9.28 Thévenin equivalent circuit.

the user, an event is added to the list. An event is processed by enumerating all nodes that may be affected by the change by following the paths of source–drain transistors for which the node being changed is a gate terminal. For each such node affected by the change, two calculations are made: a charge-sharing calculation and a final-value calculation. The charge-sharing computation is very similar to one used in the global-logic simulation. All node voltages are normalized in the range [0, 1], with the following quantized values: 0, voltages in the range $[0, V_{low}]$; 1, voltages in the range $[V_{high}, 1]$; and X, voltages in the range $[V_{low}, V_{high}]$. The idea is to determine the final voltage when a set of nodes having different capacitances are connected together. The minimum and maximum "common voltages" can be written as

$$V_{share} : \min = \frac{C_{high}}{C_{total}}$$

$$V_{share} : \max = \frac{C_{high} + C_x}{C_{total}}$$

The quantities C_{high}, C_x and C_{total} denote the sum of capacitance of high nodes, nodes in undefined states and all nodes, respectively. Since the nodes in X states contribute an undetermined amount of charge, the two voltages above are actually intervals of voltage in practice. This interval is then compared with the logic threshold when determining the final logic state as follows:

$$\text{charge-sharing state} = \begin{cases} 0 & \text{if} \quad V_{share \cdot \max} \leq V_{low} \\ 1 & \text{if} \quad V_{share} : \min \geq V_{high} \\ X & \text{otherwise} \end{cases}$$

The final-value calculation is made to determine the node's final steady-state value. If $R_{drive} < \inf.$ and the node is not an input node, the final value of the node is given by

$$\begin{array}{lll} 0 & \text{if} & V_+ \leq V_{low} \\ 1 & \text{if} & V_- \geq V_{high} \\ X & & \text{otherwise} \end{array}$$

If the value differs from the charge-sharing value, another event is added to the event list after a lapse of time which is determined by the effective time constant of the circuit [see Terman (1983) for further details].

9.6 Routing Methods

The problem of interconnecting a large number of circuit elements in a chip is taking up an increasing amount of design time and chip area. Typically, 30% of total design time and about 60% of the chip area are expended merely to interconnect the circuit elements. The *interconnection* or *routing* problem consists of a set of *modules* having

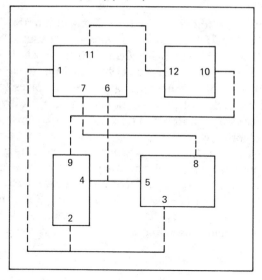

Signal nets

$\{1, 2, 3\}, \{4, 5, 6\}, \{7, 8\}$
$\{9, 10\}, \{11, 12\}$

Figure 9.29 General channel routing problem.

a rectangular bounding box and an associated set of *pins*, and a set of *signal nets* which specify which pins have to be electrically connected by wires. A typical routing problem is illustrated in Fig. 9.29. The signal nets could be *two-point* or *multipoint*. The routing region is assumed to be gridded with all pins and wires aligned on the grid, and the wires could be drawn on only two layers of metal in the horizontal direction and on polysilicon in the vertical direction. If a wire has to change direction, a *via* is used at the turning point. Layout organizations that allow both the size and placement of the modules to be specified are called *custom layout systems*. The *gate arrays* have fixed module size and the modules are placed in linear arrays of rows separated by constant size routing regions called *channels*, as shown in Fig. 9.30. In this chapter we are concerned primarily with routing of rectangular regions. If fixed pins are located on all four sides of the region, the problem is called the *switchbox* or *area routing* problem. If the pins are located on the top and bottom sides, the problem is called the *channel routing* problem. Notable works on area routers are the "sequential" algorithms, such as the Lee algorithm (Lee, 1961) and its extensions (Akers, 1967; Rubin, 1974; Korn, 1982; Soukup, 1978), the Hightower

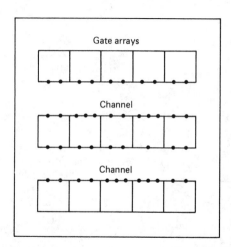

Figure 9.30 Gate array layout.

algorithm (Hightower, 1969), and the LRS and LUZ algorithm (Smith, 1983). Channel routing algorithms are "parallel." Some examples are the simple "left edge" algorithm (Hashimoto and Stevens, 1971), dogleg algorithms (Deutsch, 1976; Persky et al., 1976), the algorithm by Yoshimura and Kuh (1982), and the greedy algorithm (Rivest and Fiduccia, 1982). The interregion routing problem involves the interaction between adjacent regions. Relevant past works on this area are by the "PI" project and MIT, and the works by Baratz (1981), LaPaugh (1980), and others. We will present some of these algorithms in this chapter. We also discuss a set of new algorithms developed by Eustace (1983) for both channel and area routing.

The most difficult important measure of a routing algorithm is its completion rate in relation to the difficulty of the problem. The traditional complexity measures such as storage and computation time are, of course, important. Since most useful routing problems are "intractable," that is, NP-complete, acceptable solutions are based on good heuristics that yield reasonable time and storage complexities but guarantee completion.

9.6.1 Area Routers

The Lee algorithm. The Lee algorithm finds the shortest path if it exists between any two points, called the source and target, in a gridded routing region with any number of obstacles. The empty cells adjacent to the source are first labeled 1, indicating the cost of reaching cells from the source. Next, a cost of 2 is placed in each empty cell adjacent to cells labeled 1, and so on. Thus a diamond-shaped expanding *wavefront* of increasing numbers propagates until the target pin is labeled or there are no remaining empty cells adjacent to any labeled cells. If the target pin has been labeled, the algorithm backtracks from the target, moving at each step to

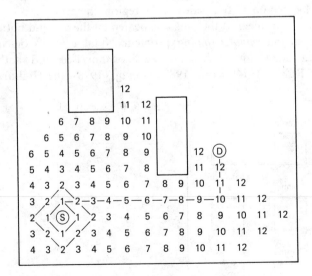

Figure 9.31 Waveform propagation in the Lee algorithm.

any cell with a lower cost. Figure 9.31 illustrates the algorithm with a backtrack path. Quite often, there will be alternative choices for the backtrack path.

Although the Lee router guarantees a path if its exists, it requires an excessive amount of storage. To store the maximum cost C, $\log_2(C + 1)$ is a bare minimum (label 0 is used to indicate empty cell). The cell must also store information about backtrack direction (a maximum of 2 bits). In large design, several million grid points are not uncommon. Akers (1967) suggested an improvement in the cost of labeling that reduces the number of bits necessary at each grid point to two. Note that a cell labeled X has neighbors which are marked either $X - 1$ or $X + 1$, denoting the predecessors and successors, respectively. The sequence $1, 1, 2, 2, 1, 1, 2, 2, \ldots$ also has the property that the predecessor of any label is different from its successor. Thus the wavefront can be expanded with two consecutive symbols 1, followed by two expansions using the symbol 2. Note that the symbols 1 and 2 no longer represent cost. The Akers algorithm is illustrated in Fig. 9.32. Moore (1959) suggested a similar algorithm earlier, even earlier than Lee's algorithm, which is also sometimes called the Lee–Moore algorithm.

Another disadvantage of the Lee algorithm is that its execution time is large since the wavefront extends in all directions irrespective of the target directions. As a result, most of the label computations are discarded when the path is found. The Lee algorithm has a time complexity of $O(n^2)$, where n is the minimum distance between the source and the target. Several modifications of Lee's algorithm have been made to improve its performance. One approach is to propagate the wavefronts from both the source and the target until they meet. This simple modification reduces the computation time by 50%. Rubin (1974) suggested that the wavefront should be expanded using a key which is the sum of the distance from the source to the current cell plus the estimated aerial distance to the target. This technique places a penalty on moves away from the target and expands moves toward the target

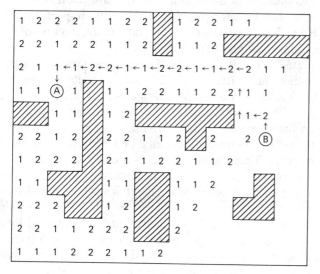

Figure 9.32 Akers algorithm.

direction unless that direction is blocked. Soukup (1978) developed an *arrow* router, which achieves a similar effect by manipulation of stack pointers representing wavefront entries. Korn (1982) suggested a router similar to that proposed by Rubin. The technique is to *overpull* the path toward the target with heavy penalties placed on moves away from the target. The algorithm, however, does not yield the shortest path. The resulting improvement in execution time incorporating these modifications is by a factor of 20 to 50.

The Lee algorithm has also been modified to work in more than one layer by assigning variable costs to grid points. These costs allow the algorithm to establish a preferential direction in a given layer, with penalties for crossing and zigzagging. For example, a variable-cost model in two layers could be as follows. Each edge has a cost of 1 in any layer. Whenever the layer has to be changed, a via is used with a cost of 3. This will encourage paths using the same layer and discourage the use of vias.

Finally, since the Lee algorithm is essentially a shortest-path algorithm, Dijkstra's shortest-path algorithm (1959) will lead to a better implementation. In Dijkstra's algorithm, by the k th step the shortest paths to the set of nodes N closest to the source are found. At the $(k + 1)$th step, a new node is added to N whose cost to the source is the minimum of the remaining nodes outside N. This process is repeated until the target node has been added to the set N.

The Hightower algorithm.   Hightower (1969) suggested a line routing algorithm that produces completion rates similar to Lee's but reduces execution time and memory requirements. The fundamental difference between this router and the Lee is that it does not store the entire plane in a matrix. Instead, only lines and points (as zero-length lines) are stored. A line segment is represented as a 3-tuple (x, y, z) where (x, y) is the coordinate of the leftmost or bottommost point on the line and z is the x- or the y-coordinate of the rightmost or topmost point in the line. The line segments can be stored lexicographically sorted with respect to their x-, y-coordinate values. This yields a fast algorithm to find the intersection of two lines or a sequence of line segments connecting two points.

The algorithm is illustrated in Fig. 9.33. The brief description presented here is based on a description by Breuer and Carter (1983). The source A and the target B have to be connected. The *blockages* are the line segments x, y, z, w, v and the boundary lines are l, r, u, d. The algorithm begins by constructing vertical and horizontal escape lines from the source and target terminals. We need a couple of definitions. A blockage is a *cover* of a point if a horizontal or a vertical line from the blockage passes through the point. Thus the blockage x is a cover for point A. An *escape line* passing through a point is a line that is perpendicular to the two nearest blockages on both sides (left–right or up–down) of the point. Thus lines a and b are escape lines through A and lines c and h are escape lines through B. Given a point m and its escape line L, an *escape point* on L is a point that is not covered by at least one cover of m. Thus point p is an escape point for the escape line b passing through A. Similarly, point q is an escape point over the escape line c passing through B. For each escape point, find the longest escape lines. If there is a choice,

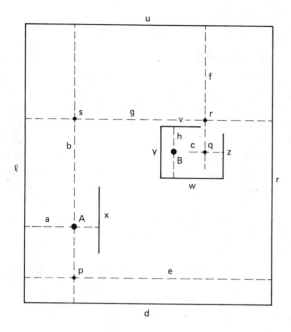

Figure 9.33 Hightower algorithm illustration.

the escape lines nearest the starting terminal are taken. Thus the escape lines e and f are drawn through the escape points p and q, respectively. Now the escape line e does not have an escape point, but the escape line f has an escape point r. The process is repeated until the escape lines originating from source and target intersect. In this example, the escape line g will intersect at point s with the escape line b for A, yielding a solution $AsrqB$. The Hightower algorithm does not perform well when the routing area is large or when the routing area has many connections. For small areas, the algorithm tends to use paths with the minimum number of vias.

Other area routing algorithms.   Finding optimal solutions to area or channel routing problems is an NP-complete problem (LaPaugh, 1980). Routing algorithms rely on heuristics to provide reasonable solutions with a good *completion rate*, which is the most important measure of effectiveness. The completion rate is usually determined empirically by the ratio of successful routing to the total number of routing problems attempted. A poor algorithm will have a high completion rate on very simple problems; a good algorithm will yield a high completion rate over more difficult problems.

For an area router, Smith (1983) proposed the Manhattan Area Measure (MAM) as a measure of difficulty of the routing problem. MAM is the ratio of the minimum theoretical area (the sum of the areas needed to route each signal net in the absence of all other nets) to the total available net area. The completion rate of the Lee algorithm has to be measured and it drops off dramatically when MAM reaches 11%, indicating that the Lee really works well only for a very sparse layout. The poor performance of the Lee is attributable to the fact that it chooses a path at random with little or no regard for nets yet to be routed and might use more vias than are needed, causing what is called *pin blocking*. Note that each via prevents any additional vertical or horizontal wires from passing through the point where the via has been placed. Agrawal (1977) showed that vias created by the Lee caused more than 35% blockage, a point at which the routeability of a net falls off dramatically. Tim Saxe at Stanford suggested the use of *reservations* to guarantee that each pin has access to the routing region. Reservations are "tails" on each pin extending across 50% of the routing region, which are placed initially. Once a net is completed, any unused section of the reservation is released. Smith showed that the use of reservations increased the completion rate of the Lee by up to 60%—an order-of-magnitude difference. The other technique to avoid pin blocking is to use judicious *net ordering*, which minimizes the number of vias. We discuss this concept later in some detail. One other technique to minimize the use of vias is to use a variable-cost Lee with a large penalty associated with a layer change or the use of a via. With the background above, we will now present some new area routers that have shown large improvements over the Lee in completion rate measurements.

Smith (1983) proposed a *loop routing scheme* (LRS) algorithm. The algorithm has a flavor of channel routers except that the tracks are square or rectangular in shape and centered around the area of the routing region. The initial tracks are opened near the center; as more tracks are needed, new tracks are assigned. If the initial area is unknown, as newer tracks are assigned near the center, the routing region is expanded outward together with all the tracks that have already been assigned. A part of the track is assigned to a net if all the pins of the net can be connected to the part by connectors, which are horizontal or vertical lines as

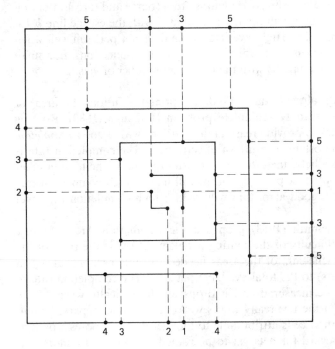

Figure 9.34 Illustration of the LRS algorithm.

illustrated in Fig. 9.34. As usual, two layers are assumed (i.e., tracks in metal, connectors in polysilicon). If more than one net can be assigned to a track, there will be a choice. Also, a net can be assigned to a part of the track or to its sort of "complementary" part, leading to a choice of which part can be left unused. The authors do not specify clear strategies to exploit these choices. Some improvements are suggested: breaking up multiple-pin nets into a collection of two-point nets, packing more than one net assignment on a single track when the nets are nonoverlapping in the radial sense, and using outer tracks for nets that have pins on one side only, which minimizes wire lengths.

Smith also implemented an algorithm called the *LUZ algorithm*, which derives its name from the three basic path shapes it uses: L, U, and Z, as shown in Fig. 9.35(a). The algorithm attempts to minimize the number of vias by routing simplest paths first using these basic path shapes, called a LUZ level 1 path. LUZ level 2 paths are formed by adding two L-shaped corners to a LUZ 1 path [Fig 9.35(b)], and similarly for higher level paths. If all the paths cannot be routed by using a level 1 path, the algorithm will use level 2 paths, and so on. Using randomly generated routing problems, Smith showed that 75% of the net can be routed using level 1 paths only, about 20% using level 2 paths, and the remaining paths are completed using level 3 paths. The analysis also shows that almost no improvement comes from routing at anything higher than LUZ level 3. The results also show that the LUZ using only level 2 paths had completion rates equal to that of the Lee. The algorithm with level 3 paths had a better completion rate than the Lee with reservations. For test cases with uniform pin distribution on all sides, a 100% completion rate was achieved up to 55% MAM. One disadvantage of the LUZ algorithm is that it does not have any means to determine when to halt—it keeps trying more and more

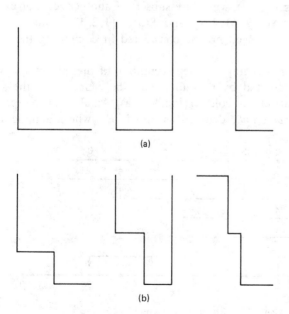

(a)

(b)

Figure 9.35 Illustration of LUZ path shapes: (a) level 1 paths; (b) level 2 paths.

complex paths. A modified LUZ algorithm proposed by Eustace (1984) uses a "degrees of freedom" concept to formalize the path shape definitions.

9.6.2 Channel Routers

The channel routing problem can be defined as follows. A rectangular region, called a *channel*, with pins located on the top and bottom edges aligned on the grid, is given. Horizontal and vertical grid lines are referred to as tracks. The objective of channel routing is to electrically connect all the nets using a minimum number of horizontal tracks. Two layers are assumed again, metal for horizontal tracks and polysilicon for vertical tracks. Figure 9.36 shows an example of a channel routing problem. The problem can also be stated as a "net list," which gives the net numbers to be connected to the top and bottom columns from left to right. A 0 means that the pin is not connected to any net. For this problem, the net list will look as follows:

$$3\ 1\ 3\ 0\ 0\ 5\ 6\ 0\ 3\ 0\ 0\ 0$$
$$1\ 2\ 4\ 2\ 4\ 1\ 5\ 7\ 0\ 7\ 6\ 0$$

The horizontal segment of a net is determined by its leftmost and rightmost terminal connections. Let $S(i)$ be the set of nets whose horizontal segments intersect column i. The number of elements in each set $S(i)$ is called the *local density*. Since horizontal segments of distinct nets must not overlap, the horizontal segments of any two nets in any set $S(i)$ must not be placed in the same horizontal track. Clearly, the set $S(i)$ with the most elements will be a lower bound on the number of horizontal tracks necessary to route the channel. This lower bound is called the *channel density*.

The computation of the maximum density can be simplified if we eliminate from consideration those sets, $S(i)$, which are subsets of another set. For example, in Fig. 9.36 $S(1) = \{1, 3\}$, $S(2) = \{1, 2, 3\}$, and $S(3) = \{1, 2, 3, 4\}$. Since $S(1)$ and $S(2)$ are subsets of $S(3)$, they need not be considered in computing the channel density.

A *zone representation* is a graphical representation of the maximal sets $S(i)$. Each maximal set is represented by a column and the elements of the set are represented by line segments in that column [Fig. 9.37(a)]. An alternative representation of zones is an *interval graph*, denoted as $HG(V, E)$, where a node $v_i \in V$

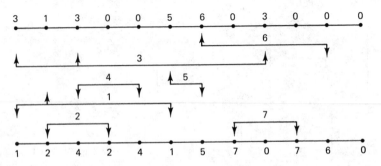

Figure 9.36 Channel routing problem.

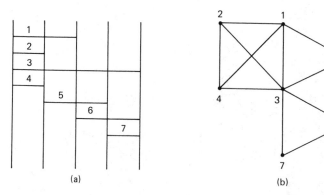

Figure 9.37   (a) Zone representation and (b) interval graph.

represents net n_i and an edge $(v_i, v_j) \in E$ if the horizontal segments of net n_i and net n_j cross a common column. Figure 9.37(b) illustrates the interval graph for the net list in Fig. 9.36. This graph is also referred to as the *horizontal constraint graph*. A zone is simply a maximal *clique* [see Harary (1969)] of this graph and the clique number is the channel density.

The "left edge" channel router of Hashimoto and Stevens (1971) attempts to maximize the placement of horizontal segments in each track. The edges by the left endpoint of each segment are sorted. The sorted list for Fig. 9.36 is $(1, 2, 3, 4, 5, 6, 7)$. The algorithm selects the first edge, 1, and places it in the lower left corner of the routing region. Net 1 is deleted from the sorted list. The algorithm then scans through the remaining list for the first net that does not overlap net 1. In Fig. 9.36, this would be net 6. This process is repeated until no more nets can be placed on track 1. The algorithm starts again using the remaining unplaced nets in the list and filling track 2, and so on. The track selections made by the left edge algorithm are shown in Fig. 9.38.

The track assignment discussed above has some serious problems. Net 3 has to be connected to the top of the channel crossing directly over the end of the net 3 horizontal segment. Since net 4 will have to place a via at that position, this will short nets 4 and 3. Thus net 4 must be above net 3. Similarly, net 1 must be placed on top of net 2. Constraints such as these are called *vertical constraints*.

The vertical constraints can be depicted in the form of a graph called a *vertical constraint graph*. Let $VG(V, E)$ be a directed graph where each node corresponds to

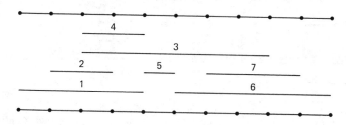

Figure 9.38   Track selections by left edge algorithm.

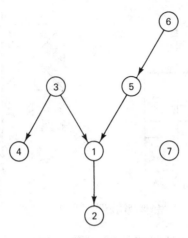

Figure 9.39 Vertical constraint graph.

a net. For each vertical column, let $(n_i, n_j) \in E$ if and only if net i has a pin on the top and net j is a pin on the bottom of the channel. An edge between nodes n_i and n_j indicates that the horizontal segment of net i must be placed above the horizontal segment of net j. The vertical constraint graph for Fig. 9.32 is shown in Fig. 9.39. The "left edge" algorithm finds the optimal solution if there are no vertical constraints.

A *constrained left edge* algorithm is reported by Perskey et al. (1978). This algorithm again places horizontal segments from the lower left corner of the routing region. The algorithm will place a horizontal segment for a net only if it does not have any descendants in the vertical constraint graph. For example, using the same example as in Fig. 9.36, net 2 can be placed since it has no descendant in the vertical constraint graph. If a net is assigned to a track, the corresponding node is removed from the vertical constraint graph. The algorithm then finds the first edge that does not overlap the previous net in the remaining sorted list of horizontal segments. Net 7 is the next net that fulfills both these requirements. This process is repeated, filling each track from left to right until all horizontal segments have been placed. Figure 9.40 illustrates the completed routing. In this case, the constrained left edge algorithm produced an optimal solution; the number of tracks used is equal to the channel density plus one.

The preceding algorithm will fail if there is a cycle in the vertical constraint graph. Cycles in the vertical constraint graph are called *vertical constraint loops*. Consider the channel routing problem and vertical constraint graph shown in Fig.

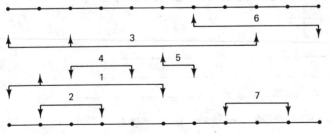

Figure 9.40 Constrained left edge track assignment.

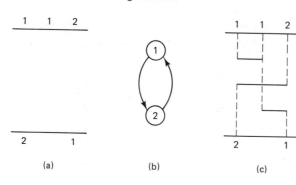

Figure 9.41 Vertical constraint loop: (a) problem; (b) graph; (c) solution using dogleg.

9.41(a) and (b). In this example, the constrained left edge algorithm will not be able to route either net 1 or net 2, since both have descendants in the vertical constraint graph. Figure 9.41(c) shows that a solution to this problem is possible only if we allow the horizontal segment of a net to be split. Splitting nets in this fashion is called *doglegging*.

The *restricted* channel routing algorithm does not allow doglegging. The dogleg algorithm proposed by Deutsch (1976) uses doglegging both to avoid vertical constraint loops and to decrease the density of the channel. Doglegging is also sometimes useful to reduce the number of horizontal tracks, particularly for channels with multipoint signal nets. This is illustrated in Fig. 9.42. The statement of Deutsch's algorithm given below has been modified (Eustace, 1984) based on an understanding of the vertical constraint graph. The algorithm also assumes that the vertical constraint graph has no cyclic loop.

Deutsch's algorithm takes each multiple-pin net and breaks it up into individual horizontal segments. A break occurs only in columns that contain a pin for that net. Figure 9.43 illustrates the horizontal segment definition and the corresponding vertical constraint graph for both the "restricted" and the "unrestricted" algorithms.

Using the new vertical constraint graph, the dogleg is very similar to the constrained left edge. Horizontal segments are sorted in increasing order of their left endpoint. The first segment in the list that has no descendants in the vertical constraint graph is placed in the channel. The node corresponding to this section of the net is removed from the vertical constraint graph. Then the next net in the list that does not overlap the first segment and has no descendants is placed. This process continues for each track, from left to right, until all segments have been completed.

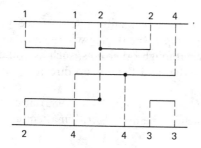

Figure 9.42 Use of dogleg to reduce the number of tracks.

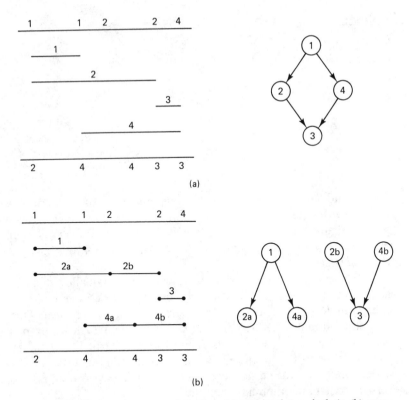

Figure 9.43 Horizontal segment definition: (a) restricted (no dogleg); (b) unrestricted (using dogleg).

The original dogleg algorithm suggested by Deutsch has one difference from the algorithm presented above. Deutsch chose segments as suggested above for the first track, placing horizontal segments from left to right in the track. Then his algorithm switched to the top track in the channel and placed horizontal segments in it, from right to left; then the second-from-the-bottom track; and so on, until all segments have been placed. This technique can easily be incorporated into the algorithm above by selecting a horizontal segment that has no descendants for placement in the bottom tracks and segments that have no ancestors for placement in the top tracks. Deutsch claimed that this symmetric alternating format produces routing with a smaller total vertical length than routing all nets from the bottom of the channel to the top.

The dogleg algorithm considers only a few of the possible candidates for merging. In Fig. 9.36, for example, the dogleg would consider only nets 2 and 7 for placement on the first row, although other nets, such as 1 and 7, could also share a track, since they do not overlap and there is no directed path between them in the vertical constraint graph.

Two algorithms by Yoshimura and Kuh (1982) attempt to exploit alternate pairings of horizontal segments. The first algorithm attempts to minimize the longest

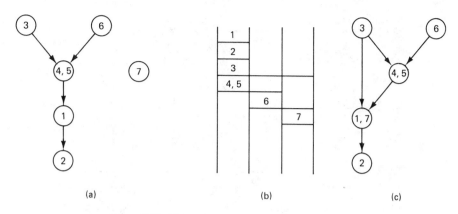

Figure 9.44   Illustration of Yoshimura–Kuh algorithm.

path in the vertical constraint graph by attempting to combine those tracks that minimize the path through the vertical constraint graph. The algorithm uses the zone representation of the horizontal constraint graph and the vertical constraint graph. The algorithm defines a merging operation as follows. If two nets i and j have no horizontal constraint or vertical constraint, the nets can be merged; that is, they can be placed on the same horizontal track. The process then yields a modified vertical constraint graph in which nodes i and j are merged into node (i, j). The zone representation of the routing problem is also updated by replacing net i and net j by a single net (i, j) which occupies the set of all consecutive zones, including the zones of both nets i and j. The merging process is illustrated in Fig. 9.44 with respect to the channel routing problem of Fig. 9.36. Nets 4 and 5 satisfy the merge criteria. If these nets are merged into one net $(4, 5)$, the modified vertical constraint graph and the updated zone representation will be as shown in Fig. 9.44(a) and (b), respectively. Note that the merge operations do not create any cycle in the modified vertical constraint graph. The algorithm uses a list L of nets which terminate at zone 1 and a list R of nets which begin at zone 2. Pairs of nets from these two lists are merged, if permitted by the constraint graph, such that the length of the longest path in the vertical constraint graph is minimized. The merged nets are now dropped from list L and then L is appended by nets that terminate at zone 2 and list R is modified to be the nets that begin at zone 3. The process is repeated until L corresponds to the last-but-one zone. A new vertical constraint graph and a new horizontal constraint graph in the form of a zone representation are now built and the entire algorithm is repeated. The algorithm is finished when no further merging takes place. Each set of merged nodes is then assigned a track of the channel. If the algorithm is applied to our example of Fig. 9.36, the final result after merging will be as shown in Fig. 9.44(c). Each set of merged nodes is now assigned a track of the channel. The second algorithm by Yoshimura and Kuh achieves longest-path minimization through matching techniques on a bipartite graph. Both techniques report better results than for the dogleg.

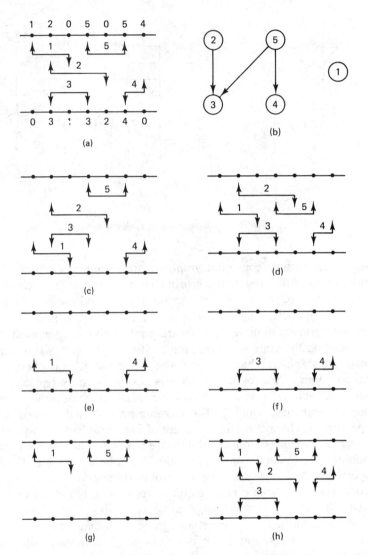

Figure 9.45 Eustace's revised dogleg algorithm.

Eustace (1984) investigated the alternative choices for net assignments to a track and developed two new channel routing algorithms: the *revised dogleg* and the *least-cost path algorithm*, which are described next.

Basic to these algorithms is the concept of *density reducing* track assignment. Define a density vector $D = (d_1, d_2, \ldots, d_l)$, when d_i is the local density in column $i, 1 \le i \le l$, l being the total number of columns. For example, the channel routing problem of Fig. 9.45(a) has the density vector $D = (1, 3, 3, 3, 2, 2, 1)$; its vertical constraint graph is shown in Fig. 9.45(b). An assignment of nets to a track, when permissible by the constraint graphs, is said to be *density reducing* if removal of the selected nets for the track results in the reduction of the channel density of the remaining routing problem by 1. For our example, the dogleg will select nets 1 and 4 for assignment to track 1. If we removed nets 1 and 4 from the problem, the density vector of the remaining problem would be $D^1 = (0, 2, 2, 3, 2, 1, 0)$. The selection is thus not density reducing since the channel density of the remaining problem is still 3. After this selection, the dogleg will choose net 3 for track 2, net 2 for track 3, and net 5 for track 4. The final solution is shown in Fig. 9.45(c). On the other hand, if the dogleg used the right ends of the nets rather than the left ends, it would have assigned nets 3 and 4 to track 1. The density vector of the remaining problem would have been $D^1 = (1, 2, 2, 2, 2, 1, 0)$. This selection is thus density reducing since the channel density of the remaining problem has been reduced from 3 to 2. In fact, for this problem, two more tracks will route the channel as shown in Fig. 9.45(d).

A theorem of Eustace (1984) proves that if an arbitrary channel routing algorithm routes the channel using $d + k$ tracks, where d is the channel density, then the algorithm must have used d density-reducing track assignments and k track assignments which are not density reducing. Obviously, if all the track assignments are density reducing, the algorithm is optimal. A good channel routing algorithm is the one that minimizes k over a large number of channels. A few more definitions: A column in the channel is said to be *critical* if its local density equals the channel density. Obviously, a track is density reducing if the nets assigned to the track cover all the critical columns. A *penalty p* associated with a track assignment equals the number of critical columns that are not covered by the selection of nets in the track. The *revised dogleg* algorithm of Eustace can now be stated as follows: For each track assignment, the following four alternatives are attempted: (a) "left edge" on a bottom track, (b) "right edge" on a bottom track, (c) "right edge" on a top track, and (d) "left edge" on a top track, so as to maximize the density-reducing choices. If more than one choice is density reducing, the one with minimum penalty is chosen. If more than one choice has the same penalty, the algorithm chooses the track that covers the most columns. Returning to the example of Fig. 9.45, the four initial choices are depicted in parts (e), (f), and (g). Of these, only (f) and (g) are density reducing and have equal penalty value. But choice (g) is better since it covers the most columns, leading to the solution shown in (h). Although this solution differs from the one shown in (c), it is one of the optimal solutions.

The revised dogleg algorithm does not always guarantee finding density-reducing track assignments. Consider the channel shown in Fig. 9.46(a); its vertical constraint graph is shown in part (b), and a solution using the revised dogleg is shown in part (c), which is not optimal since part (d) shows a better solution. Eustace (1984) proposed a new algorithm called the *least-cost path* (LCP) algorithm which guarantees selection of a density-reducing track assignment if it exists. The algorithm starts by considering two acyclic directed graphs GT and GB. Graph GT has the set of vertices corresponding to the nets that have no precedents in the vertical constraint graph, that is, the set of nodes that have no incoming arcs, and two special vertices called "source" and "sink." Two nodes corresponding to nets N_i and N_j are connected by a directed edge from N_i to N_j if they can be placed in the

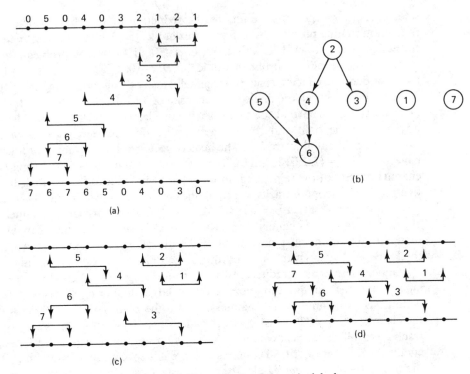

Figure 9.46 Nonoptimality of the revised dogleg.

same track (i.e., no horizontal constraint) and N_i can be placed to the left of N_j. By definition, the "source" node can be placed to the left of all other nodes in the same track, and similarly the "sink" node can be placed to the right of any other node in the same track. The arc connecting N_i and N_j has an associated cost function which equals the number of critical columns that are not covered between the right endpoint of N_i and the left endpoint of N_j. The graph GT is shown in Fig. 9.47(a) for the example of Fig. 9.46. Note that the arc connecting source and sink has weight 6, since there are six nodes with channel density 3 in the density vector $(1, 3, 3, 3, 2, 2, 3, 3, 3, 1)$. The weight connecting net 7 to net 1 is 2 since columns 6 and 7 are not covered by nets 7 and 1, and so on. Graph GB is very similar to graph GT except that it is drawn with respect to nodes in the vertical constraint graph that has no descendants. This graph for our example is shown in Fig. 9.47(b). It is obvious that the nets in any path in GT (except for the path connecting source to sink, which trivially corresponds to no track selection at all) correspond to a valid track assignment on the top part of the channel, and similarly a path in GB corresponds to a track assignment at the bottom part of the channel. Furthermore, the minimum weighted paths in these graphs correspond to track selections that maximize the number of columns being covered at the channel density. If there is more than one such path, the choice could be arbitrary. The LCP algorithm works in several passes: It computes GT and GB, selects a minimum weighted path, and then

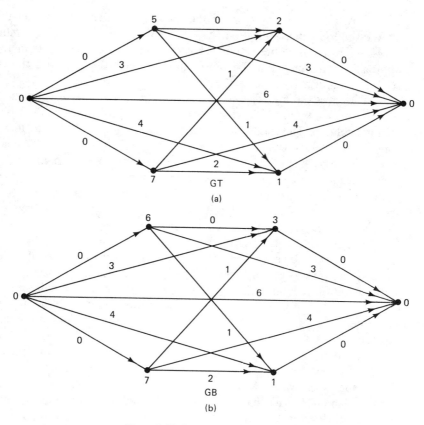

Figure 9.47 Least-cost path algorithm.

repeats the entire process for the remaining channel routing problem, after removing the nets corresponding to the least weighted path. Note that if there is a density-reducing choice, LCP will find it. But if there is no density-reducing choice at a particular point in the algorithm, it makes an arbitrary choice that may not result in an optimum solution. A final heuristic of the algorithm is invoked when there is more than one minimum weighted path in GB and GT. This consists of analyzing the possible impact of the current choice on future local densities. Intuitively, the choice that covers the maximum number of columns with higher densities will improve the chances of density-reducing future choices. This is incorporated by computing the sum of local densities in columns covered by the paths and selecting the path with the maximum value of the sum, if one exists. If more than one path has the same local density sum, an arbitrary choice is made.

The *greedy channel router* of Rivest and Fiduccia (1982) takes a totally different approach to the problem. The channel routing is done on a column-by-column basis (rather than row by row) beginning at the left side of the channel. In each column, the router tries to maximize the number of tracks available in the next vertical column by using a sequence of heuristics. It will frequently allow a net to occupy two different tracks until the heuristic decides to combine the net. This allows good channel utilization, but the number of vias and the amount of wire necessary to connect the channel may be significantly greater than for the other approaches presented thus far. Since the approach does not use the vertical constraint graph, it avoids the vertical constraint loop problem.

The greedy router begins with an *initial channel width* equal to the channel density. A new track is added whenever it is not possible to route a pin at a given channel column. The choice of the initial channel width affects the final routing. In some cases, the router may have to use a spillover area beyond the channel, which can be minimized by iterating the entire algorithm with a higher value of initial channel width. One difference between the greedy router and other channel routers is that it does not use the vertical or horizontal constraint; all decisions are made locally on a column and the algorithm handles routing problems even with cycles in the vertical constraint graph, possibly with a spillover area.

The algorithm classifies each net which has a pin to its right as either *rising*, *falling*, or *steady*. A net is rising if the next pin is at the top edge of the channel and there are no pins located at the bottom edge within a window size of columns equal to the *steady-net constant*. Falling nets are defined similarly. Finally, a net is steady if it is not a rising or a falling net. A larger window minimizes the number of times that a net changes track. A value of 10 has typically been chosen for the steady-net constant.

Another given parameter of the algorithm is the minimum convenience *jog length*. The router may place a vertical wire in order to jog a net to a track closer to the channel edge, where the next pin connection for the net is located. No jogs are made which are shorter than the minimum jog length. This value is usually taken to be channel density divided by 4. This parameter affects the number of vias and tracks in the final routing. A higher value of minimum jog length reduces the number of vias, and a lower value minimizes the number of tracks in the routing.

Finally, the algorithm handles easily the pins entering from the left side or emanating from the right side of the channels. Such *endpins* can also be handled in dogleg and other channel routers with a little effort.

We will now present the algorithm. The tracks are identified by number 1 through the maximum number of nets. Sometimes more than one track may be occupied by one net, in which case it is called a *split net*. A split net is eventually collapsed when the last pin for the net has been passed by a connection jog. The rising nets are put above the steady nets, which are placed above the fallling nets, and the nets are generally placed near the center of the channel. Each step of the algorithm is illustrated by "before–after" diagrams describing what happens in a column. An arrow indicates that the net extends to the right of the column. The net numbers are placed on pins on the top or bottom edges as well as on tracks allocated to these nets. The steps of the algorithms are illustrated in Fig. 9.48. [We refer to them as Figs. (a) through (m) in the following discussion.] Each figure has two parts —the left part depicting "before" situations and the right part showing the "after" situations.

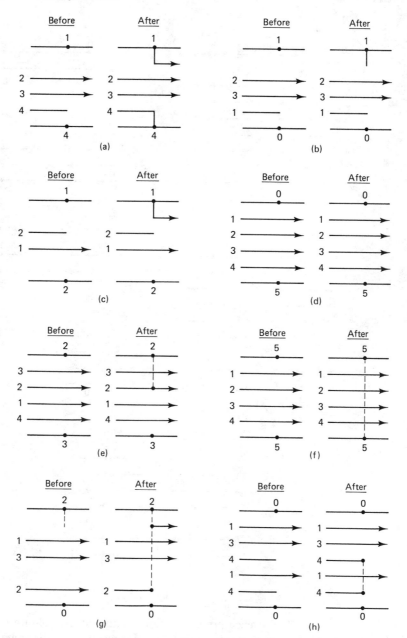

Figure 9.48 Greedy channel router steps.

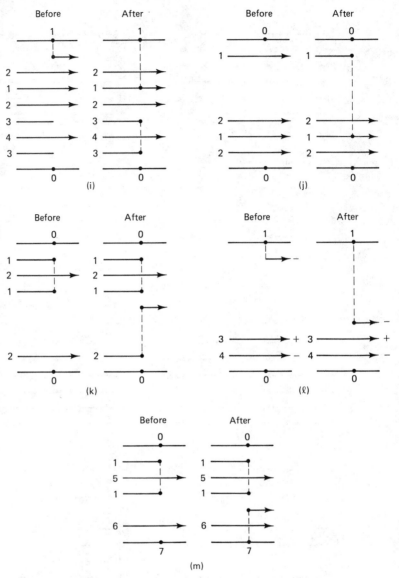

Figure 9.48 (*Cont.*)

Step 1: For each pin connection existing at the column being processed, connect it to an empty track or to a track occupied by the same net, whichever uses the least vertical wire [Figs. (a) through (c)]. If the channel is fully occupied, bringing a new net is deferred until step 5 [Fig. (d)]. If two nets, one from the bottom and one from the top, create a conflict due to overlap, bring the one that uses the least wire, deferring the other until step 5 [Fig. (e)]. If there are no empty tracks, a vertical straight-through connection is permissible [Fig. (f)].

Step 2: Collapse as many split nets as possible. This is an important step in the algorithm since it makes more tracks available to nets arriving at the channel to the right. A *collapsing segment* is a piece of vertical wire that connects two adjacent tracks occupied by the same net. A *pattern* consists of a set of collapsing segments where the segments for different nets do not overlap and no segment overlaps the routing placed in step 1. Each collapsing segment has a weight of either 1 or 2, depending on whether or not the net continues to the right beyond the current column. The weight represents the number of tracks freed due to collapse. The "winning" pattern is found by a combinatorial search that maximizes the weighted sum [Figs. (g) and (h)]. If there is a tie, the pattern that leaves the outermost uncollapsed split net as far as possible from the channel edge is chosen. In Fig. (i), split net 3 is collapsed because it has weight 2; but a tie exists between nets 1 and 2, which is broken by selecting net 1, since net 2 is the outermost uncollapsed net farthest from the channel edges. The idea is to keep the free area as close to the edges as possible. If necessary, the second outermost net is considered, and so on. If there are still remaining ties, use the pattern that maximizes the amount of vertical wire [Fig (j)]. The idea is to minimize the adverse effects on the future pattern due to large collapsing segments. The lists of tracks occupied by a net are updated to reflect the track merging for the pattern selected. If the collapsed track continues to the right, it will do so along the track that is closest to the target edge (the side of the channel where the net has its next terminal connection).

Step 3: Add jogs to reduce the range of split nets. For each uncollapsed split net, additional jogs are added so that the track in the highest level goes as far down as possible and the one at the lowest level goes as far up as possible if such jogs are permissible. No jogs can be shorter than the minimum jog length discussed earlier [see Fig. (k)].

Step 4: Add jogs to raise rising nets and lower falling nets if such jogs are permissible and if the length of such jogs is greater that the minimum jog length [Fig. (l)].

Step 5: If the nets in the current column could not be routed in step 1, add new tracks and bring them to these tracks. Such new tracks must be placed as near the center of the channel as possible if they do not conflict with existing wiring [Fig. (m)].

Step 6: For each unsplit net that ended in the current column, delete the list of tracks occupied by the net. Extend all tracks occupied by unfinished nets and split nets to the next column.

An example of the greedy router is given below. The input to the problem is a "net list" which is given below.

| Top: | 0 | 1 | 4 | 5 | 6 | 1 | 7 | 0 | 4 | 9 | 10 | 10 |
| Bottom: | 2 | 3 | 5 | 3 | 5 | 2 | 6 | 1 | 3 | 1 | 7 | 9 |

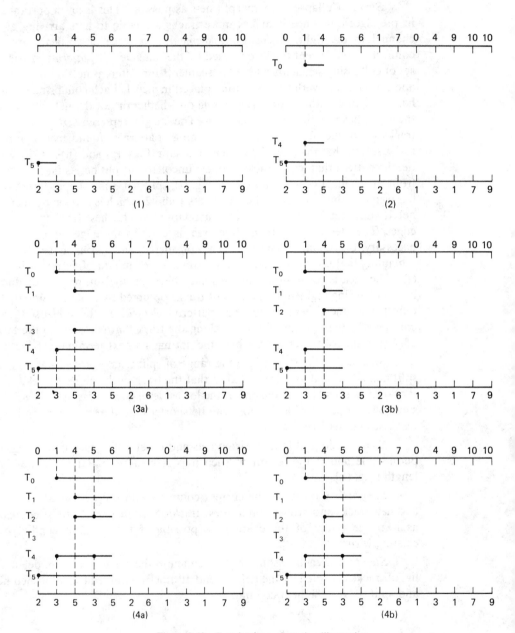

Figure 9.49 Greedy channel routing illustration.

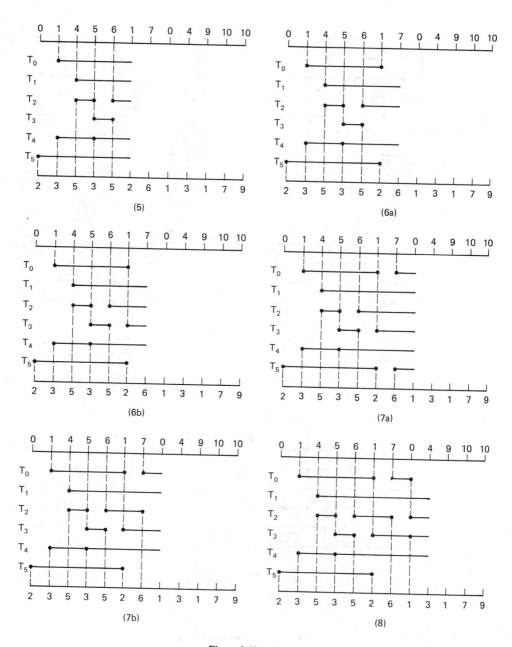

Figure 9.49 (*Cont.*)

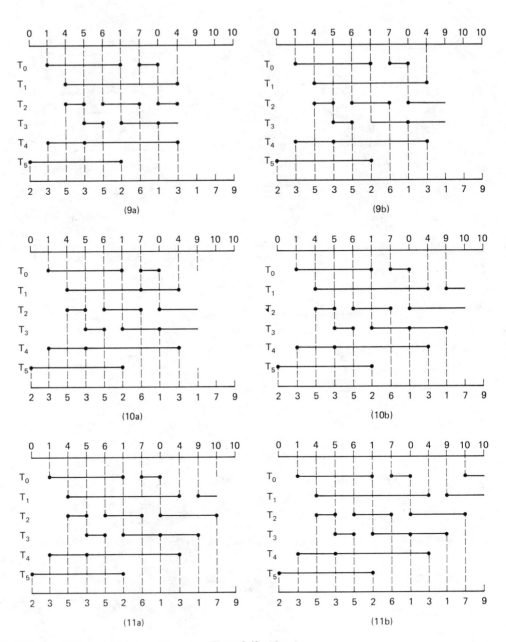

Figure 9.49 (*Cont.*)

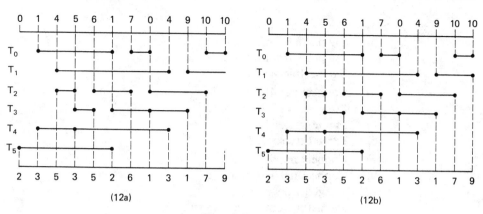

Figure 9.49 (*Cont.*)

The steps in the algorithms are as follows and are illustrated in Fig. 9.49. The step numbers are appended with parts of the figure. Some steps are illustrated in two parts, part (a) and part (b) in the figure.

1(a)   Connect net 2 to track 5.
  (b)   Extend track 5 to next column.
2(a)   Connect net 1 to track 0.
  (b)   Connect net 3 to track 4.
  (c)   Extend tracks 0, 4, 5.
3(a)   Connect net 4 to track 1.
  (b)   Connect net 5 to track 3.
  (c)   Jog net 5 from track 3 to track 2.
  (d)   Extend tracks 0, 1, 2, 4, 5.
4(a)   Connect net 5 to track 2.
  (b)   Connect net 3 to track 4.
  (c)   Jog net 5 from track 2 to track 3.
  (d)   Extend tracks 0, 1, 3, 4, 5.
5(a)   Connect net 6 to track 2.
  (b)   Connect net 5 to track 3.
  (c)   Extend tracks 0, 1, 2, 4, 5.
6(a)   Connect net 1 to track 0.
  (b)   Connect net 2 to track 5.
  (c)   Jog net 1 from track 0 to track 3.
  (d)   Extend tracks 1, 2, 3, 4.
7(a)   Connect net 7 to track 0.
  (b)   Connect net 6 to track 5.
  (c)   Merge tracks 2 and 5.
  (d)   Extend tracks 0, 1, 3, 4.

8(a)   Connect net 1 to track 5.
 (b)   Jog net 7 from track 0 to track 2.
 (c)   Jog net 1 from track 5 to track 3.
 (d)   Extend tracks 1, 2, 3, 4.

9(a)   Connect net 4 to track 1.
 (b)   Connect net 3 to track 5.
 (c)   Merge tracks 4 and 5.
 (d)   Extend tracks 2, 3.

10(a)   Connect net 9 to track 0.
 (b)   Connect net 1 to track 5.
 (c)   Merge tracks 3 and 5.
 (d)   Jog net 9 from track 0 to track 1.
 (e)   Extend tracks 1, 2.

11(a)   Connect net 10 to track 0.
 (b)   Connect net 7 to track 5.
 (c)   Merge tracks 2 and 5.
 (d)   Extend tracks 0, 1.

12(a)   Connect net 10 to track 0.
 (b)   Connect net 9 to track 5.
 (c)   Merge tracks 1 and 5.

Eustace (1984) measured the performance of the channel and area routers using a variety of randomly generated channels of two- or multiterminal nets. For channel routers, the best performance was obtained from the least-cost path algorithm, followed by the revised dogleg, the greedy algorithm, and the dogleg. For example, using a number of tracks 5% over the density, the completion rates for these algorithms were found to be 98%, 95%, 80%, and 40%, respectively. With 10% over density, these figures are 99%, 99%, 98%, and 82%. With 20% over density, all showed 100% completion. For area routers, the Lee and the revised LUZ performed equally well up to a MAM of about 0.5, beyond which the completion rate fell sharply. The modified LUZ performed slightly better than the Lee in this range.

9.7 Summary

This chapter provided a foundation for VLSI design tools and algorithms. VLSI design tools are evolving rapidly and some of the ideas presented in this chapter are destined soon to be obsolete. The chapter could be greatly expanded to include a description of several other timing and logic simulators, circuit simulators such as SPICE, schematic entry and circuit isomorphism programs, layout language tools, silicon compilers, and graphic-assisted layout systems. The description of the tools presented also reflects the author's interests, with a certain amount of bias. The reader can get an idea of the amount of new knowledge accumulated every year by looking at the proceedings of several conferences dedicated to CAD (Computer-Aided Design) tools such as design automation conferences and several journals dealing with the same topic. A comprehensive text in this area is very much needed.

REFERENCES

Agrawal, P., "Routing of Printed Circuit Cards: Density Analysis and Routing Algorithms," *USCEE Rept. 495*, University of Southern California, Los Angeles, Calif., June 1977.

Akers, S. B., "A Modification of Lee's Path Connection Algorithm," *IEEE Trans. Electron. Comput.*, Feb. 1967, pp. 97–98.

Arnold, M. H., "Specifying Design Rules for Lyra" in *1983 VLSI Tools: Selected Works by the Original Artists* (Eds. R. N. Mayo, W. S. Scott, and J. K. Ousterhout), *Rept. UCB/CSD/83/115*, University of California at Berkeley, Berkeley, Calif., Mar. 1985.

Arnold, M. H., and J. K. Ousterhout, "Lyra: A New Approach to Geometric Layout Rule Checking," *Proc. 19th Design Automation Conference*, Las Vegas, June 14–16, 1982, p. 530.

Baird, H. S., "Design of a Family of Algorithms for Large Scale Integrated Circuit Mask Artwork Analysis." M.S. thesis, Rutgers University, New Brunswick, N.J., June 1976.

Baker, C., "Artwork Analysis Tools for VLSI Circuits," *Tech. Rept. MIT/LCS/TR-239*, Massachusetts Institute of Technology, Cambridge, Mass., 1980.

Baker, C., and C. Terman, "Tools for Verifying Integrated Circuit Designs," *Lambda*, 4th quarter 1980, pp. 22–30.

Baratz, A., "Algorithms for Integated Circuit Routing." Ph.D. thesis, Massachusetts Institute of Technology, Cambridge, Mass., Aug. 1981.

Bentley, J. L., and T. A. Ottmann, "Algorithms for Reporting and Counting Geometric Intersections," *IEEE Tran. Comput.*, Vol. C-28, No. 9, 1979, pp. 643–647.

Breuer M. A., and H. W. Carter, "VLSI Routing," Chapter 15 in *Hardware and Software Concepts in VLSI* (Ed. G. Rabbat). New York: Van Nostrand Reinhold, 1983.

Bryant, R., "An Algorithm for MOS Logic Simulation," *Lambda*, Fall 1980.

Bryant, R., "A Switch-Level Model and Simulator for MOS Digital Systems," *IEEE Transactions on Computers*, Vol. C-32, No. Feb. 1984, pp. 160–177.

Bryant, R., "A Switch-Level Simulation Model for Integrated Logic Circuits." Ph.D. dissertation, Massachusetts Institute of Technology, Cambridge, Mass., 1981.

Deutsch, D. N., "A Dogleg Channel Router," *Proc. Thirteenth Design Automation Conference*, June 1976, pp. 425–433.

Dijkstra, E. W., "A Note on Two Problems in Connection with Graphs," *Numer. Math.*, Vol. 1, 1959, pp. 285–292.

Eustace, R. A., "A Deterministic Finite State Automata Approach to Design Rule Checking for VLSI." M.S. thesis, University of Central Florida, Orlando, Fla., Aug. 1981.

Eustace, R. A., "Intra Region Routing." Ph.D. dissertation. Computer Science Department, University of Central Florida, Orlando, Fla., Aug. 1984.

Eustace, R. A., and A. Mukhopadhyay, "A Deterministic Finite Automation Approach to Design Rule Checking for VLSI," *Proc. 19th Design Automation Conference*, Las Vegas, June 14–16, 1982, p. 712.

Harary, F., *Graph Theory*. Reading, Mass.: Addison-Wesley, 1969.

Hashimoto, A., and S. Stevens, "Wire Routing by Optimizing Channel Assignment Within Large Apertures," *Proc. Eighth Design Automation Conference*, 1971, pp. 155–169.

Hayes, J. P., "A Logic Design Theory for VLSI," *Second Caltech Conference on VLSI*, California Institute of Technology, Pasadena, Calif., 1981, pp. 455–476.

Hayes, J. P., "A Unified Switching Theory with Applications to VLSI Design," *Proc. IEEE*, Vol. 70, No. 10, 1982, pp. 1140–1151.

Hightower, D., "A Solution to Line-Routing Problems on the Continuous Plane," *Proc. Design Automation Workshop*, 1969.

Korn, R. K., "An Efficient Variable-Cost Maze Router," *Proc. 19th Design Automation Conference*, June 1982, pp. 425–431.

LaPaugh, A., "Algorithms for Integrated Circuit Layout: An Analytic Approach." Ph.D. thesis, Massachusetts Institute of Technology, Cambridge, Mass., Dec. 1980.

Lee, C. Y., "An Algorithm for Path Connection and Its Applications," *IRE Trans. Electron. Comput.*, Sept. 1961, pp. 346–365.

McCreight, E. M., "Efficient Algorithms for Enumerating Intersection Intervals and Rectangles," *Rept. CSL-80-9*, Xerox Palo Alto Research Center, Palo Alto, Calif., 1980.

Mead, C., and L. Conway (Eds.), *Introduction to VSLI Systems*. Reading, Mass.: Addison-Wesley, 1983.

Moore, E. F., "The Shortest Path Through a Maze," *Proc. Iternational Symposium on Switching Theory*, Harvard University Press, Vol. 1, 1959, pp. 285–292.

Ousterhout, J. K., G. T. Hamachi, R. N. Mayo, W. S. Scott, and G. S. Taylor, "A Collection of Papers on Magic," *Rept. UCB.CSD 83/154*, University of California at Berkeley, Berkeley, Calif., Dec. 1983.

Persky, G., D. Deutsch, and D. Schweikert, "LTX—A System for the Directed Automatic Design of LSI Circuits," *Proc. 13th Design Automation Conference*, June 1976.

Rivest, R., "The PI (Placement and Interconnect) System," *Proc. 19th Design Automation Conference*, June 1982, pp. 475–481.

Rivest, R. L., and C. M. Fiduccia, "A Greedy Channel Router," *19th Design Automation Conference*, June 1982 (Las Vegas), pp. 418–424.

Rowson, L. A., "Understanding Hierarchical Design." Ph.D. dissertation, California Institute of Technology, Pasadena, Calif., Apr. 1980.

Rubin, F., "The Lee Connection Algorithm," *IEEE Trans. Comput.*, Vol. C-23, 1974, pp. 907–914.

Smith, L., "An Analysis of Area Routing." Ph.D. thesis, Stanford University, Stanford, Calif., Feb. 1983.

Soukup, J., "Fast Maze Router," *Proc. 15th Design Automation Conference*, June 1978, pp. 100–102.

Supowit, K. J., "A Minimum Impact Routing Algorithm," *Proc. 19th Design Automation Conference*, June 1982, pp. 104–112.

Terman, C. J., "Simulation Tools for Digital LSI Design," Laboratory for Computer Science, Massachusetts Institute of Technology, Cambridge, Mass. (MIT/LCS/TR-304), Sept. 1983.

Whitney, T., "A Hierarchical Design-Rule Checker." Master's thesis, California Institute of Technology, Pasadena, Calif., 1980.

Whitney, T., "A Hierarchical Design-Rule Checking Algorithm," *Lambda*, 1st quarter 1981, p. 40.

Yamin, M., "XYTOLR: A Computer Program for Integrated Circuit Mask Design Checkout," *Bell Syst. Tech. J.*, Vol. 51, No. 7, 1972, pp. 1595–1610.

Yoshimura, T., and E. S. Kuh, "Efficient Algorithms for Channel Routing," *IEEE Trans. CAD Integr. Circuits Syst.*, Vol. CAD-1, Jan. 1982, pp. 25–35.

Appendix A

The Physics of Semiconductor Devices

A.1 Semiconductor Conduction

The term *very large scale integration* (VLSI) reflects the capabilities of the semiconductor industry to fabricate a circuit consisting of thousands of components on a single silicon substrate. To understand how these circuits work, it is necessary to delve into the basic physical processes that take place in the semiconductor material —silicon.

A model of the silicon atom is shown in Fig. A.1. Silicon has an atomic number 14; the nucleus consists of protons and neutrons. The nucleus is surrounded by a "cloud" of electrons which have discrete energy levels and are depicted to reside in different *shells*, the energy level increasing from inner to outer shells. The outermost shell has 4 *valence electrons* and is not completely filled since it could hold a maximum of 8 electrons. As such, a silicon atom is unstable because of the unfilled outer shell, but in solid silicon, a stable crystal lattice is formed by *covalent bonding*, in which silicon atoms share their valence electrons with neighboring atoms as shown in Fig. A.2. The sharply defined energy levels of electrons are now modified to *energy bands*, which define the allowed energy states of the electrons as drawn in Fig. A.3. This shows the two highest energy bands, the *valence band* and the *conduction band*, separated by a forbidden band. At very low temperatures, almost all the electrons reside in the valance band solidly attached to the crystal structure by the covalent bonds. Under ordinary room temperature and/or under external stimulus, some of the electrons acquire enough energy (1.1 electron volts) to break loose as *free electrons* to ascend to the conduction band, leaving a positively charged ion in the crystal lattice. The absence of an electron, called a *hole*, however, can

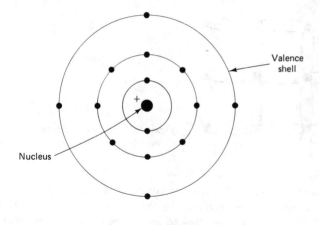

Figure A.1 Model of a silicon atom. The first shell has 2 electrons, the second 8, the third (not shown) 18 electrons. The valence shell has 4 electrons.

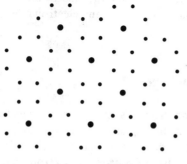

Figure A.2 Covalent bonding in silicon crystal.

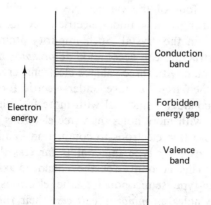

Figure A.3 Energy bond diagram for electrons in semiconductors.

move in the sense that if a neighboring atom contributes an electron to neutralize the positive charge of the ion, it becomes a positively charged ion in the process. A state of equilibrium is achieved at a given temperature when the rate of production of electron–hole pairs is equal to the rate of recombination of electron–hole pairs. When a voltage is placed across the semiconductor, the electrons in the conduction band and the holes in the valence band move in opposite directions to constitute a

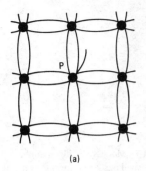
(a)

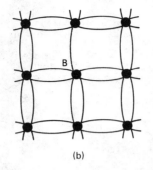
(b)

Figure A.4 *n*-type and *p*-type semiconductors. The lines denote covalent bonding.

total current flow. The energy bands of metals or conductors show no forbidden zone, signifying that free electrons are produced in the conduction band with no expenditure of energy, whereas in insulators this forbidden band is so large that a really high voltage is necessary to produce any electron conduction.

A.2 *n*-Type and *p*-Type Semiconductors

The production of holes and electrons can be greatly enhanced in a pure semiconductor by introducing controlled quantities (1 part in 10 million to 100 million silicon atoms) of impurities or *doping* material. Elements such as phosphorus, arsenic, and antimony which have five valence electrons can be placed in the silicon crystal structure and only four of the valence electrons will be used as covalent bonds, leaving a free electron to move under electric force, as shown in Fig. A.4(a), and a positive ion bound in the crystal. Such impurity atoms are called *donors*. Similarly, silicon crystal structure can accommodate *acceptor* atoms such as boron, indium, gallium, or aluminum with three valence electrons, creating electron vacancies or holes which are then free to move under electric force as shown in Fig. A.4(b). A region of semiconductor material with more free electrons than holes is called *n-type* and an area with more holes than free electrons is called *p-type*. The energy expenditure to move free electrons to conduction bands in *n*-type material and holes to valence bands in *p*-type material is far less than that used in an undoped semiconductor. This accounts for the enhanced conductivity of doped semiconductors. In an *n*-type semiconductor, the electrons are referred to as *majority carriers* and the holes as *minority carriers;* their roles are reversed in a *p*-type material.

Both *n*-type and *p*-type regions can be created in the semiconductor wafer by *selective doping.* A *p*-type semiconductor can be changed to an *n*-type semiconductor by a process called *counterdoping,* in which an increased number of donor atoms are placed in the region to neutralize the holes and to produce an excess of free electrons. Similarly, counterdoping can be used to change an *n*-type semiconductor to a *p*-type semiconductor.

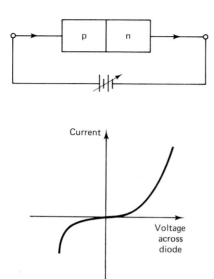

Figure A.5 Diode and its current–voltage characteristics.

The electron has a charge of 16×10^{-19} coulombs. At room temperature, the concentration of free electrons and holes in pure silicon is about 1.5×10^{10} per cubic centimeters. A typical doping dose is about 10^{16} atoms/cm^3. Silicon has about 10^{23} atoms/cm^3. The ratio of doping concentration to free electrons or holes is about 1 million to 1.

A.3 The Diode

If a p-type semiconductor adjoins an n-type semiconductor, a diode is formed, as shown in Fig. A.5. Figure A.5 also indicates that for positive current there is a voltage drop across the diode, which increases with the diode current. The plot also shows that the current–voltage relationship is nonlinear and that a small amount of reverse current exists when the diode is reverse biased. To understand the character-istic curve, we need to understand the basic physical processes going on at the junction. The first process, called *diffusion*, takes place by the movement of carriers from an area of high carrier concentration to an area of lower carrier concentration. Because of diffusion, holes from the p-type material move through the junction to the n-type material, leaving behind negative ions, and the electrons from n-type materials move through the junction to the p-type material. The region on either side of the junction from which electrons or holes have disappeared due to recombination is called the *space-charge* or *depletion* region. The ions left behind give rise to a built-in electric field of about 0.6 V which opposes the diffusion process that caused it. This field, called a *barrier potential* or *contact potential*, causes a drift of holes and electrons away from the junction, resulting in a drift current. An equilibrium state is achieved if the diffusion current is equal and opposite to the drift current, resulting

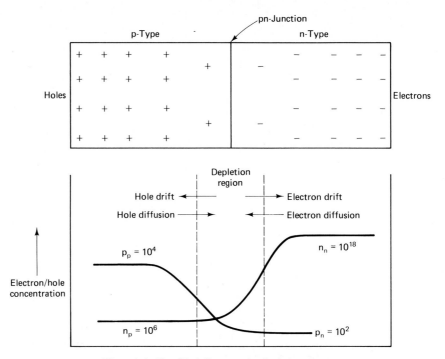

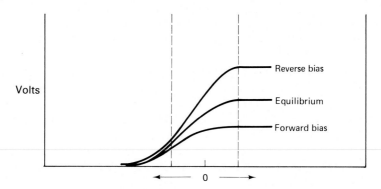

Figure A.6 Simplified diagram of a diode junction.

in a zero net current as shown in Fig. A.6. The symbols p_p, n_p denote hole and electron densities, respectively, in the p-type material. Similarly, p_n, n_n denote the same quantities in an n-type material.

An external voltage is applied across the diode terminal that will decrease the built-in electric field and will cause the holes and electrons to diffuse rapidly across the junction as shown in Fig. A.7. This will result in a large increase in diffusion current, resulting in a positive current flow as shown in the characteristic curve in

Figure A.7 Potential profiles of a diode junction.

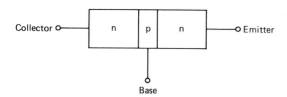

Figure A.8   Bipolar (*npn*) transistor.

Fig. A.5. The junction is now said to be *forward biased*. On the other hand, if the external voltage increases a potential barrier across the junction, the drift current will be favored against the diffusion current and the junction is said to be *reverse biased*. The drift current is due primarily to the flow of a few minority carriers on both sides of the junction, and is small since the rate of diffusion of minority carriers is practically independent of the applied voltage. If, however, the reverse voltage is increased to a large value, the minority carriers will be sufficiently energized to "knock" electron–hole pairs from the atoms near the junction, causing them to collide with the crystal lattice to produce an avalanche multiplication of electron–hole pairs. This results in a large negative current, as shown in the characteristic curve.

A junction or bipolar transistor consists of a thin layer of *p*-type material between two layers of *n*-type material. This forms two diodes, one reverse biased and another forward biased, as shown in Fig. A.8. However the forward current is mostly diffused to the collector region since the *p* region is made very thin. For an *npn* junction, if the base is negative with respect to the emitter, the current stops. Between the maximum and the zero values of the base current, the collector current is linearly proportional to the base current and could be 100 to 1000 times larger than the base current.

A.4 The MOS Transistor

A second basic type of transistor utilizes the propagation of a single kind of charge carrier. This is the MOSFET (metal-oxide-semiconductor field-effect transistor) or MOS transistor, of which there are two basic types: *n*-channel and *p*-channel. The structure of an *n*-channel MOS transistor has been described in Chapter 2 and is sketched in Fig. 2.1. It consists of two islands of *n*-type diffusions embedded in a *p*-type substrate, which are connected to external metallic conductors called the *source* and the *drain*. On the surface, a thin layer of silicon dioxide (SiO_2) is formed and on top of this a conducting material made of polysilicon called a *gate* is deposited. If the substrate material is *n*-type and the diffused islands are *p*-type, the same structure will represent a *p*-channel MOS transistor. The region between the two diffused islands under the oxide layer is called the *channel* region. In the most common mode of operation of the transistor, the source and substrate are grounded and the drain is connected to a supply voltage V_{dd}, which is positive for an *n*-channel transistor and negative for a *p*-channel transistor.

The terminal characteristics of the device are given by a plot of drain-to-source current I_{ds} against drain-to-source voltage V_{ds} for different values of gate-to-source voltage V_{gs}. All voltages are referenced with respect to the source voltage, assumed to be at ground potential. The source and substrate are assumed to be connected together. The characteristic curves have been described in Fig. 2.3. The curves show that the drain-to-source current flows only when the magnitude of the gate-to-source voltage exceeds a minimum value called the *threshold voltage*, V_{th}, that is, $|V_{gs}| > |V_{th}|$. The characteristic curve can be divided into two regions, separated by the dashed *saturation curve* corresponding to the value of I_{ds} for $|V_{ds}| = |V_{gs}| - |V_{th}|$. The region to the left (right for a *p*-channel) of this curve is called the *linear region*, where the device behaves like a voltage-controlled resistor. The region to the right (left for a *p*-channel) of the curve is called the *saturation region*, where I_{ds} remains practically constant with increasing V_{ds}. We also have noted that the amount of current flowing through the transistor is much greater for the *n*-channel transistor than for the *p*-channel transistor for given $|V_{ds}|$ and $|V_{gs}|$ and that the magnitude of the threshold voltage is higher for the *p*-channel transistor then for the *n*-channel transistor.

An explanation of the characteristic curves is rather complex. In the following, we will present a simplified explanation with respect to an *n*-channel device.

The basic device essentially consists of two reverse-connected diodes (the substrate–source and the substrate–drain) with a gate electrode spaced close to them, which can control the space-charge depletion regions of the diode. Let us start with a negative gate voltage. The negative voltage will increase the potential barrier across the junctions near the gate. To compensate for the negative charge, an *accumulation* of holes will take place under the gate [Fig. A.9(a)]. If the gate voltage is raised to zero and then to slightly positive, a negative charge will be induced in the semiconductor near the surface. Initially, this will be caused by the repulsion of the holes away from the surface, leaving behind negatively charged dopant ions and a few minority carrier electrons under the surface. This phenomenon is called *depletion* because the surface charge density near the gate is reduced below that of the bulk *p*-type substrate, resulting in the formation of a *depletion region* [Fig. A.9(b)]. If the gate voltage is increased, the width of the depletion region will increase and the strength of the total electrostatic potential across the gate surface will increase, causing a reduction of the potential barrier of the diodes near the gate ends. This will pull electrons away from the source and drain islands, resulting in a sharp rise in electron concentration near the surface in a thin layer. When this happens, an *inversion layer* is said to have been formed, which constitutes a *channel* [Fig. A.9(c)]. The gate voltage at which the inversion layer begins to form is called the *threshold voltage*, V_{th}. Once an inversion layer is formed, the width of the depletion region attains a maximum value. Any increase in gate potential will now result in increased electron concentration in the channel and the role of the dopant atoms will now be taken over by the electrons drifted into the region. This is called a *strong inversion condition* and the electron concentration level at this point exceeds the concentration of the dopant ions.

The electrons in the channel can now be withdrawn by applying a voltage between the source and the drain, producing a current I_{ds}. Let us now keep the gate voltage fixed and vary the drain-to-source potential. For a small drain voltage, the rate of withdrawal of electrons is proportional to the applied voltage and the

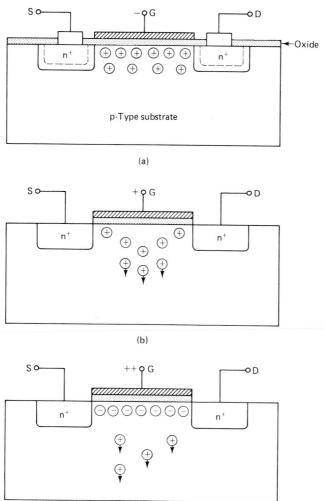

Figure A.9 State of affairs:
(a) accumulation; (b) depletion;
(c) inversion.

induced channel behaves essentially like a resistor with a lateral voltage gradient along the length of the channel. The electric field in the channel is much more complex; a transverse component of the field supports the inversion layer, which tapers to the drain end since the transverse field in excess of V_{th} is smaller as one proceeds to the drain end. The channel potential V_{ch} is 0 at the source end and is V_{ds} at the drain end. The potential difference V_{chg} between the channel and the gate is V_{gs} at the source end and drops to $V_{gs} - V_{ds}$ at the drain end. The channel support voltage V_{cs} at any point in the channel is $V_{chg} - V_{th}$, which decreases linearly along the length of the channel. The channel support voltage must be nonzero at all points under the gate for the channel to exist. It is $V_{gs} - V_{th}$ at the source end and drops to $V_{gs} - V_{ds} - V_{th}$ at the drain end. When $V_{ds} \geq V_{gs} - V_{th}$, the channel support voltage at the drain end becomes zero or less than zero and the channel near the drain disappears. The device now enters the *current-limited* or *saturation mode*. The drain voltage at this point is called the *saturation voltage* or *pinch-off* voltage $V_{d\text{sat}} = V_{gs} - V_{th}$, depicted by the dashed curve in Fig. 2.3. The gate voltage at the drain end is now V_{th}, the threshold needed to start inversion. A further increase in the drain voltage will now move the channel away from the drain, effectively reducing the channel length. The drain voltage being of the same polarity as that of the gate voltage opposes the induction process that produces channel electrons at the drain end, and the surface near the drain is not inverted any more; it is simply depleted. The potential at the end of the channel will remain constant at $V_{d\text{sat}}$, although the channel length will vary with V_{ds}. The excess of V_{ds} over $V_{d\text{sat}}$ will actually produce a lateral voltage drop in the depletion region at the surface between the channel endpoint and the drain. The drain current is now due to the injection of electrons which flow down the channel into the drain depletion region and are swept across the region by the large drain voltage. The operation of the device is now very similar to a junction transistor, where the saturation current is due primarily to the injection of carriers in the collector depletion region. The magnitude of the current depends on the rate of injection, which in this case is independent of V_{ds} and is only dependent on $V_{d\text{sat}}$, the channel end voltage at saturation. One can now easily see the effect of the increased gate voltage. First, the induced channel will have more charge, which will result in a larger current; second, the saturation voltage $V_{d\text{sat}}$ will be higher, resulting in a large saturation current.

The operation of the *p*-channel MOS transistor is very similar to that of the *n*-channel transistor except that a hole channel is formed and the polarities of the relevant voltages are exactly opposite. There is, however, one important difference: I_{ds} is much smaller for the *p*-channel than for the *n*-channel due to the fact that electron mobility is about two to three times the hole mobility. Early *p*-channel processes used a higher supply voltage (25 V) to compensate for their low hole mobility. Also, the threshold voltage of an *n*-channel MOS device is about half of that for a *p*-channel device.

One of the assumptions we made is that neither the drain–substrate nor the source–substrate diodes are forward biased. The primary reason for reverse biasing the diodes is to make the gate-induced charge the dominant phenomenon for the device. For this reason, this kind of MOS device is called an *enhancement-type* device, since the formation of the channel has been enhanced by the presence of the gate voltage. Another kind of MOS device, called a *depletion-mode* device, has a thin continuous *n*-type channel built under the gate by ion implantation. This is accomplished early in the fabrication process by using an ion acceleration gun to shoot phosphorus ions into the channel areas of the substrate, which lodge themselves near the surface up to an accurately controlled depth. The net effect of this is to change the threshold voltage V_{th} to a negative value. The characteristic curves are similar to those of the enhancement-type transistor except for the V_{th} value, which is -3 to -4 V for an *n*MOS. A source-to-drain current will flow when $V_{gs} = 0$. A negative gate voltage will deplete the electrons away from the surface and will stop the current. The magnitude of the negative voltage must be greater than a minimum voltage in order that the depletion phenomenon can occur. This negative voltage is called the *depletion threshold voltage* V_{dep}. Typically, $V_{dep} = -0.8V_{dd}$ for an nMOS. Thus with $V_{dd} = 5$ V, $V_{dep} = -4$ V. Finally, it must be noted that we made the assumption that the source and substrate are connected together to ground. If a MOS transistor is used as a transmission gate (see Section 2.3), the source terminal is at a higher potential with respect to substrate. The substrate potential V_B, which is negative with respect to the source, appears as a reverse bias or a *back bias*, tending to turn off the transistor. This necessitates a higher threshold voltage V_{thb} for inversion as given by the approximate relation $V_{thb} = V_{th} + 0.5\sqrt{-V_B}$. Thus, normally $V_B = 0$ and $V_{th} = 0.2V_{dd}$, which is 1 V with $V_{dd} = 5$ V. However, if $V_B = -4$, V_{thb} becomes 2 V. This phenomenon of changing of threshold voltage due to back bias is called the *body effect*.

A.5 Current Equations

We will now give a simple expression for the current I_{ds} for an n-channel transistor with width W and length L. The current equals the amount of charge crossing the channel per unit of time. The total induced surface charge, which must be equal and opposite in polarity to the gate charge, has two components: the charge induced in the depletion region due to the dopant ions and the charge Q due to the presence of excess electrons resulting from the inversion process, which contributes to the ohmic conduction current in the n-channel. As a first approximation we can say that charge Q equals the gate capacitance C_g times the average channel support voltage. The channel support voltage is $V_{gs} - V_{th}$ at the source and $V_{gs} - V_{ds} - V_{th}$ at the drain. Thus the average channel support voltage is $(V_{gs} - 0.5V_{ds}) - V_{th}$. Therefore,

$$Q = C_g\left[(V_{gs} - 0.5V_{ds}) - V_{th}\right] \tag{A.1}$$

The current I_{ds} is given by

$$I_{ds} = \frac{Q}{\tau} \tag{A.2}$$

where τ is the transit time of the electron across the channel. The velocity v of electrons in the channel is given by

$$v = \mu \times \text{electric field} = \mu \times \frac{V_{ds}}{L} \tag{A.3}$$

where μ is the average electron mobility in the channel. Thus

$$\tau = \frac{L}{v} = \frac{L^2}{\mu V_{ds}} \tag{A.4}$$

Substituting Eqs. (A.1) and (A.4) in Eq. (A.2), we get

$$I_{ds} = \mu\left(\frac{C_g}{L^2}\right)\left[(V_{gs} - V_{th})V_{ds} - 0.5V_{ds}^2\right] \tag{A.5}$$

The gate capacitance C_g equals $\varepsilon A / T_{ox}$, where ε is the permittivity of the oxide material; A is the area of the gate, which equals WL; and T_{ox} is the thickness of the oxide. Thus

$$I_{ds} = \left(\frac{\mu\varepsilon}{T_{ox}}\right)\left(\frac{W}{L}\right)\left[(V_{gs} - V_{th})V_{ds} - 0.5V_{ds}^2\right] \tag{A.6}$$

The quantities μ, ε, and T_{ox} are fixed for a given MOS structure and for a given fabrication process. It is seen from Eq. (A.6) that if the channel ratio W/L is high, I_{ds} is high for a given V_{ds}, whereas if the ratio is small, the channel will have a small current behaving as a high resistance. For saturation operation, the effective length of the channel is reduced to L' and the channel end voltage $V_{dsat} = V_{gs} - V_{th}$ determines the electron transit time. Substituting these values in Eq. (A.6), we get

$$I_{ds} = \frac{1}{2}\left(\frac{\mu\varepsilon}{T_{ox}}\right)\left(\frac{W}{L}\right)(V_{gs} - V_{th})^2 \tag{A.7}$$

The actual value of L' depends on V_{ds}, but as an approximation we have taken $L = L'$.

A.6 MOS Transistor as a Switch

When the transistor is operating in the linear region, the channel resistance R_{ch} can be computed as

$$R_{ch} = \frac{V_{ds}}{I_{ds}} = 1 \bigg/ \left(\frac{\mu\varepsilon}{T_{ox}}\right)\left[(V_{gs} - V_{th}) - 0.5V_{ds}\right]\frac{W}{L} \tag{A.8}$$

If V_{ds} is much smaller than $V_{gs} - V_{th}$, the effect of the term $0.5V_{ds}$ can be ignored and R_{ch} simplifies to

$$R_{ch} = 1 \bigg/ \left(\frac{\mu\varepsilon}{T_{ox}}\right)\left(\frac{W}{L}\right)(V_{gs} - V_{th}) \tag{A.9}$$

Thus the device acts as a linear resistance under gate voltage control. In this mode, the transistor can be used as an on–off switch, as symbolized in Fig. 2.5. The operation of the MOS switch is fully described in Chapter 2 and is thus omitted here.

A.7 Tutorial Discussion

In this section we compute typical values of electrical and physical parameters in order to obtain an idea of the magnitude of the different quantities involved.

Transistor dimensions are usually expressed in units of *microns* (μm); a micron or micrometer equals 10^{-6} meters or 10^{-3} cm. It is also expressed in units of λ (see Chapter 5) and we will take $\lambda = 2.5$ μm. The oxide thickness T_{ox} is expressed in *angstroms* (Å), where 1 μm = 10,000 Å. A typical value of T_{ox} is 700 Å. The quantity ε is called the *permittivity* of the silicon material and is given by $\varepsilon = 3\varepsilon_0$, where $\varepsilon_0 = 8.85 \times 10^{-14}$ farads/cm is the permittivity of free space, which is the capacitance of parallel conductors of area 1 cm² separated in free space by a thickness of 1 cm. The gate capacitance C_g is given by

$$C_g = \frac{\varepsilon}{T_{ox}}(WL) = \frac{3 \times 8.85 \times 10^{-14}}{7 \times 10^{-5}}(WL) = 3.8 \times 10^{-9}(WL) \qquad \text{farads}$$

(A.10)

Taking $W = L = 2\lambda = 5 \times 10^{-4}$ cm, $C_g = 0.01$ pF [where 1 pF (picofarad) = 10^{-12} farads] is the gate capacitance of the smallest transistor. The electron mobility μ is the velocity of electrons in the channel per unit electric field and its typical value is 8×10^{-2} m²/volt · sec. The quantity $\alpha = \mu\varepsilon/T_{ox}$ is called the *process gain factor* and its value is

$$\alpha = \frac{(8 \times 10^{-2}) \cdot (3 \times 8.85 \times 10^{-12})}{7 \times 10^{-8}} = 30 \times 10^{-6}\,\text{A/V}^2 = 30 \times 10^{-3}\,\text{mA/V}^2$$

(A.11)

The current I_{ds} for the linear region is thus

$$I_{ds} = 30 \times 10^{-3}\left(\frac{W}{L}\right)\left[(V_{gs} - V_{th})V_{ds} - 0.5V_{ds}^2\right]$$

Taking $V_{gs} = 5$ V, $V_{th} = 1$ V, and $V_{ds} = 1$ V, we have

$$I_{ds} = 0.1\left(\frac{W}{L}\right) \qquad \text{mA}$$

(A.12)

The saturation current is given by

$$I_{ds} = 15 \times 10^{-3}\left(\frac{W}{L}\right)(V_{gs} - V_{th})^2 = 0.24\left(\frac{W}{L}\right) \qquad \text{mA}$$

(A.13)

Note that the quantity $V_{gs} - V_{th}$ is the same for both enhancement-mode and depletion-mode transistors. In the later case, $V_{gs} = 0$ and $V_{th} = 0.8V_{dd} = -4$ V. The linear channel resistance R_{ch}, according to Eq. (A.9), is

$$R_{ch} = 1/30 \times 10^{-6}\left(\frac{W}{L}\right)(4) = 8.3\left(\frac{L}{W}\right) \qquad \text{k}\Omega$$

(A.14)

The velocity of electrons in the channel, v, is 8×10^5 cm/s, and the transit time τ is 0.625 ns for $V_{ds} = 5$ V and $L = 5$ μm.

REFERENCES

References 1 to 4 are classic texts on basic semiconductor physics, devices, and technology. Reference 5 is a relatively recent monograph on CMOS technology and devices. Reference 6 and 7 are texts on MOS devices and circuits written primarily for electrical engineering students. Reference 8 is, of course, the seminal text on VLSI. Reference 9 is an excellent introduction to semiconductor physics and is very clearly written. Reference 10 is a recent text on semiconductor technology.

1. Grove, A. S., *Physics and Technology of Semiconductor Devices*. New York: Wiley, 1967.

2. Penny, W. M., and L. Lau (Eds.), *MOS Integrated Circuits*. New York: Van Nostrand Reinhold, 1972.

3. Richman, P., *MOS Field Effect Transistors and Integrated Circuits*. New York: Wiley-Interscience: 1973.

4. Sze, S. M., *Physics of Semiconductor Devices*. New York: Wiley, 1969.

5. Melen, R., and H. Garland, *Understanding CMOS Integrated Circuits*. Indianapolis, Ind.: Howard W. Sams, 1975.

6. McCarthy, O. J., *MOS Device and Circuit Design*. New York: Wiley, 1982.

7. Mavor, J., M. A. Jack, and P. B. Denyer, *Introduction to MOS LSI Design*. Reading, Mass.: Addison-Wesley, 1983.

8. Mead, C., and L. Conway (Eds), *Introduction to VLSI Systems*. Reading, Mass.: Addison-Wesley, 1980.

9. Meindl, J. D., "Microelectronic Circuit Elements," *Sci. Am.*, Vol. 237, No. 3, 1977, p. 70.

10. Glaser, A. B., and G. E., Subak-Sharpe, *Integrated Circuit Engineering*. Reading, Mass.: Addison-Wesley, 1979.

Appendix B

MOSIS
Scalable and Generic
CMOS Design Rules

Instructions for Designers

Draw in units of lambda.

For all layers except metals, stay on a lambda grid. For metals, stay on a half-lambda grid.

What you draw will be very close to what you get. MOSIS will tell you the differences.

Scale your designs to centimicrons. Never submit a design in centilambda.

What Values Lambda?

Now, Lambda = 1.5 microns for 3-micron fabricators. Lambda = 0.8 microns for 1.2-micron fabricators. The scalable and generic CMOS rules are optimized for 1.2 micron process, but they are also applicable to the 2 micron processes. MOSIS also has a set of design rules for 3 micron p-well process.

*Courtesy of the USC Information Sciences Institute, Marina del Rey, California.

TECHNOLOGIES AND REQUIRED LAYERS

Process	Technology	Required layers
p well and n subs twin tub	SCP	CWP, CSP
n well and p subs twin tub	SCN	CWN, CSN
All*	SCG	CWG, CSG
All**	SCE	CWP, CWN CSP, CSN

*For a p well or n subs twin tub process, MOSIS sets CWP = CWG and CSP = CSG. For an n well or p subs twin tub process, MOSIS sets CWN = CWG and CSN = CSG.

**For a p well or n subs twin tub process, MOSIS ignores CWN and CSN. For an n well or p subs twin tub process, MOSIS ignores CWP and CSP.

Questions to Fabricators

MOSIS

- merges contact-to-active and contact-to-poly layers to generate the contact mask
- can apply bloats or shrinks to all layers
- can adjust active overlap of contact independently of bloat to active
- can adjust poly overlap of contact independently of bloat-to-poly

FABRICATORS

- What value of lambda can you support? What bloats or shrinks do you require?

MOSIS CMOS SCALABLE RULES

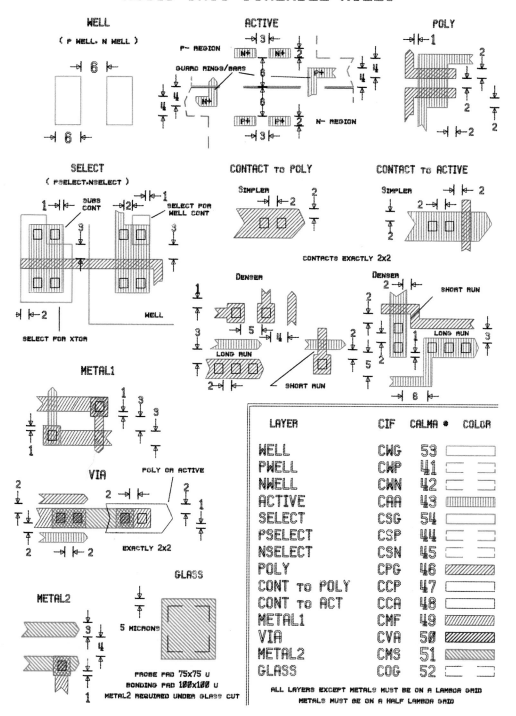

LAYER	CIF	CALMA #	COLOR
WELL	CWG	53	
PWELL	CWP	41	
NWELL	CWN	42	
ACTIVE	CAA	43	
SELECT	CSG	54	
PSELECT	CSP	44	
NSELECT	CSN	45	
POLY	CPG	46	
CONT to POLY	CCP	47	
CONT to ACT	CCA	48	
METAL1	CMF	49	
VIA	CVA	50	
METAL2	CMS	51	
GLASS	COG	52	

ALL LAYERS EXCEPT METALS MUST BE ON A LAMBDA GRID
METALS MUST BE ON A HALF LAMBDA GRID

LAYER NAMES AND COLORS

LAYER	CIF	CALMA #	COLOR
WELL	CWG	53	
PWELL	CWP	41	
NWELL	CWN	42	
ACTIVE	CAA	43	
SELECT	CSG	54	
PSELECT	CSP	44	
NSELECT	CSN	45	
POLY	CPG	46	
CONT to POLY	CCP	47	
CONT to ACT	CCA	48	
METAL1	CMF	49	
VIA	CVA	50	
METAL2	CMS	51	
GLASS	COG	52	

1. WELL
(NWELL,PWELL)

		LAMBDAS
1.1	WIDTH	6
1.2	SPACE	6

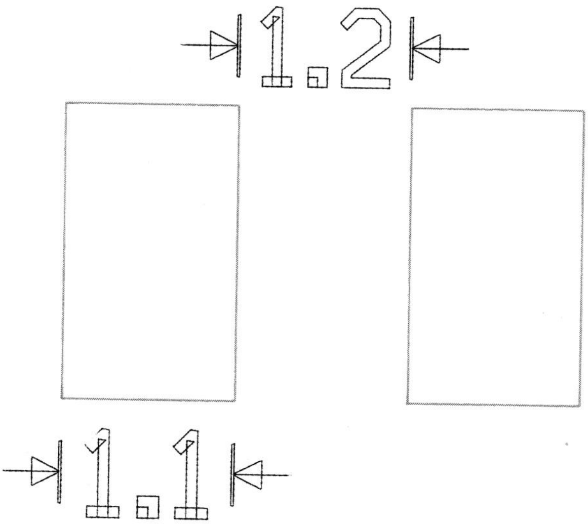

NOTE: IF P AND N WELLS
SUBMITTED, THEY MAY NOT
OVERLAP BUT THEY MAY BE
COINCIDENT

2. ACTIVE

		LAMBDAS
2.1	WIDTH	2
2.2	SPACE	3
2.3	SOURCE/DRAIN ACTIVE TO WELL EDGE	6
2.4	GUARD RING/BAR OVERLAP OF WELL	4

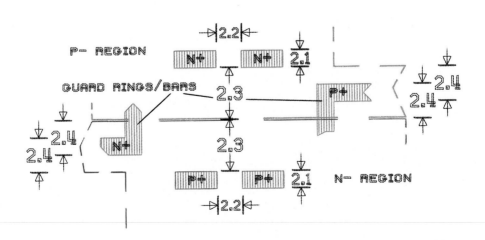

3. POLY

4. SELECT
(PSELECT,NSELECT)

		LAMBDAS
4.1	SELECT SPACE (OVERLAP) TO (OF) GATE	3
4.2	SELECT SPACE (OVERLAP) TO (OF) ACTIVE	2
4.3	SELECT SPACE (OVERLAP) TO (OF) CONTACT TO WELL OR SUBSTRATE	1

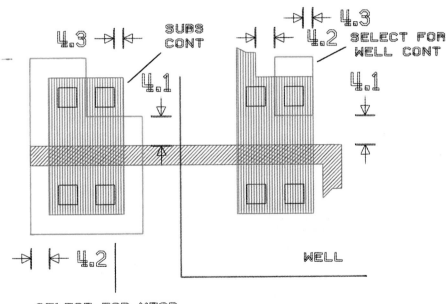

SELECT FOR XTOR

NOTE: IF BOTH PSELECT AND NSELECT SUBMITTED, THEY MAY BE COINCIDENT BUT MUST NOT OVERLAP

5A. SIMPLER CONTACT TO POLY

LAMBDAS

5A.1 CONTACT SIZE
      EXACTLY   2x2

5A.2 POLY OVERLAP   2

5A.3 SPACING   2

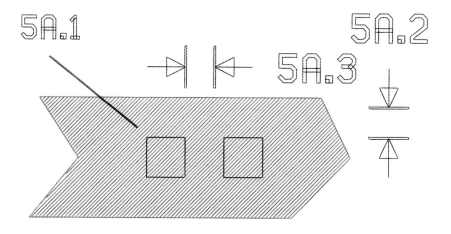

5B. DENSER CONTACT TO POLY

		LAMBDAS
5B.1	Contact size, exactly	2x2
5B.2	Poly overlap of contact	1
5B.3	Spacing on same poly	2
5B.4	Spacing on diff poly	5
5B.5	Space to other poly	4
5B.6	Space to act, short run	2
5B.7	Space to act, long run	3

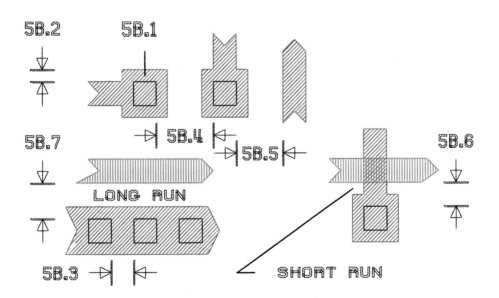

Note: your associating contacts with poly or active allows MOSIS to independently bloat the layer and the layer overlap of the contact

6A. SIMPLER CONTACT TO ACTIVE

		LAMBDAS
6A.1	CONTACT SIZE EXACTLY	2x2
6A.2	ACTIVE OVERLAP	2
6A.3	SPACING	2
6A.4	SPACE TO GATE	2

6B. DENSER CONTACT TO ACTIVE

		LAMBDAS
6B.1	CONTACT SIZE, EXACTLY	2x2
6B.2	ACTIVE OVERLAP	1
6B.3	SPACING ON SAME ACTIVE	2
6B.4	SPACING ON DIFF ACT	6
6B.5	SPACE TO DIFF ACT	5
6B.6	SPACE TO GATE	2
6B.7	SPACE TO FIELD POLY, SHORT RUN	2
6B.8	SPACE TO FIELD POLY, LONG RUN	3

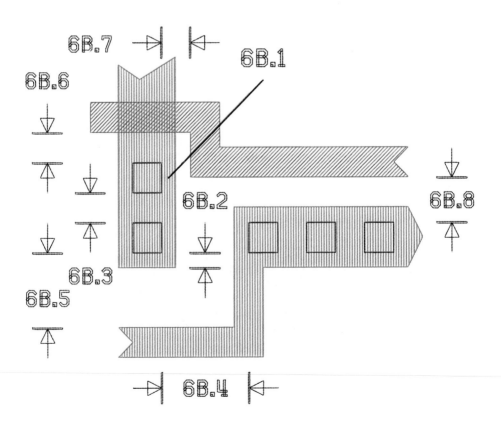

360

7. METAL1

7.3 7.2 7.1

7.4

361

8. VIA

		LAMBDAS
8.1	SIZE, EXACTLY	2x2
8.2	SEPARATION TO VIA	2
8.3	OVERLAP BY METAL1	1
8.4	SPACE TO POLY OR ACTIVE EDGE	2
8.5	SPACE TO CONTACT	2

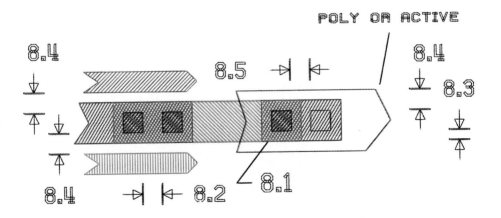

NOTE: OBJECTIVE IS VIA ON A FLAT SURFACE. VIA STACKED OVER CONTACT NOT ALLOWED.

9. METAL2

9.1

9.2

9.3

10. GLASS

MICRONS

10.1	Bonding Pad	100x100
10.2	Probe Pad	75x75
10.3	Pad to Glass Edge	5

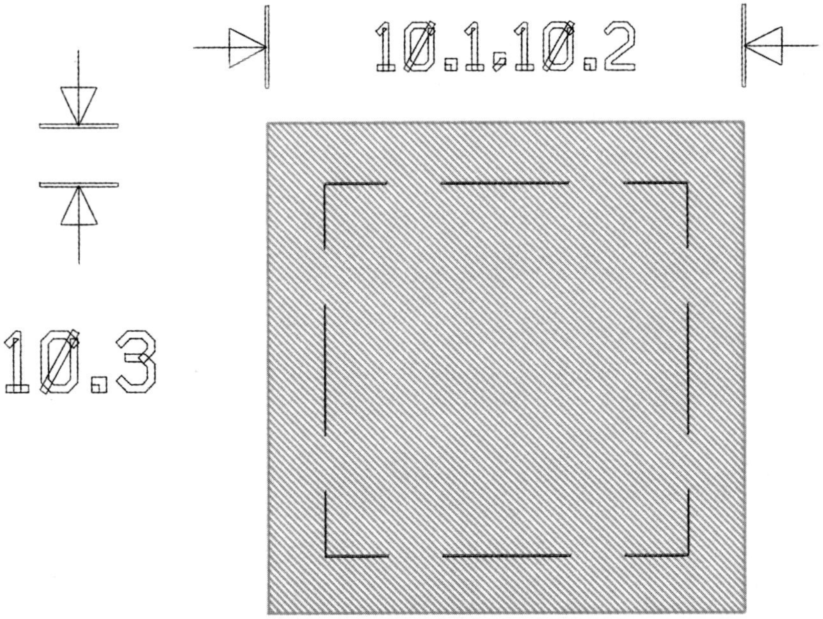

10.1,10.2

10.3

Note: There Must Be Metal2 Under a Glass Cut

Index